Applied Image

**Macmillan New Electronics Series**
*Series Editor: Paul A. Lynn*

G. J. Awcock and R. Thomas, *Applied Image Processing*
Rodney F. W. Coates, *Underwater Acoustic Systems*
M. D. Edwards, *Automatic Logic Synthesis Techniques for Digital Systems*
P. J. Fish, *Electronic Noise and Low Noise Design*
W. Forsythe and R. M. Goodall, *Digital Control*
C. G. Guy, *Data Communications for Engineers*
Paul A. Lynn, *Digital Signals, Processors and Noise*
Paul A. Lynn, *Radar Systems*
R. C. V. Macario, *Cellular Radio – Principles and Design*
A. F. Murray and H. M. Reekie, *Integrated Circuit Design*
F. J. Owens, *Signal Processing of Speech*
Dennis N. Pim, *Television and Teletext*
M. Richharia, *Satellite Communications Systems – Design Principles*
M. J. N. Sibley, *Optical Communications*, second edition
P. M. Taylor, *Robotic Control*
G. S. Virk, *Digital Computer Control Systems*
Allan Waters, *Active Filter Design*

# Applied Image Processing

G. J. Awcock and R. Thomas

*Department of Electrical and Electronic Engineering*
*University of Brighton*

**Macmillan New Electronics**
**Introductions to Advanced Topics**

MACMILLAN

© G. J. Awcock and R. Thomas 1995

All rights reserved. No reproduction, copy or transmission of this publication may be made without written permission.

No paragraph of this publication may be reproduced, copied or transmitted save with written permission or in accordance with the provisions of the Copyright, Designs and Patents Act 1988, or under the terms of any licence permitting limited copying issued by the Copyright Licensing Agency, 90 Tottenham Court Road, London W1P 9HE.

Any person who does any unauthorised act in relation to this publication may be liable to criminal prosecution and civil claims for damages.

First published 1995 by
MACMILLAN PRESS LTD
Houndmills, Basingstoke, Hampshire RG21 2XS
and London
Companies and representatives
throughout the world

ISBN 0-333-58242-X

A catalogue record for this book is available from the British Library

10  9   8   7   6   5   4   3   2   1
04  03  02  01  00  99  98  97  96  95

Printed and bound in Great Britain by
Biddles Ltd, Guildford and King's Lynn

# Contents

| | | |
|---|---|---|
| *Series Editor's Foreword* | | x |
| *Preface* | | xi |
| *Acknowledgements* | | xvi |

**1 Design of Industrial Machine Vision Systems**    1
    1.1   The importance of a visual sense for machines    1
    1.2   The role of understanding in vision    1
    1.3   Machine vision in context    2
    1.4   System design methodology    4
          1.4.1   The generic machine vision system model    6
          1.4.2   Emergent properties of the machine vision system    14
          1.4.3   Soft-systems thinking    15
    1.5   How can industrial machine vision be justified?    16
          1.5.1   Improvements in the safety and reliability of the manufacturing process    18
          1.5.2   Improvements in product quality    19
          1.5.3   An enabling technology for a new production process    20
          1.5.4   Economic and logistic considerations    21
    References    24

**2 Scene Constraints**    25
    2.1   General principles    25
          2.1.1   Exploitation of scene constraints    25
          2.1.2   Effective imposition of constraints    30
          2.1.3   Systemic considerations    34
    2.2   Lighting techniques    34
          2.2.1   Front-lighting    34
          2.2.2   Back-lighting    36
          2.2.3   Structured lighting    39
    2.3   Lamps    44
          2.3.1   General considerations    44

|  |  |  |  |
|---|---|---|---|
|  |  | 2.3.2 Summary of characteristics of some common sources | 46 |
|  | 2.4 | Optics | 48 |
|  |  | 2.4.1 Basic lens formulae | 48 |
|  |  | 2.4.2 Focal length | 50 |
|  |  | 2.4.3 Depth of focus and depth of field | 51 |
|  |  | 2.4.4 Lens aberrations | 51 |
|  |  | 2.4.5 Practical lenses | 53 |
|  |  | 2.4.6 Specialised optics | 55 |
|  | 2.5 | Achieving robust solutions | 56 |
|  | References |  | 58 |
| 3 | **Image Acquisition** |  | **59** |
|  | 3.1 | Representation of the image data | 59 |
|  |  | 3.1.1 What are the issues in image representation? | 59 |
|  |  | 3.1.2 Binary images | 62 |
|  |  | 3.1.3 Array tessellation | 63 |
|  | 3.2 | Image transduction (or sensing) | 66 |
|  |  | 3.2.1 What is an image? | 66 |
|  |  | 3.2.2 Principal photosensitive mechanisms | 67 |
|  |  | 3.2.3 Principal imaging architectures | 72 |
|  |  | 3.2.4 Comparison of vacuum-tube and solid-state area arrays | 82 |
|  | 3.3 | Digitisation | 83 |
|  |  | 3.3.1 Image capture | 83 |
|  |  | 3.3.2 Image display | 86 |
|  |  | 3.3.3 General system performance issues | 88 |
|  | References |  | 89 |
| 4 | **Image Preprocessing** |  | **90** |
|  | 4.1 | Introduction and theoretical background | 90 |
|  |  | 4.1.1 Digital convolution | 93 |
|  | 4.2 | Point operations | 99 |
|  |  | 4.2.1 Image brightness modification | 99 |
|  |  | 4.2.2 Contrast enhancement | 101 |
|  |  | 4.2.3 Negation | 101 |
|  |  | 4.2.4 Thresholding | 103 |
|  | 4.3 | Global operations | 106 |
|  |  | 4.3.1 Histogram equalisation | 106 |
|  | 4.4 | Neighbourhood operations | 108 |
|  |  | 4.4.1 Image smoothing | 109 |
|  |  | 4.4.2 Image sharpening | 113 |
|  | 4.5 | Geometric operations | 118 |
|  |  | 4.5.1 Display adjustment | 119 |

|  |  |  | |
|---|---|---|---|
|  | 4.5.2 | Image warping, magnification and rotation | 120 |
| 4.6 | | Temporal (frame-based) operations | 123 |
| References | | | 125 |

## 5 Segmentation 126
|  |  |  | |
|---|---|---|---|
| 5.1 | Introduction | | 126 |
|  | 5.1.1 | Formal definition | 127 |
|  | 5.1.2 | Towards good segmentation | 128 |
| 5.2 | Pixel-based or local methods | | 130 |
|  | 5.2.1 | Edge detection | 131 |
|  | 5.2.2 | Boundary detection | 136 |
| 5.3 | Region-based or global methods | | 141 |
|  | 5.3.1 | Region merging and splitting | 141 |
|  | 5.3.2 | Thresholding | 144 |
| References | | | 147 |

## 6 Feature Extraction 148
|  |  |  | |
|---|---|---|---|
| 6.1 | Image features | | 148 |
|  | 6.1.1 | Design of the feature extraction process | 149 |
| 6.2 | Image codes | | 150 |
|  | 6.2.1 | Run code | 150 |
|  | 6.2.2 | Chain code | 152 |
|  | 6.2.3 | Crack code | 155 |
|  | 6.2.4 | Signatures and skeletons | 155 |
| 6.3 | Boundary-based features | | 157 |
|  | 6.3.1 | Perimeter, area and shape factor | 157 |
| 6.4 | Region-based features | | 160 |
|  | 6.4.1 | Topology and texture | 161 |
|  | 6.4.2 | Moments | 162 |
| 6.5 | Mathematical morphology | | 165 |
|  | 6.5.1 | Basic definitions | 166 |
|  | 6.5.2 | The hit–miss transform | 167 |
|  | 6.5.3 | Erosion and dilation | 169 |
|  | 6.5.4 | Opening and closing | 171 |
| References | | | 175 |

## 7 Pattern Classification 176
|  |  |  | |
|---|---|---|---|
| 7.1 | Introduction | | 176 |
| 7.2 | Statistical methods | | 177 |
|  | 7.2.1 | Template matching | 178 |
|  | 7.2.2 | Template matching and optical character recognition | 179 |
|  | 7.2.3 | Feature analysis | 181 |
|  | 7.2.4 | Decision theoretic approaches | 182 |

|  |  |  |  |
|---|---|---|---|
|  | 7.2.5 | Probabilistic approaches – Bayes classifier | 184 |
| 7.3 | Syntactic methods | | 188 |
| 7.4 | Neural network approaches | | 191 |
|  | 7.4.1 | Neural network nodes | 192 |
|  | 7.4.2 | Network topologies | 194 |
|  | 7.4.3 | Training strategies | 196 |
|  | 7.4.4 | A memory-based classifier | 197 |
| References | | | 202 |

## 8 Image Understanding: Towards Universal Capability — 204

| | | | |
|---|---|---|---|
| 8.1 | Image formation and the visual processes | | 205 |
|  | 8.1.1 | Monocular images | 205 |
|  | 8.1.2 | Binocular images | 207 |
|  | 8.1.3 | Visual perception | 208 |
| 8.2 | Marr's computational theory of vision | | 212 |
|  | 8.2.1 | Depth from stereo | 213 |
|  | 8.2.2 | Depth from motion | 214 |
|  | 8.2.3 | Depth from texture | 215 |
|  | 8.2.4 | Depth from shading | 216 |
| 8.3 | Image representations | | 216 |
|  | 8.3.1 | Generalised cylinders | 218 |
|  | 8.3.2 | Volumetric elements | 218 |
|  | 8.3.3 | Surface representation | 219 |
|  | 8.3.4 | Wire-frame representation | 220 |
| 8.4 | Pragmatic modelling and matching strategies | | 221 |
|  | 8.4.1 | Two-dimensional modelling and matching | 222 |
|  | 8.4.2 | Three-dimensional modelling and matching | 223 |
|  | 8.4.3 | Three-dimensional modelling and matching to two-dimensional image data | 224 |
| 8.5 | The beginning of the end . . .? | | 224 |
| References | | | 225 |

## 9 Image Processing Case Histories — 227

| | | | |
|---|---|---|---|
| 9.1 | Industrial machine vision applications | | 228 |
|  | 9.1.1 | Automated visual inspection | 229 |
|  | 9.1.2 | Process control | 230 |
|  | 9.1.3 | Parts identification | 232 |
|  | 9.1.4 | Robotic guidance and control | 233 |
| 9.2 | Space exploration | | 235 |
| 9.3 | Astronomy | | 236 |
| 9.4 | Diagnostic medical imaging | | 238 |
|  | 9.4.1 | Medical image processing | 238 |
|  | 9.4.2 | Medical image reconstruction | 240 |

| | | |
|---|---|---|
| 9.5 | Scientific analysis | 243 |
| 9.6 | Military guidance and reconnaissance | 244 |
| 9.7 | Remote sensing | 250 |
| | 9.7.1 The nature of remote sensing | 250 |
| | 9.7.2 Applications of remote sensing | 251 |
| | 9.7.3 Image processing for remotely sensed data | 253 |
| 9.8 | Information technology systems | 256 |
| | 9.8.1 The nature of DIP | 256 |
| | 9.8.2 Applications of DIP | 257 |
| | 9.8.3 Image processing for DIP | 259 |
| 9.9 | Telecommunications | 264 |
| 9.10 | Security, surveillance and law enforcement | 270 |
| 9.11 | Entertainment and consumer electronics | 275 |
| 9.12 | Printing and the graphic arts | 283 |
| References | | 287 |

*Appendix: The Video Image Format* — 292
- A.1 The TV line format — 293
- A.2 Textual displays — 294

*Index* — 296

# Series Editor's Foreword

The rapid development of electronics and its engineering ensures that new topics are always competing for a place in university and college courses. But it is often difficult for lecturers to find suitable books for recommendation to students, particularly when a topic is covered by a short lecture module, or as an 'option'.

*Macmillan New Electronics* offers introductions to advanced topics. The level is generally that of second and subsequent years of undergraduate courses in electronic and electrical engineering, computer science and physics. Some of the authors will paint with a broad brush; others will concentrate on a narrower topic, and cover it in greater detail. But in all cases the titles in the Series will provide a sound basis for further reading in the specialist literature, and an up-to-date appreciation of practical applications and likely trends.

The level, scope and approach of the Series should also appeal to practising engineers and scientists encountering an area of electronics for the first time, or needed a rapid and authoritative update.

Paul A. Lynn

# Preface

Images are a vital and integral part of everyday life. On an individual, or person-to-person basis, images are used to reason, interpret, illustrate, represent, memorise, educate, communicate, evaluate, navigate, survey, entertain, etc. We do this continuously and almost entirely without conscious effort. As man builds machines to facilitate his ever more complex lifestyle, the *only* reason for NOT providing them with the ability to exploit or transparently convey such images is a weakness of available technology.

*Applied Image Processing*, in its broadest and most literal interpretation, aims to address the goal of providing practical, reliable and affordable means to allow machines to cope with images while assisting man in his general endeavours.

By contrast, the term 'image processing' itself has become firmly associated with the much more limited objective of modifying images such that they are either:

(a) corrected for errors introduced during acquisition or transmission ('*restoration*'); or
(b) enhanced to overcome the weaknesses of the human visual system ('*enhancement*').

As such, the discipline of 'pure' image processing may be succinctly summarised as being concerned with

> '*a process which takes an image input and generates a modified image output*'

Clearly then, other disciplines must be allied to pure image processing in order to allow the stated goal to be achieved. 'Pattern classification', which may be defined simply as

> '*a process which takes a feature vector input and generates a class number output*',

confers the ability to identify or recognise objects and perform sorting and some inspection tasks. 'Artificial intelligence', which may be defined as

*'a process which takes primitive data input and generates a description, or understanding or a behaviour as an output',*

confers a wide range of capability from description, in the form of simple measurement of parameters for inspection purposes, to a form of autonomy borne out of an ability to interpret the world through a visual sense.

These disciplines have been evolving steadily and independently ever since computers first became available, but only when they are all effectively harnessed together do machines acquire anything like the ability to exploit images in the way that humans do. In particular, the marriage of one, or both, of the first two disciplines with artificial intelligence has given birth to the new, image specific disciplines, namely 'image analysis', 'scene analysis' and 'image understanding'.

*Image analysis* is normally satisfied with quantifying data about objects which are known to exist within a scene, or determining their orientation, or recognising them as one of a limited set of possible prototypes. As such it is largely concerned with the development of descriptions of the 2-D relationships between regions within single images. However, for many applications, there is an undoubted need to extend this activity to the description of 3-D relationships between objects within a 2-D view of the real-world scene.

*Scene analysis* was the original term coined to describe this extension of image analysis into the third dimension. Such work flourished in the 1960s and was concerned with the rigorous visual analysis of three-dimensional polyhedra (the so-called 'blocks-world'), on the mistaken premise that it would be a trivial matter to extend these concepts to the analysis of natural scenes. The work was finally abandoned in the late 1970s when it was realised that the exploitation of application-dependent constraints was no way to research *general-purpose* vision systems.

Consequently, the term scene analysis fell into disuse only to be replaced by that of *image understanding*, which is more fundamentally based upon the physics of image formation and the operation of the human visual system. It aims to allow machines to operate with ease in complex natural environments which feature, for example, partially occluded objects or, ultimately, previously unseen objects.

A broad overview of the literature in the field of machine *perception* of images suggests the existence of two distinct 'camps' whose followers, while sharing common roots, set out to achieve fundamentally different objectives. We have chosen to label these camps as 'computer vision' and 'machine vision', and feel that they are essentially distinguished by their different approaches to the use of artificial intelligence and the degree to which it is employed. ('Robot vision' was also a popular alternative at one time, although it appears to be slowly falling into disuse, perhaps because of rather unfortunate science-fiction connotations.)

'*Computer vision*' is ultimately concerned with the goal of enabling machines to understand the world that they see, in real-time and without any form of human assistance. Thus, application-specific constraints are rejected wherever possible as the world is 'interpreted on-line'. The complexity of this task is easily under-estimated by those who take human vision for granted, but it is fraught with many immensely difficult problems, and seriously hampered by inadequate processing power.

'*Machine vision*', on the other hand, is concerned with utilising existing technology in the most effective way to endow a degree of autonomy in specific applications. The universal nature of the computer vision approach is sacrificed by deliberately exploiting application-specific constraints. Thus knowledge about the world is 'pre-compiled', or engineered, into machine vision applications in order to provide cost-effective solutions to real-world problems.

Thus one group of workers, primarily from engineering backgrounds, is application specialists, while the other group is more strongly motivated by the quest for knowledge and the desire to establish a solid research base for a 'universal' visual capability for machines. Both communities are vital for the successful development of the field, and the scope of their interests will continually converge as the performance of cost-effective computer capability improves. Clearly, the goal of producing a universal vision system which compares favourably with the human visual system is a very long way off, but progress towards that goal will continue to be absorbed in raising the level of autonomy exhibited by industrial automata etc.

However, while the labels 'computer vision' and 'machine vision' address a major sector of the field which aims to offer 'image manipulation for human advancement', they do not adequately embrace the full range of disciplines involved in meeting that aim. For example, consider the 'information technology' roles of image manipulation, such as 'document image processing' in business or 'image reconstruction' in medicine, which are critical to the effective utilisation of images in their respective application domains. Underpinning each of these and many other applications is image data compression, which is a vital part of *practical* storage and transmission of image-based information.

So, what should a book be called which aims to introduce its readership to the exciting new field of 'image manipulation for human advancement'? How should such a text be structured?

We have targeted the book at people with an engineering or general scientific background who are now in a position to *exploit* this new technology, rather than at computer scientists and AI workers who might have *developed* it. Therefore we have chosen to return to first principles and name the book *Applied Image Processing*. This deliberately uses the term 'image processing' in its broadest, colloquial sense and prefixes it with the word 'applied' to reflect the practical bias that pervades the whole book.

The book divides naturally into two parts — theory and applications — although the theory is always treated with a strongly pragmatic bias. This is reflected in the choice of industrial machine vision as a vehicle through which to investigate many of the disciplines defined above. Such an approach also allows the book to achieve a second important objective: that of providing readers with an insight into the design methodology of effective machine vision systems. This is intended to address one of the main weaknesses of the machine vision approach with respect to that of computer vision, i.e. the amount of 'bespoke' engineering that is required to realise effective solutions, coupled with the scarcity of people qualified to undertake it.

Thus the theoretical treatment is underpinned by a 'systemic' philosophy which is introduced in Chapter 1 and which aims to ensure that pragmatic and cost-effective solutions can be achieved through the well-placed application of a little forethought. This places rather uncommon emphasis upon the acquisition of good quality images (where 'good' implies *fit for purpose*), particularly through the exploitation and application of 'scene constraints' and appropriately matched image acquisition techniques (Chapters 2 and 3). Chapter 4 discusses the commonly encountered image processing techniques, but the systemic philosophy ensures that these are not used simply to compensate for poor quality image acquisition.

All image processing operations up to this point in the text have sought simply to modify the array of stored image data in order that it might better serve its intended purpose. As such, these processes are generally classified as 'low level' or pre-processing operations. 'High level' operations are concerned with the analysis, description and understanding of images, where the information representation format of an image changes from an array of numbers to symbolically meaningful strings of text. This treatment begins with discussion of segmentation of an image into meaningful regions and subsequent feature description (Chapters 5 and 6). Chapter 7 introduces the three major pattern classification strategies, including a treatment of the rapidly developing field of 'neural networks' in this context.

Until this point in the text, it is assumed that the machine is intent on deriving a two-dimensional description of the scene under investigation. While this dramatically simplifies the processing problem, and also suits an introductory text, it clearly imposes excessive constraints on many desirable applications. Thus Chapter 8 addresses the problem of helping a machine to understand and interpret a two-dimensional view of a three-dimensional world. It inevitably leans towards computer vision and includes a discussion of Marr's pivotal contribution to that field.

For undergraduate courses or self-study at any level, these first eight chapters provide a complete and thorough introduction to machine vision and through that, all the supporting disciplines. Chapter 9 uses a wide range of case studies to introduce and illustrate the breadth of application of 'image manipulation for human advancement'. It aims to fire the imagination of the readership, and inspire them to seek applications within their own sphere of influence and personal experience. It also serves as a practical introduction to many of the techniques, such as image data compression, which are not adequately addressed by the machine vision orientation of the earlier chapters.

Affordable computer performance has at last begun to become equal to the demanding nature of image processing, and the situation can only get better as the years go by. Therefore the stage is set for massive exploitation of 'Applied Image Processing'.

# Acknowledgements

The authors would like to acknowledge the support of their families and the patience of their publisher, Malcolm Stewart, throughout the gestation period of this book. A particular debt of thanks is due to David Thomas for his efforts in originating many of the line drawings in the book. Thanks are also due to the companies, and individuals within them, who supplied photographs to help illustrate the book. Finally, a big thank-you to all our friends and colleagues, in particular David Lawrence, who contributed to the book through their ideas, comments and suggestions.

# 1 Design of Industrial Machine Vision Systems

## 1.1 The importance of a visual sense for machines

There seems to be little disagreement that vision is the most valuable sense that an automaton can possess. The information that it conveys is extremely rich. It can provide absolute and relative position, range, scale, orientation etc., and all this is achieved without the need for physical contact.

The term *computer vision* serves to associate the machine world of computing and electronics with the human attribute of vision. The 'computer' aspects of such a system are related to the hardware elements of optical sensors, parallel processing architectures, computer graphics and displays, and the software elements of data manipulation and calculation. The 'vision' aspects mirror the human visual system and encompass the functional aspects of the eye, optic nerve, and brain.

Ideally machines should be endowed with the same visual sensing capabilities as humans. In attempting to define a viable computer vision system, which emulates the essential functionality of the human system, a decision must be made on which characteristics of the human system must be, or should be, included. This may seem a deceptively trivial objective to the uninformed, but this is because humans are so good at vision that we take it for granted. In fact the human visual system is extremely complex, with many stages of processing both in the eye and in the visual cortex of the brain [1].

The distinguishing feature of higher order animate vision is the perception of images rather than their simple *sensation*. Perception, coupled with the ability to *actively* investigate a scene confers the incredible flexibility which most humans take for granted.

## 1.2 The role of understanding in vision

The need to emulate perception makes image understanding a vital ingredient of any computer vision system. For many years now workers in the field of computer vision have striven to develop universal vision systems which see and understand the world in much the same way as we do. Since the pivotal contribution made by Marr [1] and others in the late 1970s, worthwhile progress has been made in understanding the processes of visual perception. However, the problem with a *universal* vision system is that it must cope with operation in unconstrained environments containing objects that have never been specifically encountered before. This highly prized human attribute of 'generalisation' requires an incredible depth and breadth of knowledge and understanding about the world to be achieved. In the absence of computers featuring the sheer power and architectural elegance of the brain, the implementation of vision systems offering the versatility of biological vision must be considered a very remote possibility.

Fortunately, despite this gloomy prediction, computer vision does not have to remain totally in the realms of science fiction. Provided that we accept sensible limitations and do not become seduced by the ideal of fully emulating the human visual capabilities, useful visual perception for machines can be brought within the bounds of realistic processing power. Throughout the rest of this book, this pragmatic approach to computer vision will be distinguished from the more generalised aims of computer vision research *per se*, by use of the term 'machine vision'.

## 1.3  Machine vision in context

Thus the term 'machine vision' is used to describe any work which aims to provide a practical and affordable visual sense for any type of machine, which works in 'real-time'. In order to satisfy this definition it is necessary to operate in relatively constrained environments. Fortunately, the modern manufacturing environment is characterised by a high degree of order and industrial automata are generally required to perform repetitive tasks on a limited range of well-defined objects. Therefore such an environment allows exploitation and imposition of application-specific constraints and it is here where machine vision has made the most progress.

Exploitation of *a priori* knowledge about the working environment of the machine considerably eases the problem of understanding that environment. For example, an assumption can often be safely made from a single visual cue without having continually to support its validity with other cues. Reliable recognition of complex objects can often be achieved by evaluating a simple set of features which have been determined to be uniquely characteristic during an 'off-line' training process.

Using structured light to actively probe the machine's environment is much like the way a human uses a torch as a tool to investigate an unknown scene. The essential difference is that the machine must be told which tool to use and when. The pattern of light is also specially formulated and used so that some important parameter within the scene, such as depth, becomes explicitly observable. It can then be easily measured in a two-dimensional view. The system designer may also use lighting or careful selection of viewpoint deliberately to suppress details within a scene, such as surface texture, which would otherwise complicate its understanding.

This is the stuff of which 'industrial machine vision' systems are made. Essentially the machine is *adapted* by its designer (or better still, by its installer and/or programmer) to its working environment so that it does not have to be *adaptive* in the extreme. Obviously this compromises the long-term goal of *fully* flexible machines, but it allows a useful level of flexibility to be achieved in the short to medium term. The aim of industrial machine vision system design is the provision of affordable and realisable solutions to the visual sensing problems of automation. These must be achieved using currently available technology and their prime purpose is to maximise profit.

An extremely simple example of industrial machine vision concerns the measurement of the diameter of steel bar as it leaves a rolling mill [2]. It works by simply causing the bar to interrupt a collimated light beam, thus casting a shadow on an image sensor. Clearly the system is not required to develop any 'on-line' understanding of its environment whatsoever, but it still performs a very important role.

By contrast, a cruise missile may be seen as being typical of functional computer vision applications. It must be able to navigate unaided across hundreds of miles of hostile territory. The inadequacy of pure dead reckoning and the hazards of low-level, high-speed flight, make it necessary to check its position periodically by viewing the terrain over which it is flying. In order to do this it must acquire radar images and attempt to fit these observations to pre-stored maps. It must operate in a three-dimensional world, in any season, at any time of the day or night and in all weather conditions, and it must accommodate errors in the maps that it uses. Therefore it must achieve a level of understanding that allows it to perform an activity akin to orienteering!

Despite these widely differing application examples, there is no hard and fast dividing line between 'industrial machine vision' and 'computer vision'. As progress is made in pure computer vision research it will be absorbed into the set of tools available to the industrial machine vision worker. One application which perhaps serves to mark the boundary is 'bin-picking'. This is concerned with using artificial vision to help retrieve components that have been randomly stored in a bin. Many existing industrial processes use this as an intermediate form of storage. However, there is much debate in the computer vision community as a whole as to whether or not it is a valid problem for industrial machine vision to solve. Some say this form of storage can and should be avoided in modern factories [3, 4], while others believe that such avoidance is an unacceptable constraint [5]. As there are several grades of bin-picking, the truth is that there are some which can be easily resolved into trivial vision tasks and some which cannot.

## 1.4 System design methodology

In the above discussion the term *system* appears frequently and is used to identify a wide variety of items whether technical or financial, optical or electronic, or even whether software or hardware. Every system however has one unique property, that is it must have a *system boundary*, see figure 1.1a. This may appear a facile statement at first but it is this unique property which must be used to define any system for the purposes of design. For example, the boundary for the vision system which inspects the diameter of rolled steel bar encompasses far less scope than that of the cruise missile 'vision' system. The designer of the former system had obviously clearly identified that the 'goal' of the system was *only* to measure diameter and *not* surface finish, velocity of travel, position within the factory, etc.

All artefacts or parameters which fall within this boundary are essential ingredients in the operation and performance of the system and must therefore be taken into account in any analysis. All artefacts and parameters falling outside this boundary do not affect the system operation and can therefore be neglected. For any given engineering problem or application it is important for the engineer to know precisely where this boundary is located.

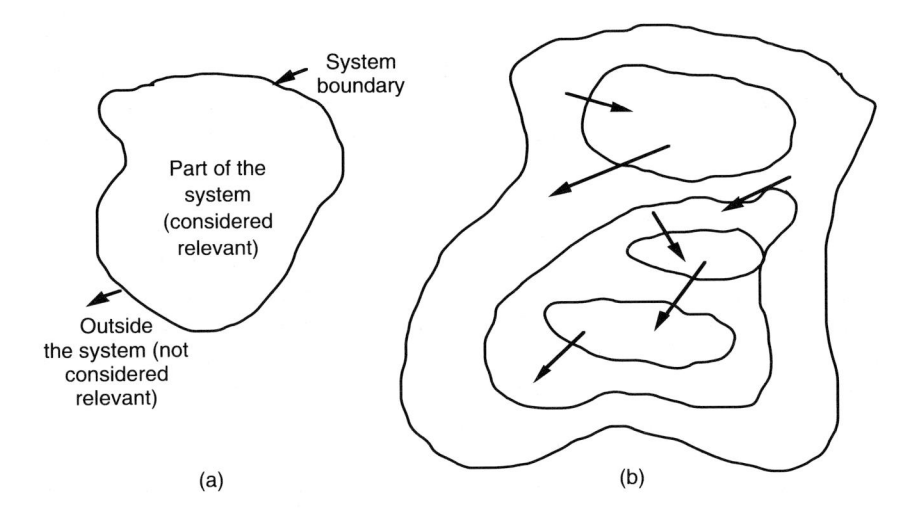

Figure 1.1 Introductory systemic concepts: (a) a system boundary; (b) a complete system, showing interdependencies

Identification of the boundary is not trivial because it has as much to do with the engineer's interpretation of the problem as it has to do with analysis of hard facts. Moreover the degree of magnitude assigned to the system is an issue. For example, the phrase 'machine vision system' could refer to a camera and a computer only, or may refer to a company's complete automated visual inspection equipment inventory and its operational schema.

One further term must be defined: namely *sub-system*. A sub-system is a smaller part of a larger system, and a system can contain many sub-systems within it. This definition requires the engineer to identify the interrelation between the various sub-systems, see figure 1.1b. Sub-systems may have varying degrees of interdependence where one sub-system is related to, or affected by, events occurring in one or more other sub-systems which cooperate with it.

Finally it must be noted that the well known statement *'the whole is greater than the sum of the parts'* has a systems interpretation. Within any complex system, containing many and varied sub-systems and sub-system interdependencies, the system may exhibit particular characteristics which are not readily apparent by an analysis of its component parts in isolation. These new characteristics, or *'emergent properties'* are often of vital interest to the successful performance of the overall system.

For example, consider two 'hi-fi' 'systems': one is made up of a CD player, an amplifier and a tape deck, each from a different manufacturer; the other consists of all of these components sourced by one manufacturer with a unified control interface. Now consider the task of recording music from CD onto tape. In the first 'system', which is really only an aggregate of components, the user must manually start the CD playing and then start the tape deck recording at precisely the right moment. In the second system the user simply selects CD as the required source and selects record on the tape deck. The system control interface then takes care of the timing – selecting record-pause and releasing it at precisely the right moment when the CD begins to play.

Furthermore, consider the case where the tape runs out before the CD finishes. The aggregate hi-fi will simply ignore this fact. However, the true system is aware that its goal is to record the CD onto tape. Therefore it will detect the tape end, pause the CD and search for the beginning of the track that was playing when the tape ended. In the meantime the tape will be 'auto-reversed' and put into record-pause at the beginning of the reverse side. Only when both sub-systems are ready will the record-pause be released so that the rest of the CD can be successfully recorded on the reverse side of the tape. Undoubtedly the second system has significantly greater useful capability than the first, because it has been designed as a *true* system where awareness of its overall goals has been encapsulated in the form of management of interdependencies between its sub-systems.

It is the job of the systems engineer to take such a *systemic* view, to be aware of all sub-system parameters and characteristics, and to be cognisant of any emergent properties that might be achieved. Such 'systems thinking' will help engineers to maximise the performance of practical industrial machine vision systems.

### 1.4.1 The generic machine vision system model

An overview of manufacturing industry readily identifies inspection and assembly tasks which traditionally would be performed by a human operator. Indeed it was the Industrial Revolution and the introduction of mechanisation which caused complex manufacturing operations to be broken down into simple step-by-step instructions which could be repeated by a machine. The need for systematic assembly and inspection operations, being performed in an appropriate sequence or being implemented in parallel, lends itself to possible solution by vision controlled automata.

As noted above, a suitable systemic approach to solving complex problems is to decompose them and thus form a hierarchical, or a distributed, system consisting of sub-systems. If the result is to be a true system rather than a simple aggregate, each module must be designed and developed with system-building in mind. Each one must provide a defined communications interface with the modules around it in the system; this communication interface ensures that relevant detail about the overall goals of the system can be propagated throughout the set of sub-system modules, in order that they are each able to make a wholly beneficial contribution to the achievement of those goals. Exploitation of this property of 'interdependence' ensures that the resulting system has the potential to be 'near-optimal' in terms of efficiency.

The level of flexibility which is necessary may be achieved by the adoption of a generic modular structure for the particular application domain; a complete range of sub-system modules, sufficient to provide all the functions essential to problem solving in that domain, must be identified. Each sub-system module must be allowed the freedom to perform its function in the way best able to realise the overall goals of the system. The term 'ordered liberty' succinctly describes this kind of control structure operating within a system architecture.

Necessary functions for a visually equipped machine can be identified as:

(a) the exploitation and imposition of environmental constraints;
(b) the capture of an image;
(c) the analysis of that image;
(d) the recognition of certain objects within it;

(e) the initiation of subsequent action taken as a result, in order to complete the task in hand.

Therefore a generic model of a machine vision system can be defined and illustrated as in figure 1.2. This generic model identifies the seven elements or sub-systems which may be identified within a typical machine vision system.

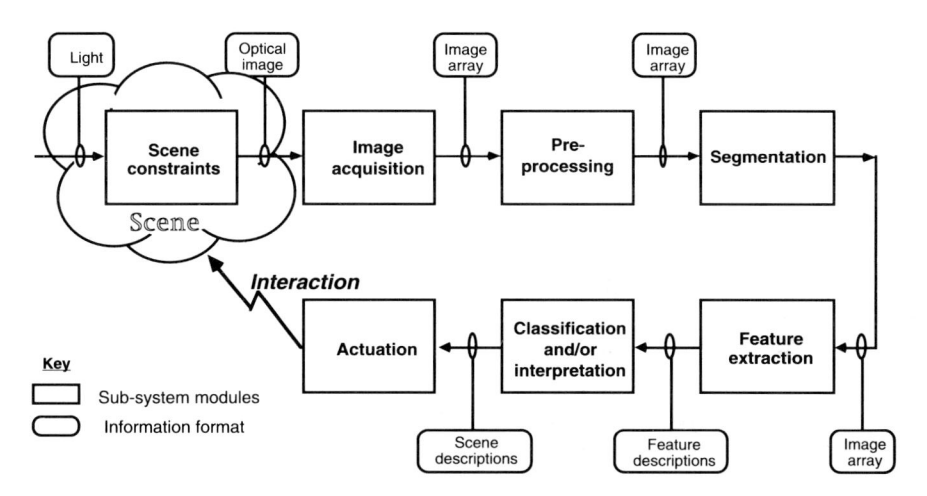

Figure 1.2   The generic model of a modular machine vision system

*Scene constraints*

The first, and arguably the most important, element is identified as scene constraints. The scene refers to the industrial environment in which the manufacturing task is to take place and into which the machine vision equipment is to be placed.

The ultimate aim of the scene constraints sub-system is to reduce the complexity of all of the subsequent sub-systems to a manageable level. This is achieved by proper exploitation of *a priori* constraints and imposition of new ones. Existing constraints might include knowledge of the limited number of objects manufactured, knowledge of their surface finish and appearance etc., while replacement of ambient light with carefully controlled lighting is generally an effective *imposed* constraint.

In order to approach the aim of the scene constraints sub-system, the desired information content should preferably be encapsulated in a form which is easy to extract. However, within the typical industrial environment many factors conspire to make the installation of cameras and special lighting conditions fraught with difficulties. In particular, most production and assembly lines have evolved to facilitate production with *human* operatives and utilise a plethora of machining, assembly, cleaning, test, rework and packaging facilities and equipment which cannot readily be replaced or reorganised when retrofitting a new machine vision system. Safety rules, operative interactions and other constraints can all dictate that the new system has to occupy a location which is somewhat less than ideal. In addition, the lighting conditions on production lines, dust and dirt of machine shops, hazardous chemicals and paints of processing and finishing shops, and even the hostile attentions of disgruntled workers all contribute to a harsh working environment for vision-based machinery.

These factors complicate the apparently simple task of presenting the objects under consideration to a camera in order to produce an optical image which contains the maximum amount of relevant information. Nevertheless, a little time and effort spent in achieving good control over the scene constraints will almost certainly result in a less complex, more reliable and less expensive vision system implementation.

*Image acquisition*

This sub-system is concerned with the process of translation from light stimuli falling onto the photosensors of a camera to a stored digital value within the computer's memory. Typically a digitised picture can be up to $512 \times 512$ pixels resolution with each pixel representing a binary, grey scale or colour value. The choice of spatial and luminance resolution parameters critically affect the workload of the computer and therefore must be minimised while ensuring that no useful information is lost.

Cameras incorporating either line scan or area scan elements are available and the correct choice will depend upon the particular application. A line scan camera, incorporating a single line of up to (exceptionally) 12 000 sensors, requires relative movement between camera and object in order to build up a complete two-dimensional image. Where necessary, this is often achieved by exploiting the movement of objects on a conveyor, for example. Area scan cameras, on the other hand, are readily available at lower spatial resolutions (generally not more than $780 \times 580$) utilising either CCD (charge coupled device) sensors or Vidicon tubes. However, they do offer the significant advantages of highly standardised interfacing to computers and being able to acquire information about the relative locations of several objects in a single 'glance'.

*Preprocessing*

The many processes within this sub-system seek to modify and prepare the pixel values of the digitised image to produce a form that is more suitable for subsequent operations. Typically contrast enhancement and adjustment, filtering to remove noise and improve quality, and correction for sensor distortion may all be desirable during this stage. These operations only change the pixel values of the digitised image within the frame store, or subsequent display memory, and do not seek to make any fundamental change in the information content of the captured image.

The digitised image, as initially acquired, has a direct relation, pixel by pixel, to the original scene and thus lies in the spatial domain. Preprocessing operators, which act on the spatial domain, are generally concerned with single pixels, or small neighbourhoods of pixels, only. Operations which are required to act globally on the image often make use of the transformation of the image from the spatial domain into the (spatial) frequency domain. This allows the image to be considered as being a series of periodic functions which can be easily and intuitively modified by standard filter functions. The most well known transformation operator is the Fourier transform, although it is generally not the most computationally efficient transform for image processing work.

Low level processing includes operations to clean up 'noisy images', and highlight particular features of interest. Histogram manipulations provide simple image improvement operations, either by grey level shifting or equalisation. An image histogram is easily produced by recording the number of pixels at a particular grey level. If this shows a bias towards the lower intensity grey levels, then some transformation to achieve a more equitable sharing of pixels among the grey levels would enhance or alter the appearance of the image. Such transformations will simply enhance or suppress contrast, and stretch or compress grey levels, without any alteration in the structural information present in the image.

Another important class of spatial domain algorithms is designed to perform pixel transformation, whose final value is calculated as a function of a group of pixel values (or 'neighbourhood') in some specified spatial location in the original image. Many filtering algorithms for smoothing (low pass) and edge enhancement (high pass) are firmly in this category. This introduces the basic principle of 'windowing operations' in which a 2-D (two-dimensional) mask, or window, defining the neighbourhood of interest is moved across the image, taking each pixel in turn as the centre, and at each position the transformed value of the pixel of interest is calculated (see chapter 4).

*Segmentation*

Segmentation is the initial stage for any recognition process, whereby the acquired image is 'broken up' into meaningful regions or segments. The segmentation process is not primarily concerned with what the regions represent, but only with the process of partitioning the image. In the simplest case (binary images) there are only two regions: a foreground (object) region and a background region. In grey level images, there may be many types of region or classes within the image; for example, in a natural scene to be segmented, there may be regions of sky, clouds, ground, building and trees. There are, broadly speaking, two approaches to image segmentation, namely thresholding and edge-based methods.

Thresholding techniques can be classified as employing either global or local methods. A global thresholding technique is one that thresholds the entire image with a single threshold value, whereas a local thresholding technique is one that partitions a given image into sub-images and determines a threshold for each of these sub-images. In general, a thresholding method is one that determines the value of threshold based on some predetermined criterion.

The classical approach to edge-based segmentation begins with edge enhancement which makes use of digital versions of standard finite difference operators, as in the first-order gradient operators (e.g. Roberts, Sobel) or in the second-order Laplacian operator (see sections 4.4.2 and 5.2.1). The difference operation accentuates intensity changes, and transforms the image into a representation from which properties of these changes can be extracted more easily. A significant intensity change gives rise to a peak in the first derivative or a zero crossing in the second derivative of the smoothed intensities. These peaks, or zero crossings, can be detected easily, and properties such as the position, sharpness, and height of the peaks infer the location, sharpness and contrast of the intensity changes in the image. Edge elements can be identified from the edge-enhanced image and these can then be linked to form complete boundaries of the regions of interest (see section 5.2).

Figure 1.3 illustrates thresholding and edge enhancement operations.

*Feature extraction*

The feature extraction sub-system is an important precursor to classification, whereby the inherent characteristics or features of the different regions within the image are identified. Relating this operation to a basic industrial visual inspection task it is clear that an object's features must be checked against a predetermined standard. Therefore any derived description of objects within the image should contain all the relevant shape and size information originally contained within the stored image.

In addition, the description should be invariant to position, orientation and, ideally, scale, of the object – the measurement of these parameters being an important function of the visual system for subsequent manipulation. A universal vision system should be able to accommodate a number of objects randomly positioned in the field of view. However, this also implies the need to solve the difficult problem of dealing with partial views of overlapping objects. Once again the controlled environment of most industrial tasks

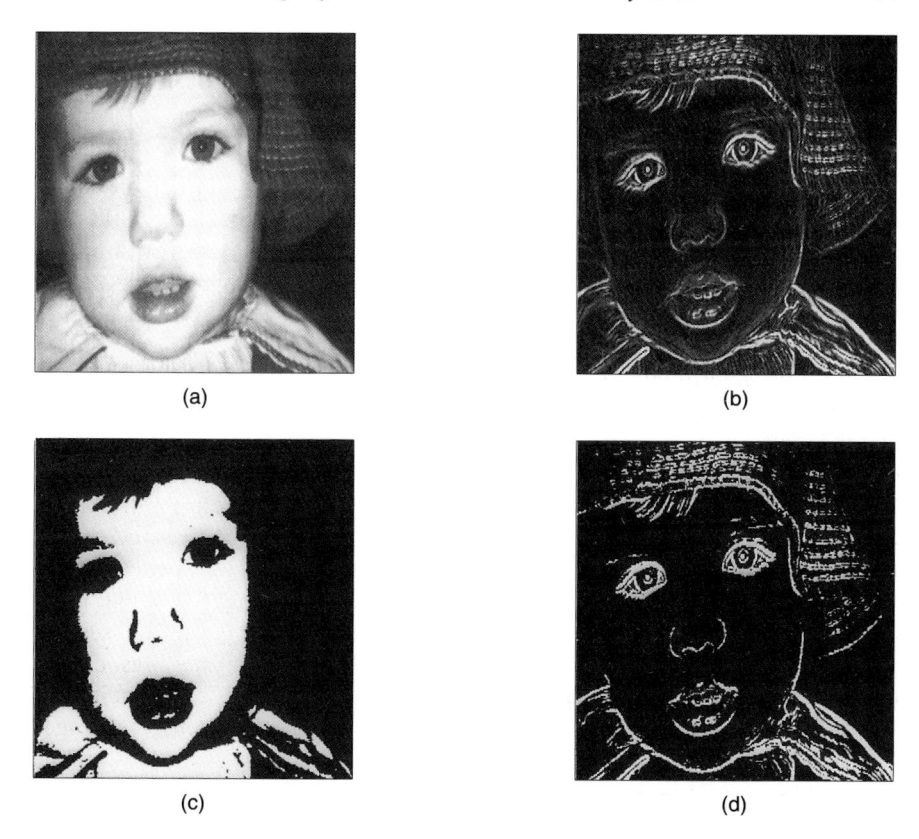

Figure 1.3 Preprocessing and segmentation operations: (a) original grey scale image (512 × 256 × 6 bits); (b) the Sobel algorithm used to edge-enhance (a); (c) binary version of (a) thresholded at grey level 47; (d) binary version of (b) thresholded at grey level 28

comes to the rescue, since it is often relatively easy to use mechanical means to ensure that only single objects are presented to the data acquisition unit.

A number of basic parameters may be derived from an arbitrary shape, to provide valuable classification and position information. These include perimeter, minimum enclosing rectangle, centre of area and information about holes (number, size, position), as illustrated in figure 1.4. Measurements of area and perimeter provide simple classification criteria, which are both position and orientation invariant. The derived 'shape factor', (perimeter$^2$/area), is also a popular parameter used in object recognition, because it is dimensionless and therefore invariant to scale.

The centre of area is a point that may be readily determined for any object, providing an object-oriented origin which allows the development of a series of feature descriptors that are independent of orientation and, with

*12*  *Applied Image Processing*

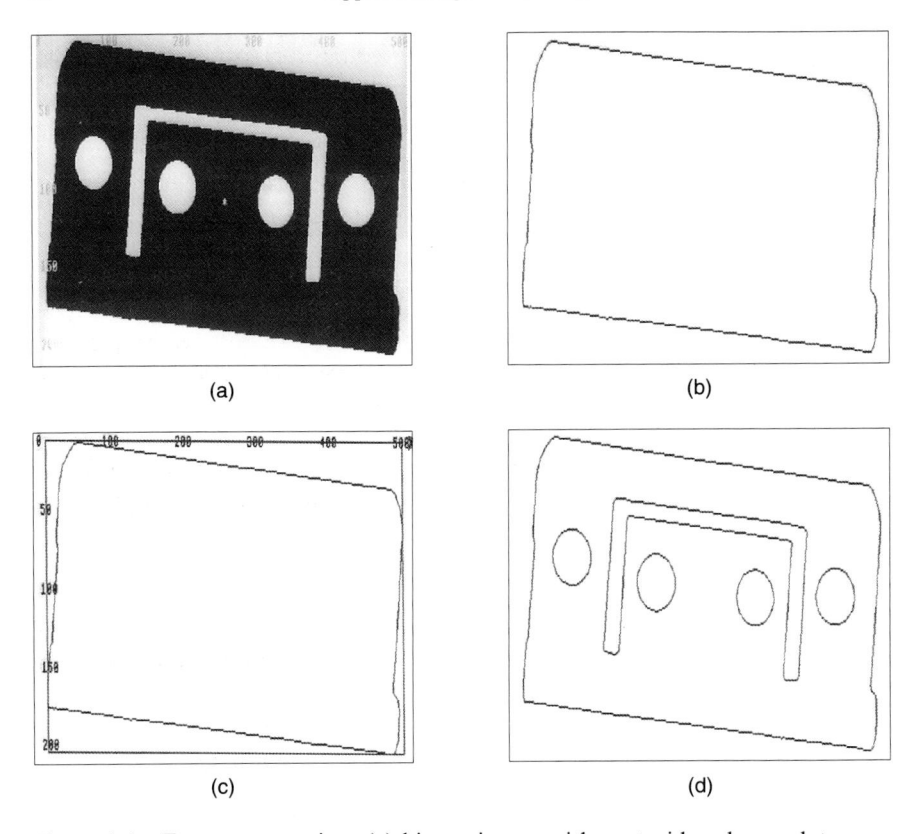

Figure 1.4  Feature extraction: (a) binary image with centroid and area data; (b) perimeter; (c) minimum enclosing rectangle; (d) number of holes identified

suitable normalisation, scale. Examples are minimum and maximum radius vectors, polar radii signatures, central moments etc. These are of considerable importance for recognition and location purposes.

*Classification*

The classification sub-system is concerned with the process of pattern recognition or image classification. In essence this process utilises some or all of the object features or descriptors that have been extracted from the image to make an accurate decision as to which category of objects (or 'prototypes') the unknown object belongs. Various techniques exist for the classification stage, including template matching and a collection of statistically based approaches (see chapter 7).

Template matching is used in situations where the objects to be identified have well defined and highly differentiated features, for example standard alphanumeric character fonts. In such cases an unknown character is compared with a set of templates or masks, each of which fits just one character uniquely. Statistical techniques can be selected to provide optimum classification performance for more varied industrial applications.

If the vision task is well constrained then classification may be made via a simple tree searching algorithm where classification proceeds by making branching choices on the basis of single feature parameters. In more complex cases, $n$ features are combined to create a 'feature vector' which places a candidate object within the $n$-dimensional feature space. Provided that the features have been properly chosen to divide the allowable range of candidate objects into well separated 'clusters', then classification merely consists of dividing the feature space with one or more 'decision surfaces', such that each decision surface reliably separates two clusters (see figure 1.5).

*Actuation*

Finally, the actuation sub-system provides a means of 'closing the loop' and allowing interaction with the original scene. From an industrial viewpoint,

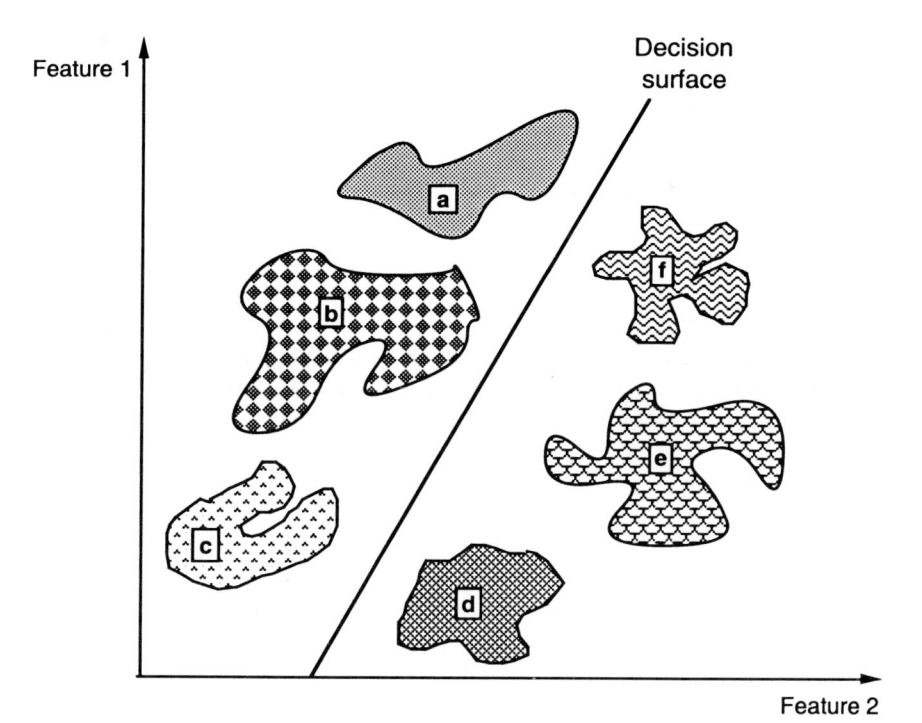

Figure 1.5 Classification via a feature space

any machine vision system must contribute to the efficiency of production, inspection or test functions within the company. Hence, the actuation module identifies any subsequent action that the vision system makes, either directly through explicit robotic interactions with the original environmental scene, or indirectly as implicit interactions (as in manual intervention, following vision system directives).

*Information format*

The generic model of figure 1.2 highlights one further attribute of interest, namely the information format available at each stage of processing. Explicit awareness of this format helps the engineer to decide where design effort should be concentrated in order to maximise the effectiveness of the machine vision system. In particular it helps to decide upon the most cost-effective partition of the system between dedicated processing hardware and general-purpose software implementation methods.

From the starting point of an appropriately illuminated industrial scene comes the optical image which is influenced by choice of lights, lenses etc. After the image is acquired, the information is in the form of an array of sampled data stored in the image frame store memory (the image array). The preprocessing and segmentation operations upon this array produce the same image format, that is, an array of data in memory (a modified image array). In an ideal world all processing up to this point would be performed by fully parallel hardware because of the spatial correlation that the data possesses. This would provide maximum performance but would naturally be very costly.

The feature extraction process, however, uses the image array data in order to produce a list of appropriate feature descriptions, this list being stored in memory. From this point on the processing requirement is better suited to the decision-taking capabilities of a software-programmable machine or possibly neural processing architectures. The classification or analysis operation continues to increase the level of abstraction of the data by producing some precise descriptions of the scene or objects within it. Therefore the information format is again changed into a new list of these 'enriched' scene descriptions.

## *1.4.2 Emergent properties of the machine vision system*

The functional blocks identified within the generic model (figure 1.2) all act upon the data they receive, before passing this modified data to the subsequent module. The machine vision system can be regarded as containing several sub-systems which, though complete in themselves within the enclosing sub-system boundary, cause interaction between modules across the functional spectrum. Moreover, it is clear that the level and degree of complexity of computing during the preprocessing stage will relate intimately to the quality of the image captured during the acquisition stage.

Although sub-system interdependence can be exploited to radically improve the cost/performance ratio of the overall system, it also tends to complicate its design, particularly if the interdisciplinary engineering nature of the system is properly addressed. Thus, the scene constraints can impinge upon the domain of the production engineer, the analysis and actuation areas can impinge upon the manufacturing and test engineers, and the rest of the system can be influenced by the hardware and software design engineers. In addition to this, the speed of operation and accuracy of any system will necessarily be governed by the financial limitations to which the system is subjected.

Consideration of the generic machine vision model in the light of a systemic philosophy has indicated the clear presence of sub-system boundaries, together with a worthwhile degree of sub-system interdependence. Each sub-system would be a candidate for application of the latest technological developments, and considered exploitation of *a priori* knowledge within the constrained industrial environment should produce 'emergent properties' from the vision system that are both predictable and beneficial. In short, when the generic model is utilised a systemic approach to optimality must be adopted, rather than the alternative option of simply optimising each module individually for its own particular sub-task.

### 1.4.3 Soft-systems thinking

The generic model has been considered as an example of a typical 'hard' system, made up of hardware and software, techniques and technologies. However, in the real world such hard systems will obviously have to interact heavily with humans, either as prospective or actual users, operatives, customers etc. Consideration of the impact of interaction across this type of system boundary enters the domain of 'soft' systems thinking [6], and is crucial to the development of practical systems which offer real benefit to humankind.

Soft systems thinking is concerned with identifying the nature of the overall goal of the total system and managing the human factors that assist or prevent its achievement. The first major problem is simply to get machine vision widely recognised as a realistic potential solution to some of the problems of manufacturing industry. Outside the electronics industry itself this is a far from trivial problem because of the highly technical nature of the machine vision systems on offer. Thus, the 'top-level' managers in industry must be subjected to a subtle process of education, in order that they can actively consider themselves as potential users.

One of the things which most easily puts off potential users is the risk factor associated with a relatively new industrial tool. When this is coupled with the expense of early, high profile, piecemeal solutions it is perhaps hardly surprising that managers are reluctant to adopt such new technology. Proper recognition of the generic model should allow system designers to select fairly standard implementations from a range available for each sub-system which grows steadily over the years. The ability simply to choose from a selection of pre-characterised sub-system implementations will reduce the overall cost of systems while maintaining their performance. There is already a wide choice of specialised, high performance, image processing boards that can be bought to plug into standard PC computers [7].

However, the total soft systems problem encompasses many non-technical domains such as financial, personal, organisational, etc. Therefore it is appropriate to take a wider industrial perspective and consider the problem of introducing machine vision into the manufacturing and assembly process for the future success and prosperity of the company. Arguably the first issue that must be tackled is that of justification.

## 1.5  How can industrial machine vision be justified?

The efficiency of its manufacturing industry is of major importance to the economic welfare of any major country. In increasingly competitive world markets it is highly desirable to employ automation to improve and/or maintain this efficiency. However, the short life-cycles and sheer variety of modern products challenge the flexibility of manufacturing automation. Conventional 'hard' automation which is dedicated to a particular sequence of actions is restricted to high volume mass production. Unfortunately, the vast majority of products are made in small batches, to customer order.

Therefore, if the use of manufacturing automation is to be significantly extended it must bridge the gap between mass production and hand-crafting of one-offs and very small batches (see figure 1.6). In order to do this it must be made more flexible, so that it can adapt to product changes as and when they occur. Much of this flexibility can be achieved by building programmability into the automation, interlinking machines and adapting the manufacturing process. However, ultimately the machinery must be provided with senses so that it can perceive changes to the product and its environment and then react accordingly. This will also allow machines to

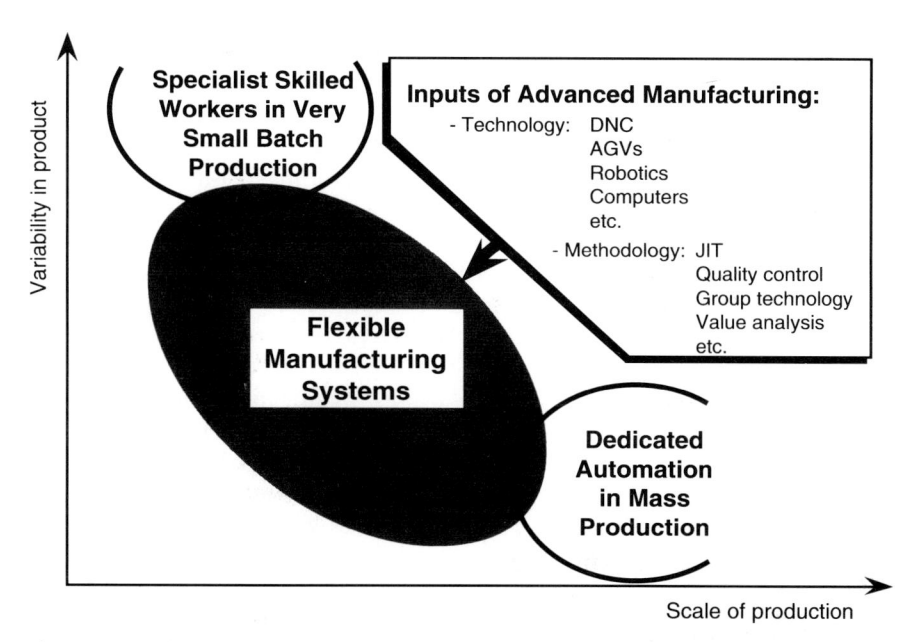

Figure 1.6  The value of flexibility in manufacturing automation – after Bessant and Heywood [9]

undertake tasks which have hitherto been difficult to automate, such as assembly.

In well designed products, assembly tasks are dominated by 'peg-in-hole insertions', 'push and twist bayonet fixings' and 'screw/bolt insertions' [8]. These can make severe demands of the precision of any machine but are greatly eased by the provision of localised sensory feedback, where the most versatile sense is vision.

When used in this way vision-assisted automation can ensure that cost savings follow from benefits such as the reduction of 'work in progress' and the adoption of 'just in time (JIT)' manufacturing methods. Savings in capital can be made through the exploitation of plant and tooling which is adaptable to many products. Of course there is also the potential of savings in direct labour costs through an overall reduction in manning and the requisite skill levels. Further savings accrue from more efficient utilisation of capital equipment through the elimination of sickness, holidays, tea-breaks and limited shift lengths. It has also been shown that improvements in throughput of as much as 50% can be made when human labour is removed from difficult and dangerous tasks [10].

Furthermore, the visual sense of the human worker features a number of operational deficiencies which can make industrial machine vision more attractive:

(a) Humans are subjective. Vision systems can make measurements with greater objectivity and repeatability than humans.
(b) Humans are fragile. Vision systems can be used in situations where humans would be uncomfortable or in danger.
(c) Humans are fallible. Vision systems need never make a mistake through boredom or inattention.
(d) Humans can be too slow. Vision systems can often operate in real-time relative to the manufacturing process, where humans cannot.
(e) Humans can be insensitive to subtle change. Vision systems can observe such changes and react to them objectively.

Naturally the human worker also brings significant advantages to an application. The performance of a 'free' visual inspection at every stage of handling by a human is one which is easily overlooked. Machines will only look for defects if they are specifically programmed to do so. There is also some considerable difficulty in getting machines to make 'fuzzy' decisions based on a number of qualitative parameters. Skilled human operators are very good at this.

Despite these potential disadvantages, there are several factors resulting from the frailties of human vision that can ensure that machine vision wins the day. In such cases the desire to improve the performance of the manufacturing process in some way can justify the development of stand-alone machine vision equipment. This may either be retrofitted to existing plant or integrated into new production lines.

### 1.5.1 *Improvements in the safety and reliability of the manufacturing process*

In some manufacturing sectors products may have a 'safety critical' status. Therefore it may be vital to perform 100% inspection, since any failure of the product in service may carry the very real threat of financially crippling litigation. However it is known that human inspectors become less reliable as defects become less common. In addition to this lack of reliability, human inspectors may not be cost-effective at normal manufacturing throughput rates.

The aerospace and defence industries are obvious examples where safety is a critical feature in manufacturing. However, many pharmaceutical products are potentially lethal and have a limited safe shelf-life, making it necessary to check every package to ensure that the product identification, instructions for use, lot code and date stamp are correct and legible.

In the automobile industry it is now common to use automated methods to build parts such as brake drum asemblies. Such automated processes suffer common failure modes including missing parts, mis-oriented parts and contamination by grease etc., which could threaten life if passed unnoticed. This is clearly a case where the elimination of the 'free inspection' given by human labour during the manufacturing process must be compensated by an explicit inspection stage at the end of the assembly. A well documented solution to this problem was implemented at Volkswagen some years ago [2]. It makes very effective use of scene constraints and *a priori* knowledge to achieve cost-effective and reliable 100% inspection of brake assemblies.

Glass products such as milk bottles are manufactured in environments which are very hostile to human labour because of extremes of heat and particularly noise. Common defects include chips, cracks and loose shards of glass which must be eliminated, but which can be difficult for the human eye to spot unaided. Furthermore, the volume production is generally so high that sufficiently detailed inspection by humans is simply not possible. Machine vision systems using carefully controlled lighting techniques can be used to fulfil this important inspection function.

## *1.5.2 Improvements in product quality*

Aside from the absolute requirement for 100% inspection for the purpose of product safety, there is a growing desire to use this technique to improve product quality. This is generally a result of the increasingly competitive marketplace.

An obvious application is the elimination of blemishes in fruit and vegetables, since this will have a direct impact upon the saleability of goods. However, discerning customers will also perceive other improvements in quality such as better performance or longer life, etc. Then there is undoubtedly the potential to increase market share by developing an enhanced reputation for high quality products.

It may also be desirable to enhance the cosmetic appearance of products even though this has no effect upon their suitability for their intended purpose. Market surveys have found that a non-uniform appearance of labels etc. can adversely influence the delicate balance of consumer preference when a product is offered for sale alongside similar items from competitors. A single event of this kind can potentially cause a lifelong switch in brand-loyalty within any one consumer. Furthermore, a really poor example could remain unsold at the front of a shelf and serve to dissuade many potential purchasers, even though there may be perfectly good examples stacked on the shelf behind it.

It is generally accepted that the cost of diagnosing and correcting faults in products rises by a factor of ten for each manufacturing stage completed after the fault was introduced. This means that there are much more tangible financial reasons for wanting to improve product quality by tightly integrating 100% inspection into the manufacturing process. These include reduction in rework costs and, even more significantly, the reduction of warranty failures in the field. Careful control of product quality means that a manufacturer can further enhance the competitiveness of his product by offering an attractive warranty period, secure in the knowledge that it will not cost too much!

The speed and reliability of machine vision systems in the role of inspection means that it is often the only way to realise these financial benefits and thus retain a competitive edge in the modern marketplace.

### 1.5.3 *An enabling technology for a new production process*

Some industrial processes would simply not be viable without machine vision in one form or another. These often involve situations where a human workforce would otherwise be placed in grave danger or could simply not survive.

One of the most commonly encountered sources of such danger is nuclear radiation. The nuclear power generation industry relies for its safe operation upon routine inspection and maintenance being carried out under conditions intolerable to human life. Robotised machines equipped with cameras are used to perform such tasks. However, even these do not have an easy time since the ionising radiation will cause lens coatings to become opaque and can easily destroy unprotected sensors and electronics. Machine vision technology has also been applied to nuclear fuel reprocessing. The UK's National Engineering Laboratory has implemented a system which can automatically identify, locate and remove fuel 'pins' from their sub-assemblies [11].

The control of certain manufacturing processes could not realistically be achieved without the aid of machine vision. Very often these involve continuous production at a rate which is too fast for the human senses to react and correct reliably. However one documented case involving the manufacture of sheet glass uses a vision system to control the process and eliminate a systematic waviness too subtle for a human to identify [12]. The system is also used to detect randomly occurring defects such as scratches and bubbles. The position of these is logged so that they can be excluded from 'first quality' sheets at the automated cutting stage.

Kodak have used a similar concept in the manufacture of photographic film and paper [2]. Even if this was not produced at such a high rate (3 m s$^{-1}$) it could not be studied by a human because of its light-sensitive properties. Therefore an infra-red laser is used to scan the material and identify scratches as small as 50 μm wide and 'spot defects' of more than 300 μm diameter, in webs up to 1.5 m wide. Once again these defects are logged and eliminated from the output as the web is cut into strips.

## 1.5.4 Economic and logistic considerations

The driving force behind the introduction of automation and robotic assembly into manufacturing industry is one of economics. The same can be said of the introduction of machine vision systems – they will only be introduced when they are economically viable. It has been noted that this will be when the vision system is able to perform tasks that cannot be performed manually for reasons of health or safety, or which need to be performed more accurately, reliably and efficiently than any company employee can achieve.

The second point needs further consideration since it is usual in medium size companies for the vision system to be considered for installation into an existing manufacturing plant where current vision tasks are performed by company operatives. Factors such as cost, speed, accuracy, reliability, flexibility and serviceability have then to be balanced against possible changes in manufacturing technique, the introductory cost of more automation, and possible changes and re-deployment of personnel that may occur. If, on the other hand, the vision system is integrated into the manufacturing process at the plant design stage, as it was with the robotic assembly lines of the automobile industry, then a more harmonious solution is often possible.

Since the application of machine vision technology to industry is a relatively new phenomenon, with few vision suppliers (in the UK) being in business for more than ten years, potential users really need to have an awareness of optics and sensors, together with computing, electronics and engineering. Furthermore, this relative immaturity necessitates that both suppliers and users of the vision system resolve the sometimes conflicting views of academics, who are making significant advances in image understanding and scene interpretation, with industrialists, who are forced to make severe compromises in terms of viable functionality and basic economy.

This potential conflict of objectives, illustrated in figure 1.7, needs to be borne in mind. An academic research programme requires innovative features and techniques to contribute to the general fund of research knowledge in the area. The industrial system, on the other hand, must focus on performing specific objectives within a predetermined and limited financial and technical resource.

It is not unusual for uninitiated machine vision users to expect much more of machine vision systems than they can realistically be expected to achieve: this stems from the fact that one's own visual capability is very much taken for granted. Introduction to the true level of performance can often dismay industrialists, particularly in small or medium size companies who might regard the machine vision system as a kind of 'magic bullet' in attaining their profitability targets.

*22*  *Applied Image Processing*

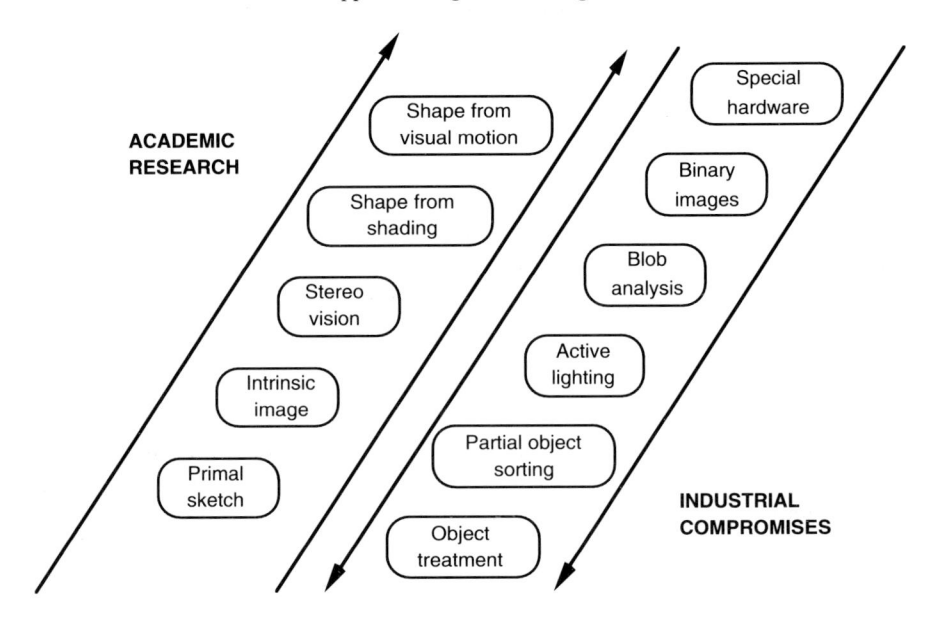

Figure 1.7  Comparison of academic and industrial objectives

In truth, costs are always relative, and the greater the technological barriers to be overcome the greater is the cost of the machine vision installation. The technological parameters to be considered include: the resolution of the sensor, the image acquisition time, whether the processing generally utilises simple two dimensional images or the two-dimensional projected image of a three-dimensional original, and the response time to process the image and then make a decision.

Cameras able to capture images up to 1024 × 1024 pixels in resolution are not that uncommon, and the adoption of 'array sensors' or 'line-scan sensors' can affect the resolution and acquisition time parameters. Binary and grey scale images can be processed with current vision systems, while the processing of colour images (with the inherent increase in data) remains too expensive and/or slow for the majority of industrial applications. The analysis of two-dimensional projected images of a scene can produce shortcomings when objects are partially occluded during an inspection process; such limitations may be overcome by the use of more than one camera, or the utilisation of different camera angles, all of which add to the cost and time in processing and merging the data. Image processing speed limitations are generally overcome by using parallel-pipeline architectures and special-purpose integrated circuits. However, this brings an obvious cost penalty, not simply because of the expense of the parallel processing hardware itself, but particularly because of the cost of bespoke engineering that is currently necessary in implementing such a solution.

Minimising system cost will depend upon the level of technical specification that the company is prepared to accept and the degree to which expertise is bought-in or home-grown. Current proprietary machine vision systems vary from basic PC-based two-dimensional measurement and inspection, at around £2000–£5000 ($3000–$7500), to complex and totally flexible three-dimensional real-time systems which retail upwards of £50 000 ($75 000).

If a completely proprietary system can be purchased then the appropriate blend of software packages may still have to be assembled. It is also necessary to ensure that appropriate cameras, lighting and optical systems are purchased and correctly positioned within the manufacturing and test facility to maximise effectiveness while minimising disruption to existing manufacturing plant. The development of a suitable maintenance strategy is also vital to the success of a machine vision system and has significant impact upon its annual running costs. Each of these activities may either be considered as part of the system vendor's responsibilities or carried out by the user's own engineers. This decision not only has a direct financial implication but also impacts upon the level of expertise currently available within the user's company or needing to be employed in the future.

So, from a purely economic viewpoint the size of the company affects the viability of a machine vision system. Small companies of less than fifty employees are least likely to invest unless it is critical to the company's existence. Medium size companies who employ up to 250 employees, and larger companies, all have to ensure that adequate financial and technical analyses are undergone in order fully to justify such purchases. Financial factors to be considered encompass the cost of raw materials and bought-in parts, the cost of production and test equipment, rent of factory premises including offices and stores, the total company salary bill, the cost of repayments on existing loans, plough-back of investment for the future, as well as day-to-day running costs and overheads.

Larger companies are often able to take a longer term view and can more easily commit themselves to in-house research and development programmes. They may also be able to benefit from incorporating their machine vision systems into an overall computer integrated manufacturing (CIM) strategy. For example, products could be designed on computer-aided design (CAD) workstations which are able to communicate in both directions with automation systems intended to manufacture those products. This can lead to the ultimate systemic control of 'scene constraints' where new product designs are checked for suitability for vision-assisted manufacture and the design database could be accessed by the vision system for inspection and/or recognition purposes.

**References**

1. D. Marr, *Vision*, W. H. Freeman, New York, 1982.
2. J. Hollingum, *Machine Vision: The Eyes of Automation*, IFS (Publications), Bedford, 1984.
3. A. Owen, *Assembly with Robots*, Chapter 10, p. 137, Kogan Page, London, 1985.
4. B. Batchelor, 'Enthusiasts debate illumination and optics at US workshop', *Sensor Review*, July 1982, pp. 157–159.
5. R. B. Kelley, J. R. Birk *et al.*, 'A robot system which acquires cylindrical workpieces from bins', *IEEE Transactions on Systems, Man and Cybernetics*, **SMC-12**(2), pp. 204–213, March/April 1982.
6. P. Checkland, *Systems Thinking, Systems Practice*, Wiley, Chichester, 1981.
7. D. Koenig, 'The boards are back in town', *Image Processing*, **3**(3), pp. 18–36, July/August 1991.
8. A. S. Kondoleon, Application of technology-economic model of assembly techniques to programmable assembly machine configuration, Masters Thesis, MIT, Cambridge, Massachusetts, 1976.
9. J. Bessant and W. Heywood, 'The introduction of flexible manufacturing systems as an example of computer integrated manufacturing', *Innovation Research Group Occasional Paper No. 1*, Fig. 5, p. 33, Brighton Polytechnic Dept of Business Management, 1985.
10. Ingersoll Engineers, *Industrial Robots*, Appendix A, p. 2, Dept of Industry/National Engineering Laboratories, Glasgow, 1980.
11. J. D. Todd, 'Advanced vision systems for computer-integrated manufacture: Part II', *Computer-Integrated Manufacturing Systems*, **1**(4), pp. 235–246, Butterworth, London, 1988.
12. H. Geisselmann *et al.*, 'Vision systems in industry: application examples', *Robotics*, **2**, pp. 19–30, Elsevier (North-Holland), Amsterdam, 1986.

# 2 Scene Constraints

## 2.1 General principles

Scene constraint was defined as the first sub-system component in the generic model of a machine vision system presented in figure 1.2. It is a key area in the application of systems engineering principles to the solution of machine vision problems. The designer of machine vision systems should have two principal aims when manipulating scene constraints;

- To maximise the use of prior knowledge of the scene, i.e. by exploiting existing knowledge
- To trivialise the problem of image analysis as far as possible, i.e. by effective imposition of constraints.

### 2.1.1 *Exploitation of scene constraints*

Exploitation of constraints refers to the use of knowledge about the objects under investigation and their environment to simplify problem-solving. The following four categories will serve to illustrate this point:

*Characteristics of materials*

Examination of the scene can reveal properties of the materials involved which influence the quality of optical images obtained. The amount of light reflecting from the surface of the objects in the scene depends upon their surface reflectivity, and will be markedly different for a highly polished or shiny surface in comparison to a flat or matt finish (see figure 2.1). This can be usefully exploited when inspecting for deep scratches on a matt finished surface, for example.

Objects made of opaque materials may lend themselves to back-lighting in cases where all of the features of interest may be efficiently represented by a silhouette (see figure 2.6). By contrast, transparent objects made of glass or plastic cannot generally be viewed in this way but, since they refract light,

(a) Smooth flat surface: - coherent reflection

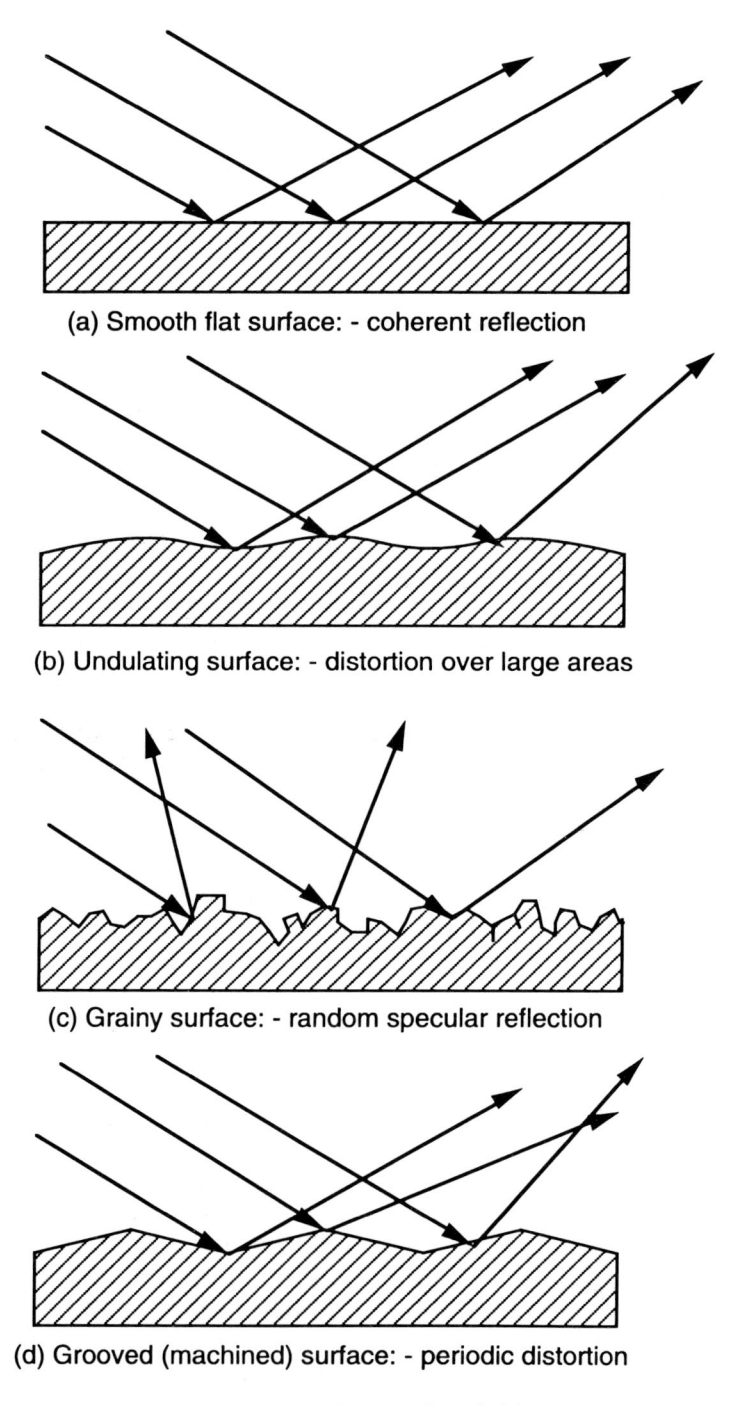

(b) Undulating surface: - distortion over large areas

(c) Grainy surface: - random specular reflection

(d) Grooved (machined) surface: - periodic distortion

Figure 2.1  Light reflection from various surface finishes

they can be highlighted very effectively by the dark-field illumination technique (see figure 2.7).

The fact that steel is magnetic is exploited in a method of crack detection which begins by immersing the part under test in magnetic UV-fluorescent ink. Any cracks cause the ink to be retained when the part is rinsed. The cracks then glow selectively when the part is illuminated by ultra-violet light. Thus the magnetic properties of the material have been exploited in an illumination technique which explicitly highlights any cracks, making their identification a trivial exercise.

Colour is an important characteristic of materials and it can often indicate the condition of a product. For example, consider the case of the banana, which is green when unripe, turning to yellow as it ripens and black when over-ripe or blemished. In this case the use of appropriate colour filters would assist in the capture of effective images, without resorting to costly and complex colour image acquisition and processing.

*Inherent features*

Careful examination of the objects under investigation may reveal particular features that can be exploited during image acquisition. For example, many industrial objects contain holes, and it is often possible to reliably distinguish parts and orientate them using these holes alone. When this is so, the silhouette lighting technique should be employed. This facilitates binary image representation which means that it is a relatively simple matter to segment the holes from the object and thus identify it by the number, or relative position, of holes.

Similarly, orientation can be determined by identifying the 'centroids' (see chapter 6) of all significant elements and then constructing vectors to join the centroid of the object to the centroid(s) of the hole(s) contained within it – see figure 2.2a [1]. In some cases common object outlines cannot be orientated easily unless there are strategically placed holes (see figure 2.2b). This may suggest that it is appropriate to include holes at some earlier stage in the manufacturing process *specifically* for the purpose of identification, provided their inclusion does not impair the normal operation of the part in any way.

Note that significant cut-outs or notches in the outline of an object can also be exploited for orientation purposes. This can be accomplished by constructing a circle centred on the centroid of the object which will only intersect the outline if a cut-out is present. In this case vectors are constructed to join the main object centroid and the intersections, and these characterise the orientation of the object (see figure 2.2c). Appropriate choice of the radii of a set of constructed circles may also allow this technique to be used for object recognition [1].

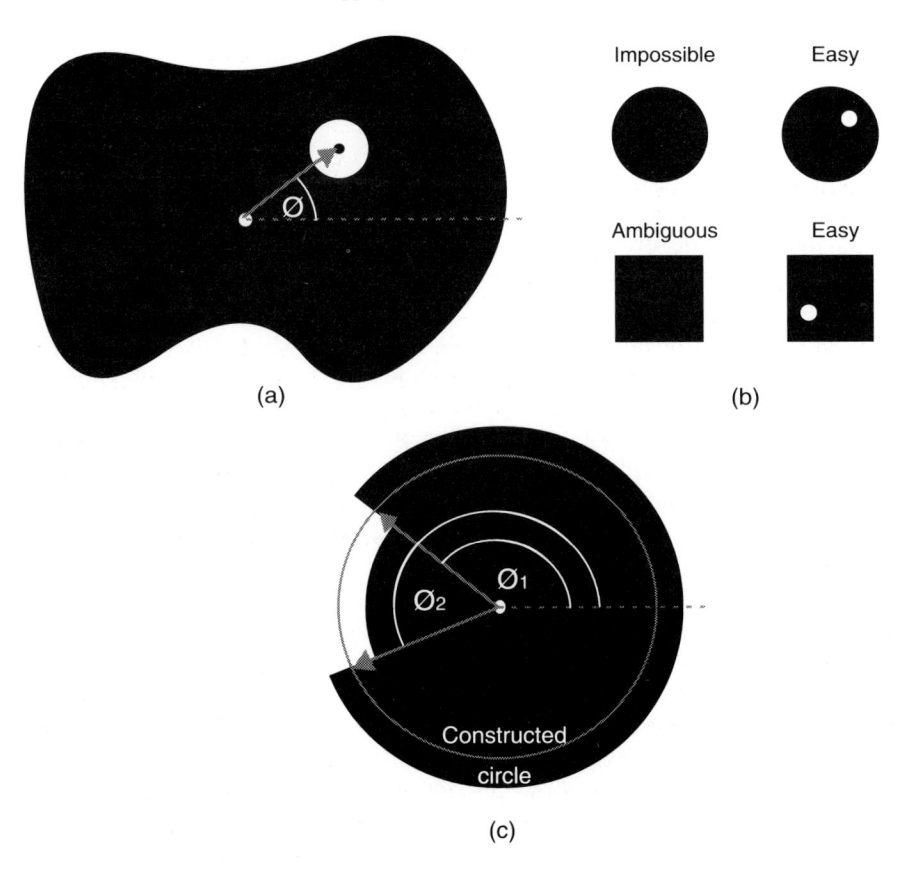

Figure 2.2 Orientation by inherent features: (a) object centroid/hole centroid method; (b) what a difference a hole makes!; (c) use of constructed circle, centred on object centroid. Note that in (c), the angle ($\phi_2 - \phi_1$) is characteristic of the object outline (for a particular radius of constructed circle, in the general case) and thus the methods may be used in object recognition as well as orientation

In some cases objects under investigation carry markings which can be exploited to simplify processing. An application that relies on this is the automatic fitting of tyres to rims at the Ford Motor Company's Cologne plant [2]. The aim is to minimise the static eccentricity of the wheel/tyre combination before balancing, by fitting the tyre so that its 'low' point coincides with the 'high' point of the rim. Fortunately, the tyre manufacturer determines the low point and marks it with a bright spot while the rim manufacturer identifies the high point with a dark adhesive label. Both of these markings contrast strongly with their local background, and are easily picked out in a binary image.

*Limitation in the range of objects*

Within the spectrum of machine vision applications it is common to have a strictly limited range of workpieces or prototypes, and knowledge of this can be exploited to focus attention on the most effective image acquisition and analysis techniques. In one particular application any one of twelve different types of truck axle is presented to a paint spraying robot in controlled orientation but random sequence [2]. Careful study of the silhouette of each of the twelve types allowed three 'windows' to be defined within the image, thus minimising the processing effort required. A simple hardware binary vision system measured the area of silhouette contained in each of the three windows and classified the axles accordingly (see chapter 7).

Optical character recognition (OCR) for printed English text is much simpler than for Chinese or Japanese text because there are only 26 letters and 10 numbers to be identified, rather than thousands of icons. In applications where only numbers are allowed the pattern recognition problem is considerably further simplified.

In some OCR applications contextual knowledge can also be very helpful. For example, in British car number-plate (licence-plate) recognition, the realisation that the letters Z, I and O are not allowed reduces the complexity of the problem because there are fewer symbols to identify. More importantly, however, the disallowed symbols are the ones which are most difficult to distinguish from the numbers 2, 1 and 0 respectively. However, this constraint was built into the system many years before machine vision systems were even thought of, simply because even the human visual system has great difficulty separating these characters under arduous conditions!

*Inherent positional limitations*

Within manufacturing industry the position of objects is often constrained by existing automation processes. As a very trivial example, the use of a conveyor-belt to transport objects requires that all objects be confined within the dimensions of that belt! This immediately defines the maximum field-of-view requirement (in one dimension) for a vision system placed at this point.

A more sophisticated example of field-of-view constraint concerns an application of vision systems to sealant spraying at the Austin–Rover Motor Company Ltd, Cowley (UK) [3]. In this case the objective is to identify the precise position of a car body which is hanging on an overhead conveyor, so that sealant can be accurately deposited on body seams. The position of the body is constrained by the conveyor system to ±30 mm. While this is clearly not accurate enough for precision spraying, it does allow effective use to be made of the available camera resolution (see chapter 3) such that key datum points on the car body can be reliably identified and located.

A less direct, but nevertheless important, use of positional information is made in another assembly operation at the Ford Motor Company at Saarlouis, Germany [2]. This time, a vision system is employed to determine the orientation of wheel hubs so that wheels may be automatically fitted. The solution implemented is critically dependent upon the prior knowledge that the degree of 'lock' on the front wheel steering is less than $\pm 5°$. Exploitation of this constraint, which is actually imposed by other, non-vision related assembly processes, means that a simple lighting and viewing arrangement can be used without prejudicing performance or reliability of the vision system.

In printed text, all alphanumeric characters are separated by a blank space. This simple fact is what makes OCR of printed text *viable* because it reduces the problem to one of identification of the limited number of symbols, as discussed above. Without this spacing, for example in cursive text, it becomes necessary to try and recognise not just symbols, but whole words or fragments of words. Obviously there are many more valid combinations of these than there are alphanumeric symbols, and the problem complexity becomes worse than that of dealing with Chinese icons.

However, even in recognition of printed English text, contextual information may be exploited to improve the confidence of decisions taken on the basis of textual input. A particularly clear example is reading of UK post-codes (equivalent to ZIP-codes in the USA) which adhere to a strict format of two letters (upper case) followed by one or two numbers, then a space. After the space comes one number and two more letters, e.g. BN2 4GJ. *A priori* knowledge of this format helps to eliminate some candidate matches in the recognition problem directly. In addition, the groups of characters can be checked for valid meaning to improve further the reliability of the recognition process: BN is the abbreviation for the Brighton postal district, whereas BZ may have no acceptable meaning, for example. This stategy is used to great effect in British postal sorting offices where around 70% of letter mail carries a type-written address and post-code.

### 2.1.2 *Effective imposition of constraints*

The imposition of constraints refers to effective control of parameters within the scene in order to maximise the information content of images, while minimising the complexity of extracting that information. The following four points may be considered:

*Control of object features*

In the design of a machine vision system the systems engineer should always be aware that it is but one component of a complete manufacturing system and that even small changes to other components could dramatically simplify its task. For example, it may be possible to arrange for salient features to be identified at minimal cost at a relevant point in the existing manufacturing processes. Although not imposed constraints in the application in question, the use of contrasting labels and paint spots in the tyre/rim assembly operation clearly illustrates the benefits of such an approach.

Another very familiar example of feature control yielding a simplified solution to the analysis problem is bar-coding of products. These markings are applied to almost all household products and many industrial ones. The bar-code reader can be regarded as a highly specialised vision system, but it is one which only works because of the existence of a predefined bar-code data format. In industrial applications one major problem with the bar code is the amount of space that it occupies. Printed circuit board (PCB) manufacturers, for example, would like to be able to add a bar-code label to identify their products but often find themselves unwilling to sacrifice sufficient surface area on the board. Philips Industrial Automation (Netherlands) have proposed a 'dot code' which addresses some of these limitations by greatly increasing the density of information storage [4].

Another example of effective addition of features exists in PCB manufacturing. This is concerned with the use of 'fiducial marks' which are added to PCB artworks to act as a datum point [5]. These marks consist of bold geometric patterns which can be easily located when reproduced on the final PCB. They are used by drilling machines, component insertion machines and surface-mount 'pick-and-place' machines etc., to 'calibrate-out' any dimensional or positional tolerance introduced during the PCB manufacturing process. Fiducial marks have been used for some time with camera-assisted optics and human operators, but the increasing throughput and precision demanded of modern electronic assembly automation are making automatic fiducial mark alignment relatively common. Proper design of the marks means that they can be accurately located using fairly simple binary vision techniques.

*Control of object position*

Here we refer to the control needed to position a piece-part in order to provide a better camera view and hence better image acquisition. If the tolerance in object position is known fairly accurately, effective use of camera resolution can be maximised since the optical magnification can be adjusted so that the field-of-view just encompasses the necessary area, and no more. Effective exploitation of *existing* positional constraints in this way has been discussed above, but in many instances it will be highly beneficial to introduce orientation and/or alignment equipment specifically to ease the vision problem.

For example, in mass production industries where a machine vision system is installed in association with a conveyor belt system, suitable mechanisms may be provided to feed the objects onto the conveyor belt one at a time so that overlapping objects, and hence complex images, can be prevented. The vibratory 'bowl feeder' is another form of mechanism which is commonly used to present discrete components to robots in a standard orientation. As such they represent an ideal location to install a vision system whose function is to monitor the integrity of such components.

A further development may be the simple repositioning of the object to offer the most appropriate surface for inspection with respect to the plane of the camera. Some applications use custom jigs and rotating platforms, etc. to achieve this *during* visual inspection operations. An extension of this thinking is the provision of more than one viewpoint to simplify the machine vision operation. This might be achieved by the use of other cameras or by splitting the field-of-view of a single camera (see figure 2.19). This approach has been exploited in a system developed for Volkswagen to inspect car brake assemblies. This whole application is a particularly good case study of the use of scene constraints in industrial machine vision and is documented in Hollingum [6].

*Control of lighting conditions*

In its simplest form, control of the lighting conditions implies the elimination of problems such as random fluctuations, unwanted gradation across the field-of-view, inadequate intensity, unhelpful colouration etc. which occur with ambient light. At a higher level it refers to the control of the position and intensity of the lighting source or sources so that desired information within the scene can be highlighted and unwanted information, which unnecessarily complicates subsequent analysis, is suppressed (see figure 2.3 and [6, 7]).

Specialised lighting can also be designed to:

- see into blind holes (e.g. using fibre-optic light guides, see figure 2.17)
- contrast holes with the surface of an object or reveal surface texture (e.g. 'grazing' illumination, see figure 2.5)
- eliminate specular reflections (see figure 2.4)
- view glass objects (see figure 2.7)

and so on.

*Structured lighting*

Structured lighting is a set of particular techniques whereby the engineer aims to make certain information *explicit* in a two-dimensional visual image by causing interference between special geometric patterns of light and the features of interest.

*Scene Constraints* 33

(a) Object with depth 'hidden' by inappropriate lighting

(b) Improved lighting reveals object as a cube with depth implied by shading

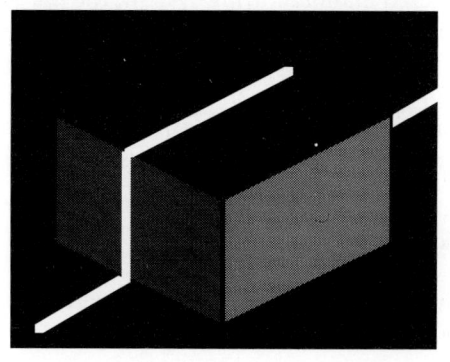

(c) Depth information explicitly revealed by distortion of a 'structured' light-stripe

Figure 2.3  Illustration of the effects of various lighting conditions

For example, the use of a fine narrow slit of light to illuminate sloping surfaces or surface indentations can produce relatively simple images for depth analysis (see figure 2.3c). Moiré fringes can also be used to determine depth information with a high degree of accuracy [8]. Polarised light is a potent tool for visualising stresses in optically-active transparent objects as well as the sugar content in some materials (see figure 2.11).

### 2.1.3  Systemic considerations

It is almost always easier and less expensive to solve a machine vision problem by manipulating scene constraints than to tolerate a loosely constrained scene and be forced to apply exotic image processing and image analysis software to overcome that lack of constraints. For example, if the salient features of an object can be recognised from its outline alone then imaging it as a high contrast silhouette through the use of back-lighting will undoubtedly be more efficient than applying edge detectors to a normal front-lit view. Naturally, not all manufacturing environments will allow a free choice of lighting conditions, or other constraints, but there is no excuse for not *investigating* the possibilities.

Where it is possible to consider the use of a vision at the planning stage of a manufacturing environment, careful systemic thought can eliminate vision systems that would otherwise be essential in retrofitted applications! 'Bin-picking' is a classic example of this. In many factories this task is necessary because isolated islands of automation store their intermediate output in bins, often in random orientation. It then becomes a very challenging problem for a vision system to look into such a bin and identify and locate individual objects for subsequent processing.

However, it is obviously valid to ask why the parts need to be stored in a bin at all in a well ordered manufacturing system; the use of 'flow-racks' as a buffer storage medium between processing cells may completely eliminate the need for bin-picking. If there must be medium- to long-term storage at least it should feature a high degree of order so that retrieval is simplified. For example, if components can be placed in matrix boxes [9] or in ordered layers on pallets [10], then relatively simple machine vision systems can be utilised.

## 2.2  Lighting techniques

When considering techniques for lighting the scene under consideration, three basic techniques can be considered.

### 2.2.1  Front-lighting

*Omni-directional illumination*

This is the most obvious form of lighting and it provides uniform, omni-directional illumination which can eliminate shadows on the objects within the scene provided it is suitably positioned and adjusted. Uniform illumination is usually easier to obtain if some kind of diffusing surface is incorporated. Figure 2.4 illustrates the use of a circular lighting source, e.g. a circular fluorescent tube, to provide omni-directional illumination. The light from the tube is prevented from directly illuminating the object by the inclusion of the light shields; the position of the diffusing screen, which in this case is almost hemispherical, ensures that scattered light from its inside surface will illuminate all aspects of the object, even surface concavities which would normally appear as dark regions.

*Directional illumination*

A variation in the front lighting technique provides for the use of directional rather than diffuse illumination to highlight surface texture

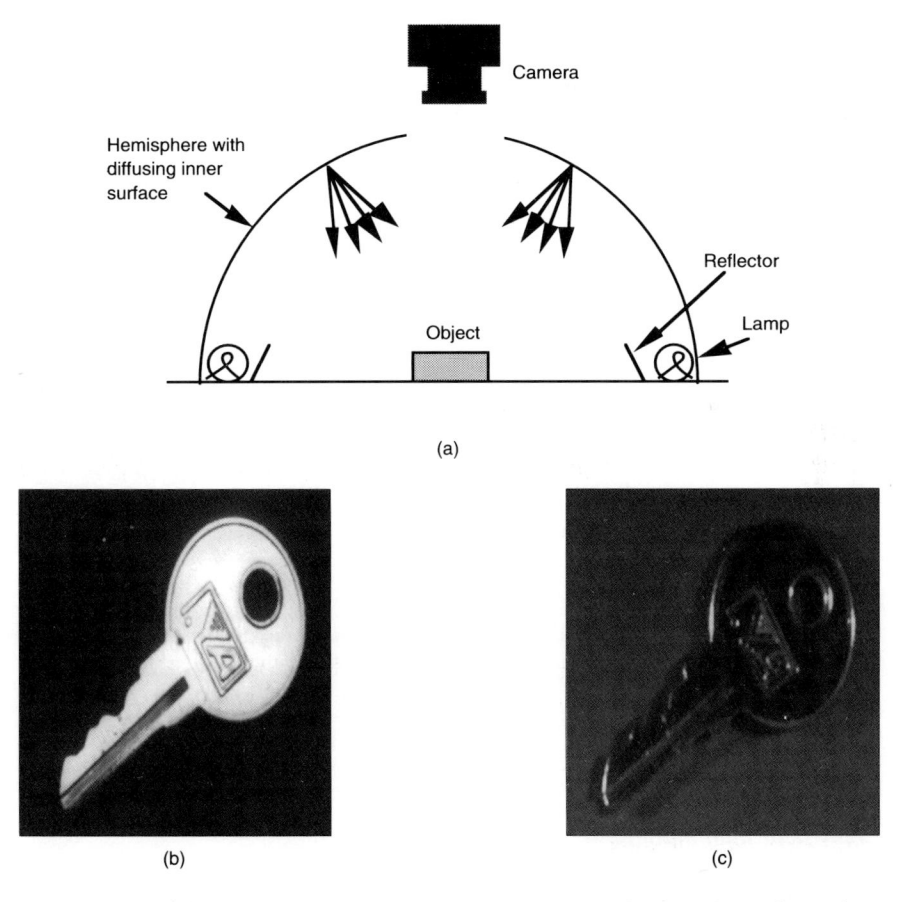

Figure 2.4 Diffuse omni-directional illumination: (a) implementation schematic; (b) key under directional illumination; (c) the same scene with uniform, diffuse, lighting

(see figure 2.5). This technique has been successfully used to identify wheel stud holes drilled into a wheel hub [2]. The holes reflect no light back to the camera, yielding a highly contrasting region on the bright image produced by the machined surface of the hub. This illumination technique simplifies subsequent image analysis but only if the number and surface area of the holes, and not the depth of the holes, are of importance.

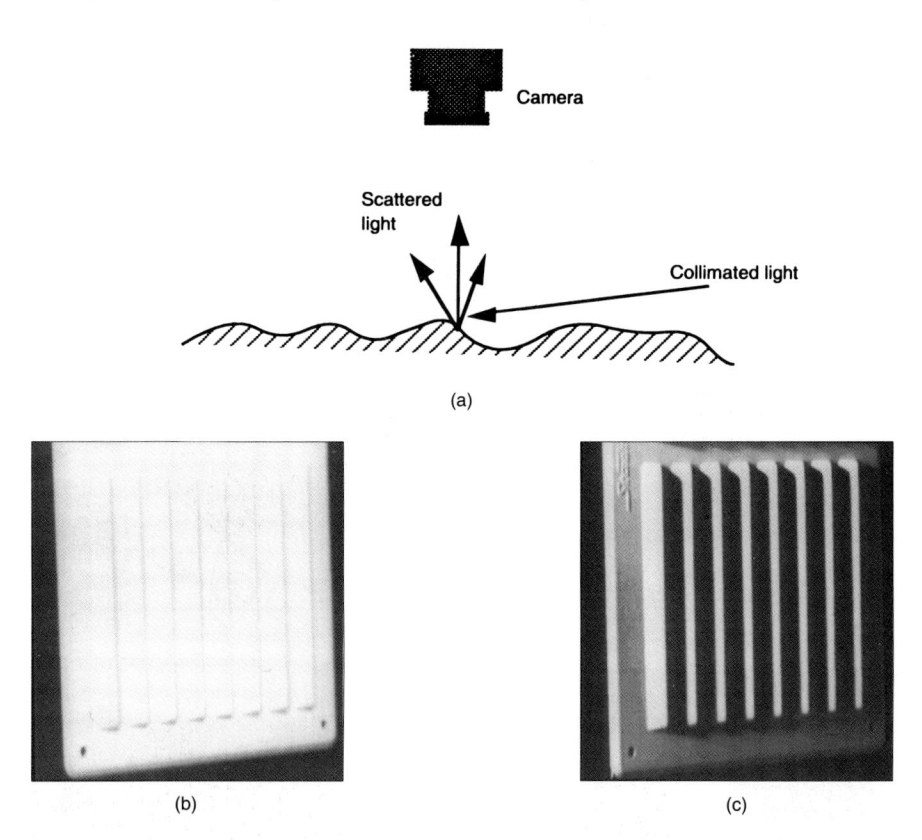

Figure 2.5 Directional lighting: (a) implementation schematic; (b) louvred surface with diffuse illumination; (c) the same scene under directional lighting

### 2.2.2 Back-lighting

*Light-field illumination*

In visual inspection applications which involve the shape analysis of stampings from sheet material, e.g. shoe leather parts, aluminium sheet assemblies, pastry shapes, etc., only the outline shapes of the objects are of interest. In this case the optimal way to illuminate the objects for maximum information acquisition is to view their silhouette.

The technique requires some form of light box with a diffusing surface at its front and the objects are viewed when positioned between the camera system and the light-emitting surface as illustrated in figure 2.6. Silhouettes may also be used to determine depth data from 3-D objects when they are rotated in front of the light source, *provided* that the object does not feature concavities.

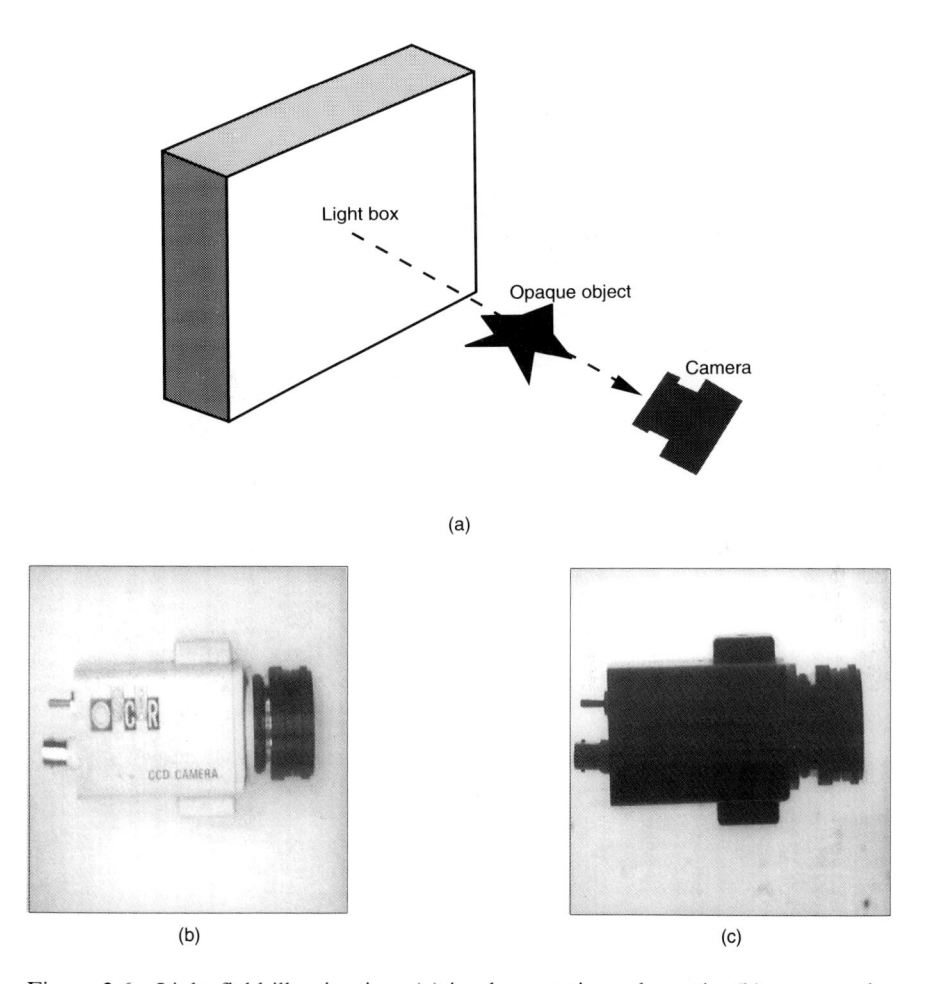

Figure 2.6  Light-field illumination: (a) implementation schematic; (b) scene under normal front-lighting conditions; (c) the same scene when back-lit. Note that in (c) all surface detail on the object is suppressed and a high contrast image which lends itself to thresholding to binary is produced. See also figure 4.8 in section 4.2.4.

## Dark field illumination

Dark field (or dark ground) illumination is the name given to a particular technique developed to view the edges of transparent or translucent materials. It can be used to highlight the walls of glass objects where the refractive properties of the material are used constructively to produce an image which is similar to the cross-section of the object. This technique is illustrated in figure 2.7.

Initially a mask is prepared which is made only slightly larger than the object in question (not necessarily the same shape as the object) and placed

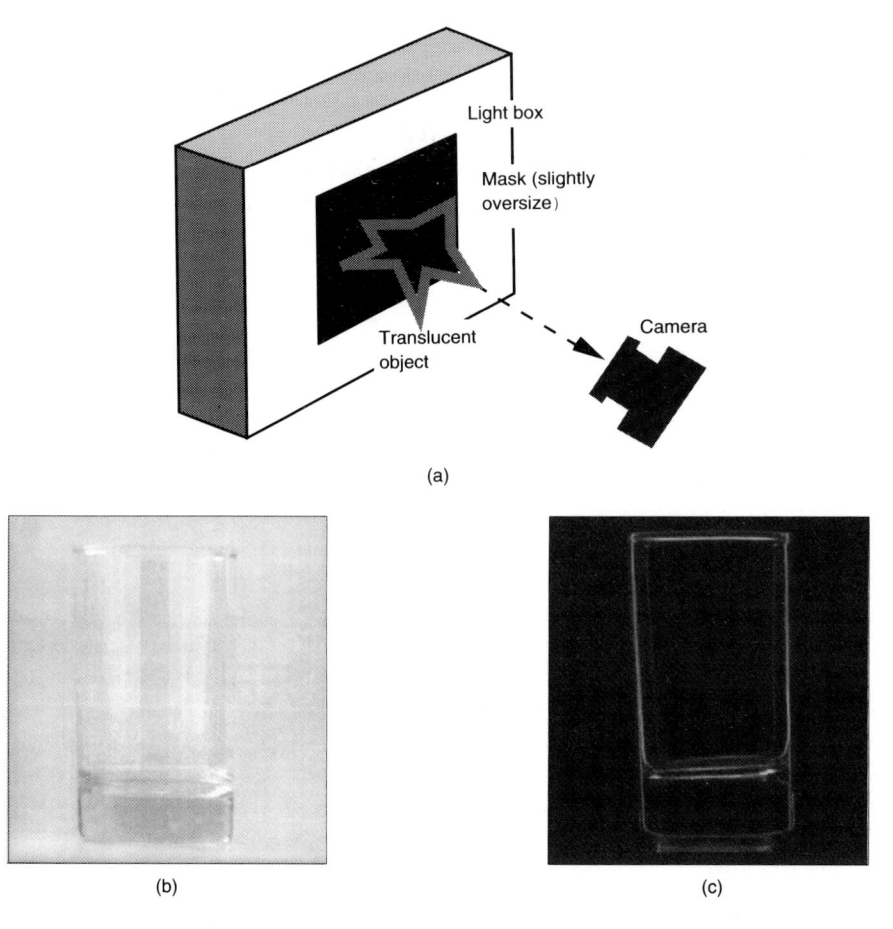

Figure 2.7 Dark-field illumination: (a) implementation schematic; (b) drinking glass under ambient light conditions on a light background; (c) effect of carefully controlled dark-field illumination. Note how the walls of this glass are sharply contrasted with the dark background. Flaws and cracks within the glass will also be highlighted.

on the surface of the light-box which is used to illuminate the object. No rays of light emerge from the light box directly behind the object because of the mask, however light rays from the rest of the light box diffusing surface pass through the object. When viewed from the front these rays are refracted by the translucent material at the edges of the object and appear in the camera field as bright areas on the dark mask area behind the object.

Varying the distance between object and mask (light box) will enable the optimal position for maximum brightness to be found. The size of the mask is important as it affects the quality and 'edge' continuity of the perceived image of the object under examination. With some objects, especially those with smoothly curved surfaces, best results are achieved if light is only emitted from a narrow 'annulus' around the mask.

The high contrast image that results from this and other lighting techniques will confuse the automatic gain control (AGC) circuitry featured by almost all video cameras. The AGC will continually strive to turn carefully contrived scenes into bland images with a mid-grey average brightness. Therefore it is critical that cameras used for such work can have their AGC circuits manually defeated.

### 2.2.3 Structured lighting

The concept of structured lighting was referred to in section 2.1.2 where it was suggested that the three-dimensional shape of an object and the orientation of its surfaces can be revealed by the use of strips of light incident upon the object under consideration. One-dimensional (1-D) and two-dimensional (2-D) object scanning techniques can thus reveal the true three-dimensional (3-D) nature of the object.

In principle a high-contrast pattern of light is directed onto the scene. The distortion of this light pattern as it is incident of the differing surfaces of the object gives rise to an image in which the surface relationships, both surface orientations and edge discontinuities, become more readily apparent to the viewer. In its simplest form this pattern of light consists of a single stripe projected in a plane set at a known angle to the camera normal (see figure 2.8).

*Polarised light*

The use of polarised light can be readily identified in engineering applications such as photo-elasticity and magneto-optics. Light can be regarded as a transverse electromagnetic wave and can be depicted as comprising two mutually perpendicular vectors, the electric vector and magnetic vector (see figure 2.9a).

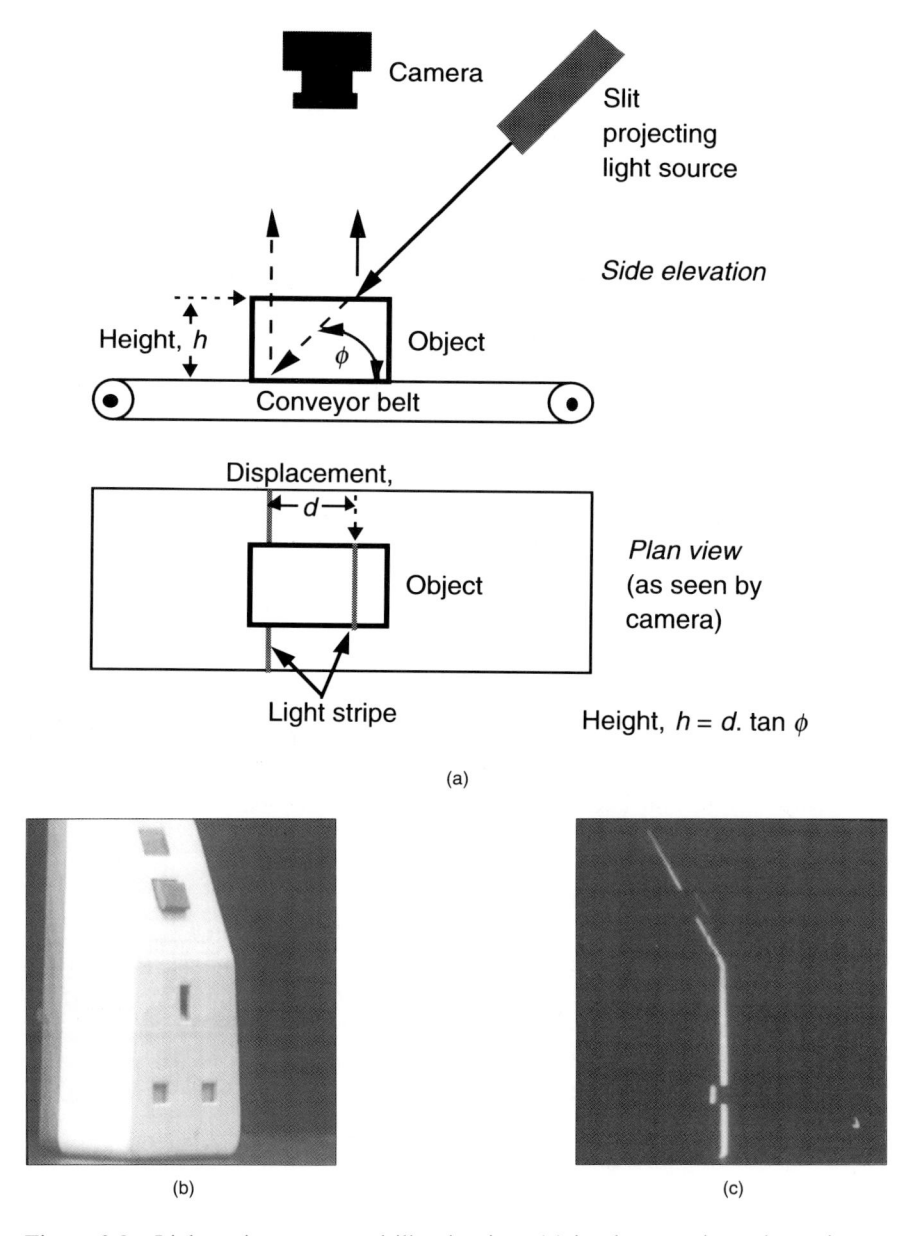

Figure 2.8 Light-stripe structured illumination: (a) implementation schematic; (b) scene under ambient light; (c) effect of light-stripe projection. Note that the sloping surface and holes are apparent. Depth data can be easily calculated, especially if $\phi = 45°$

(a) Perpendicular **E** and **H** fields

(b) Polarisation patterns

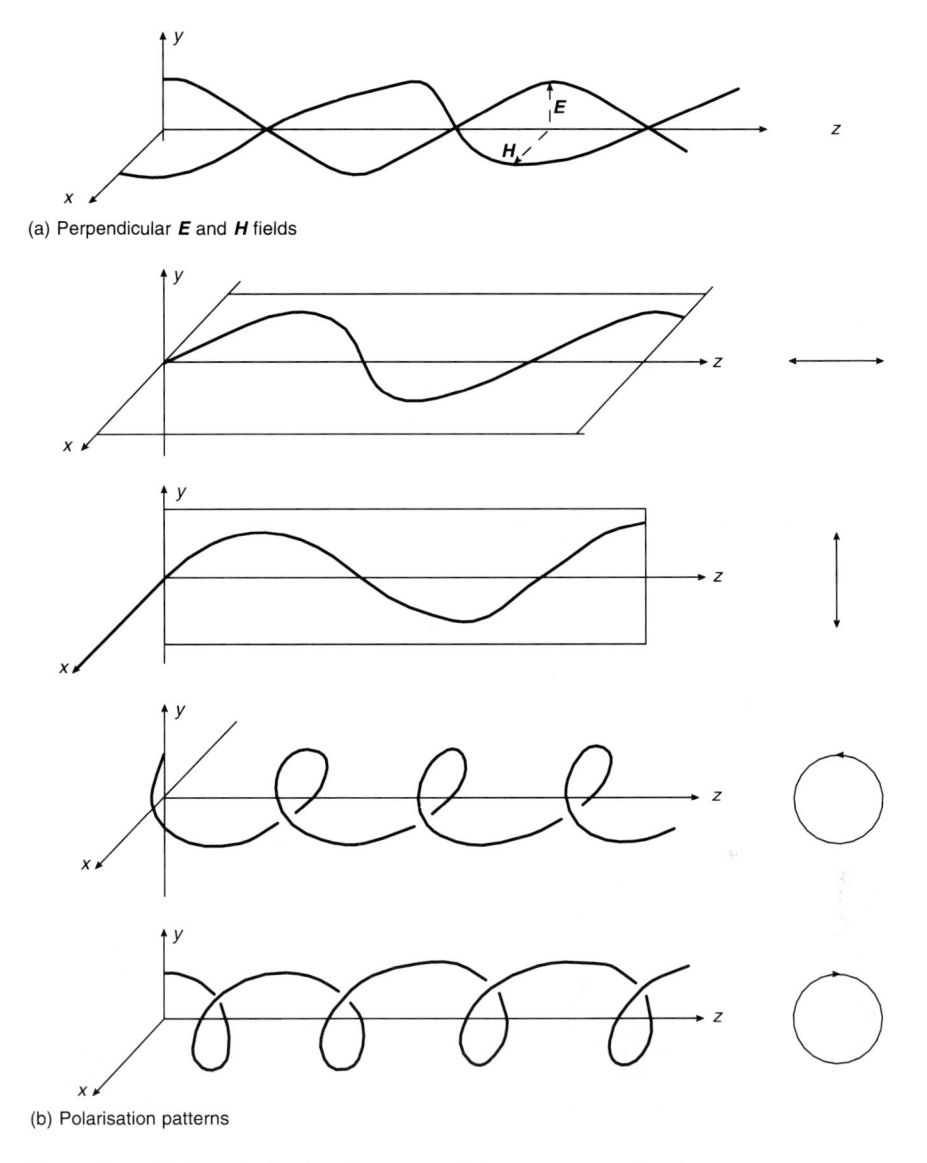

Figure 2.9 Light polarisation in terms of the transverse electric vector

Polarisation relates to the control of the plane of vibration of the transverse electric vector. Polarised light is light in which this transverse vibration has a simple pattern such as that illustrated in figure 2.9b. Here the light is assumed to be travelling along the $z$-axis, if the electric vibrations are horizontal the light is said to be polarised linearly and horizontally. When the transverse vibrations are vertical, the light is said to be polarised linearly and vertically. Some light beams have a circular sectional vibration pattern characteristic of either a right-hand or left-hand helical transverse wave. Ambient light is normally unpolarised or of random polarisation.

A polariser is a device that resolves the transverse vibrations into two components and selects one for transmission. Two most commonly found polarisers are:

(a) plastic polarising sheets which utilise the structural asymmetry within its atoms to create the transmission axis;
(b) polarising prism which utilises the polarising effects of oblique reflection within its two prism shaped calcite components.

Polarisers are well known in everyday use for reducing the amount of 'glare' from horizontal reflecting surfaces such as water and road surfaces since this is caused predominantly by light acquiring a horizontal vibration direction on reflection. By using a polariser with its transmission axis oriented vertically such glare is eliminated. Similar benefits may be obtained in industrial environments by polarising the incident light and placing a crossed polariser in front of the camera (see figure 2.10).

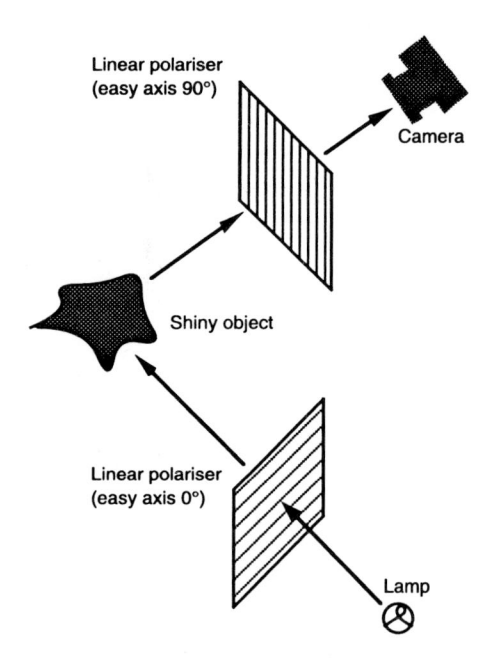

Figure 2.10  Glare-reduction crossed using polarising filters

Another well-known application of polarised light is in photo-elasticity where the effects of mechanical stress in a transparent material can be studied. Photo-elasticity is the creation of birefringence (i.e. light-splitting in two beams which travel at different velocities) in a transparent object when the object is strained. Here an incident light beam is plane polarised along one principal axis by passing through a polarising element.

The transmitted light beam is viewed through a second polariser oriented so that its transmission axis is set orthogonal to the incident beam. If the incident light beam is unaffected by transmission through the test piece then no light emerges from the second polariser. However, if the incident light beam has been depolarised owing to the stress effects in the material then the light passing through the second polariser gives a measure of the optical activity in the stressed material which can subsequently be related to the mechanical stresses present (see figure 2.11).

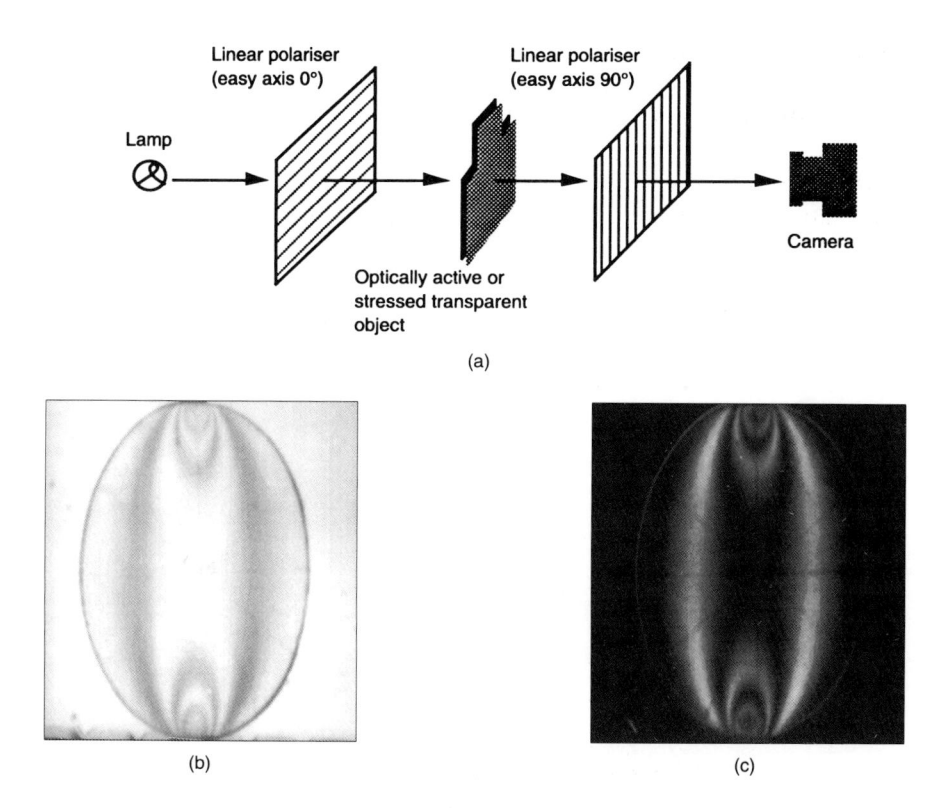

Figure 2.11  Polarised light in the assessment of photo-elasticity:
(a) implementation schematic; (b) raw image produced by polarisers; (c) enhanced image achieved by subtraction of background image (see section 4.6). The photographs reveal lines of stress in a transparent plastic disc placed under compression.

## 2.3 Lamps

### 2.3.1 *General considerations*

To ensure that high quality images will be acquired, it is essential to have good illumination of the scene. Parameters of the light source include intensity, directionality and spectral distribution, and these may influence the system designer in the selection of an appropriate light source.

Selecting the most appropriate lamp for a particular industrial task in hand requires consideration of these and other parameters, e.g. wavelengths required to match the sensors in use, area to be illuminated, reflectivity of objects in the scene, space available within the locality of the objects, heating effects, etc.

Most lamps radiate both ultra-violet and infra-red light in addition to the visible wavelengths and each has its own continuous spectral distribution, as shown in figure 2.12.

The spectral distribution of a lamp is often summarised by its 'colour temperature' (K). This corresponds to the temperature of a heated surface that would give off light of the same colour as the lamp. The higher the colour temperature the 'bluer' the light. (Curiously, sources with a low colour temperature which give off a pinky-orange light are often described as 'warm white' while bluer sources may be described as 'cool white'!). Some commonly encountered colour temperatures are given in table 2.1.

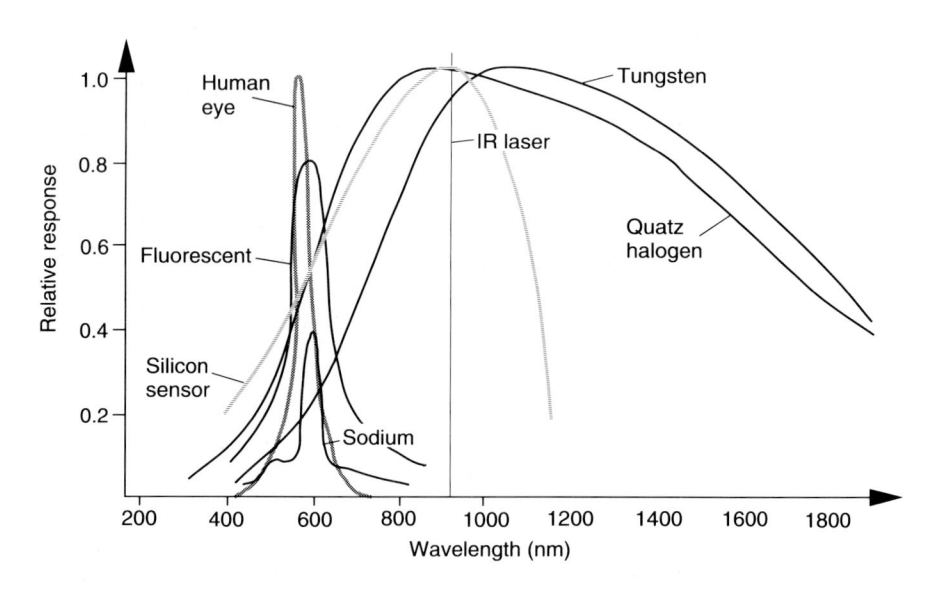

Figure 2.12 Spectral emissions of some common light sources

Table 2.1  Colour temperature of some common light sources

| Type of illumination | Colour temp. |
|---|---|
| Tungsten lamp for home use | 2 800 K |
| Tungsten lamp for photographic use | 3 200 K |
| Quartz–halogen lamp | 3 200 K |
| Blue lamp for photographic use | 5 000 K |
| Fluorescent lamp | |
|   Warm white | 3 500 K |
|   White | 4 500 K |
|   Daylight type | 6 500 K |
| Cloudy sky | 7 000 K |
| Clear blue sky | 10 000 K |

The output from a light source may be expressed in *candelas*, *lumens*, or *watts*. The candela (cd) is defined as the luminous intensity, in a given direction, of a source emitting 1/683 watts per steradian at a wavelength of 555 nm. The reason why the wavelength of 555 nm (green light) is significant is that this is the peak of the response curve of a typical human eye. Some units take this response curve into account because they are concerned with the brightness perceived by a human observer: these are called 'photometric' units. Other units are only concerned with the amount of light energy emitted or received and thus assume a flat response: these are called 'radiometric' units.

The luminous power, unit lumen (lm), is defined as 'the luminous flux in a steradian from a point source of one candela (cd sr)'. Therefore it is linked to the radiometric radiant power, unit watt (W) by the definition of the candela, given above. Some other useful units and their definitions are shown in table 2.2.

Table 2.2  Photometric and radiometric units

| Photometric term and unit | Radiometric term and unit | Definition |
|---|---|---|
| Illuminance (lm m$^{-2}$, or lux) | Irradiance (W m$^{-2}$) | The flux incident on unit area of a surface |
| Luminous intensity (lm sr$^{-1}$, or cd) | Radiant intensity (W sr$^{-1}$) | Power radiated by a point source into a unit solid angle |
| Luminance, or brightness (lm m$^{-2}$ sr$^{-1}$, or cd m$^{-2}$) | Radiance (W m$^{-2}$ sr$^{-1}$) | Power radiated from unit area into a unit solid angle |

Lighting levels or lamp output may also be specified in old imperial units. The foot-candle is a unit of illuminance equivalent to 10.76 lm m$^{-2}$ (lux). The foot-lambert (ft L) is a unit of luminance equivalent to 3.425 cd m$^{-2}$.

An indication of the illuminance of a surface for naturally occurring illumination is given in table 2.3.

Table 2.3 Naturally occurring illuminance values

| Lux rating | Typical scene description | |
|---|---|---|
| 100 000 | • Clear sky, mid-day, under sunlight | (100 000) |
| | • Clear sky, 10.00 am, under sunlight | (65 000) |
| | • Clear sky, 3.00 pm, under sunlight | (35 000) |
| | • Cloudy sky, mid-day, under sunlight | (32 000) |
| | • Cloudy sky, 10.00 am, under sunlight | (25 000) |
| 10 000 | | |
| | • By the windows during the afternoon | (3 500) |
| 2 000 | • Cloudy sky, one hour after sunrise | (2 000) |
| 1 000 | • Clear sky, one hour before sunset | (1 000) |
| 600 | • Counters at department stores | (500–700) |
| 500 | • Bowling alley | (500) |
| | • Office under fluorescent light; library | (400–500) |
| | • Direct light of torch at 1 m distance | (250) |
| | • Streetlights at night | (150–200) |
| 100 | | |
| 15 | • Cigarette lighter at 30 cm distance | (15) |
| 10 | • Candlelight at 20 cm distance | (10–15) |

Lamp power supply considerations, i.e. DC or AC supply input, may affect the quality of the captured image. Most AC light sources produce some ripple in their output and the image sensor may pick this up if the exposure time is shorter than the period of the supply frequency.

Finally, safety considerations require correct cooling of lamps and protection of operators from any high voltage, high temperatures or harmful radiation (e.g. ultra-violet).

### 2.3.2 Summary of characteristics of some common sources

*Tungsten filament lamps*

Range of power ratings up to 500 watt.

Various shapes available; cylindrical and bulbous shapes; with integrated lens; physical size 50–100 mm diameter.

Provides a non-uniform point source of moderate efficiency. Generally used in household lighting, spotlights, vehicles, etc.

*Quartz–halogen lamps*

Range of power ratings up to 1500 watt.

Often used in low voltage DC form.

Usually available in cylindrical form with integral reflectors; envelope temperature around 350°C.

Provides a non-uniform point source of moderate efficiency; more efficient than tungsten filament lamps.

Generally used in car headlights, slide projectors, shop display lighting and security applications.

*Fluorescent tubes*

Range of power ratings up to 150 watt (low pressure).

Lengths from 30 to 2400 mm.

Provides even, linear distribution; low temperature; high efficiency; long life.

Generally used in office and shop lighting.

*Sodium discharge lamps*

Selected power ratings only.

Available in cylindrical form; requires time to reach optimum operating conditions.

Provides a characteristic yellow light peaking around 590 nm.

Used in street lighting, laboratories or special industrial applications.

Incandescent lamps are readily available and commonly used in machine vision systems, however a high proportion (75%) of their energy is converted into heat with some being radiated in the infra-red part of the spectrum. Under-running the lamps to increase their life tends to shift their spectral distribution further towards the infra-red.

Quartz–halogen lamps operate at higher temperatures (300–900°C) to ensure the necessary chemical reactions occur. Some lamps employ integral dichroic reflectors that direct visible light energy forward while allowing infra-red radiation to pass backwards through the 'reflector'. This helps to keep delicate objects cool while allowing bright illumination.

Discharge lamps operate by the electrical excitation of the atoms of a gas, resulting in the release of radiation as they return to their original state. The spectrum of the radiation is characterised by the gas used, particularly when the gas is at low pressure. At high pressure the spectral lines are broadened. In some lamps the ultra-violet radiation from the discharge is used to irradiate phosphors which re-radiate (fluoresce) at the longer visible wavelengths. Most lamps identified here radiate both ultra-violet and infra-red in addition to visible wavelengths.

Note that the light output of many types of lamp falls off during their life. This means that routine calibration of optical systems is necessary in applications where the absolute level of light is critical. This is common in industrial machine vision systems.

## 2.4 Optics

Within the optical front end, an image of the object under consideration has to be formed on the surface of an image sensor by a suitable lens system. The size and resolution of the sensor will affect the design of this optical lens arrangement.

### 2.4.1 Basic lens formulae

In order to calculate the correct focal length of lens required to achieve a given magnification, or to predict the effect of a change in focal length, the simple lens formulae can be used:

$$1/f = 1/u + 1/v \tag{2.1}$$

and $\quad m = v/u \tag{2.2}$

rearranging $f = um/(m+1) \tag{2.3}$

also $\quad n = f/d \tag{2.4}$

where $f$ is the focal length of the lens, $u$ the object-to-lens distance, $v$ the image-to-lens distance, and $m$ the magnification defined as image size divided by object size. $n$ is the numerical aperture (or 'f-number') of the lens (measured in 'f-stops'), and $d$ is the diameter of the effective lens aperture. Figure 2.13 shows some illustrations of how a simple lens can be used to form a 'real' image – i.e. one that exists in space such that it can be seen on a screen. The image is always inverted, and $u + v$ is always greater than or equal to $4f$.

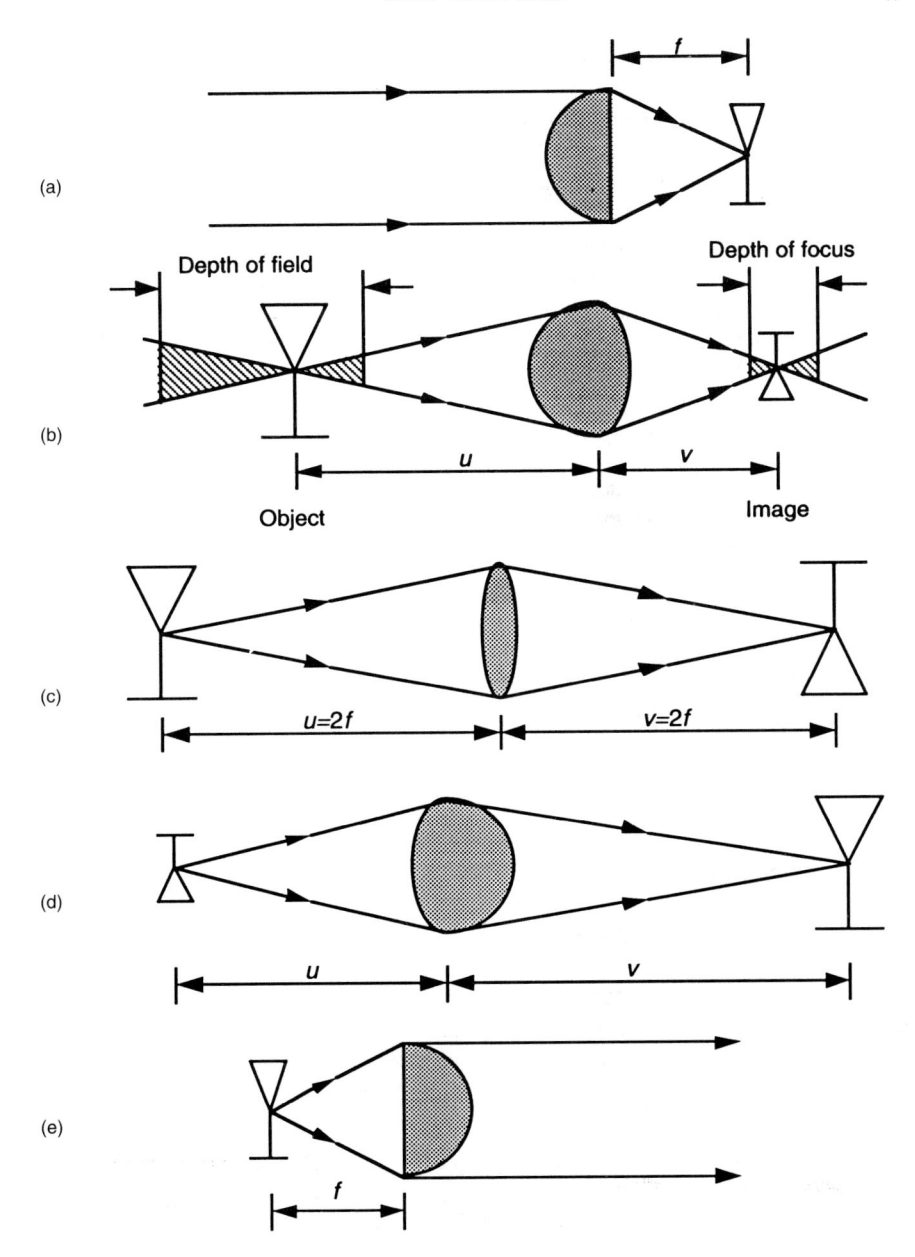

Figure 2.13  Basic image formation: (a) image of a distant object;
(b) demagnification; (c) unit magnification; (d) magnification;
(e) collimation of a light source or distant image forming. After
Batchelor *et al.* [7]

### 2.4.2 Focal length

The focal length of a lens is chosen to obtain a level of magnification which makes the best use of the resolution of the image sensor. Figure 2.14a, b shows how the image size can be increased by choosing a longer focal length lens (or vice versa).

Alternatively the object-to-lens distance can be adjusted to modify the image size – see figure 2.14c. It should be noted that the same size image can be achieved with a given size object by moving closer with a shorter focal length lens, or staying further away with a longer focal length lens. Each method has advantages and disadvantages. The short focal length method makes the optical path nicely compact but the 'wide-angle' view obtained may produce unacceptable distortion of the scene perspective – see figure 2.15. The long focal length approach eases this problem, but requires a long optical path and gives a shallow depth of field.

Whatever method is used to adjust the size of the image, the magnification achieved must be incorporated into a calibration coefficient so that the dimensions of the real object can be related to the number of pixels

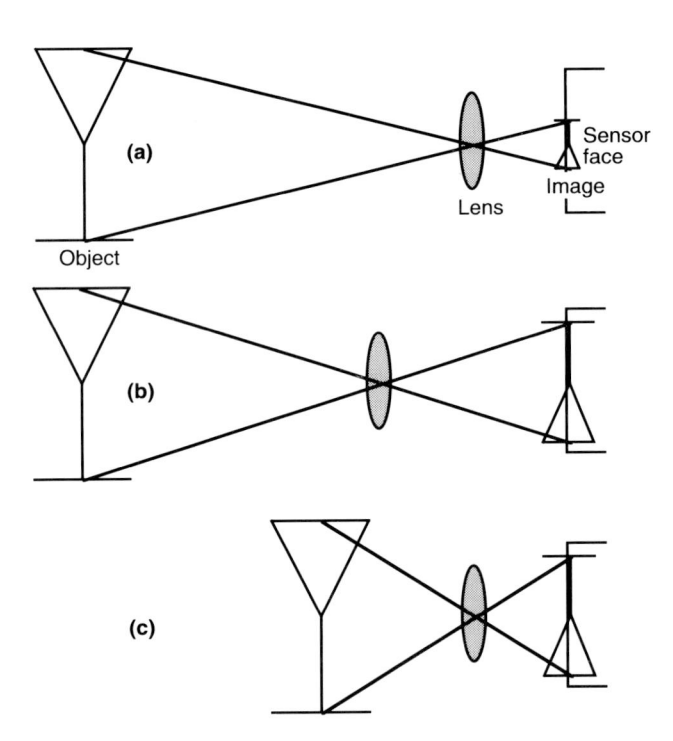

Figure 2.14 Effect of changing focal length and object-to-lens distance – after Batchelor *et al.* [7]

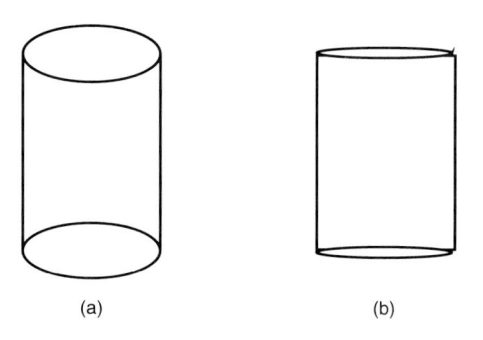

Figure 2.15 Views of a glass cylinder: (a) with a wide angle (short focal length lens); (b) with a longer focal length lens

measured on the screen. Note: it is quite common for the aspect ratio (height-to-width ratio) of screen pixels to be something other than 1 : 1. Therefore it may be necessary to have different calibration coefficients in the $x$- and $y$-axis.

### 2.4.3 Depth of focus and depth of field

The 'depth of focus' is the lateral movement on the image-to-lens distance allowable before the image begins to go out of acceptably sharp focus – see figure 2.13b. Therefore, in designing an optical system, it determines the tolerance allowable in the position of the image sensor, relative to the lens.

However, when using the system the depth of focus (or focus tolerance) can be used to accommodate object movement in the scene, or objects with significant depth. Assuming the sensor is fixed relative to the lens, then the object may move by an amount equal to the depth of focus divided by the magnification; this is termed the 'depth of field'. Note: the depth of field is not symmetrically disposed about the optimum focus position – refer again to figure 2.13b.

Both of these parameters are affected by the numerical aperture of the lens, $n$. Closing the aperture, or 'stopping down the lens', increases the depth of field etc. Note that the standard scale of f-numbers is 1.4, 2, 2.8, 4, 5.6, 8, 11, 16 etc., where each increase halves the amount of light passed by the lens.

### 2.4.4 Lens aberrations

All real lenses suffer from defects or 'aberrations', which cause sharp points and edges in the image to become blurred. When regular patterns occur, this blurring causes light and dark parts of the image to be smeared together, reducing contrast.

For monochromatic lighting the following lens aberrations can be identified:

*Spherical aberration*

Light rays passing centrally through a lens and those passing off-centre are brought to a focus at different distances from the lens. The resultant image appears blurred.

*Coma*

When light passes obliquely through a lens the off-centre rays come to a focus on one side of the central ray position. This produces comet-shaped images of point objects.

*Astigmatism*

In an image of a spoked wheel centred in the frame, spokes at different angles come to focus at different distances from the lens.

*Field curvature*

The surface of best focus is domed. Focus therefore varies across the image if a flat image sensor is used.

*Distortion*

Straight lines in the object plane become bent in the image plane. This is a particular problem with wide-angle lenses.

If white light is used then additional 'chromatic' aberrations occur:

*Longitudinal chromatic aberration*

Light of different colours comes to a focus at different distances from the lens, causing coloured fringes at the image plane.

*Transverse chromatic aberration*

Light of different colours passing obliquely through the lens strikes the image plane at different points, causing chromatic differences of magnification.

The limiting resolution of a lens is often specified as the number of line-pairs per millimetre (lppm) that it can accurately reproduce at its nominal focal plane. This amounts to a spatial frequency bandwidth limit for the lens. The figure is usually quoted at the optical axis of the lens and falls off towards the edge of the field-of-view.

All of the above aberrations get worse as the lens aperture is increased. Similarly, depth of field is reduced by wide apertures. Therefore, for best results from a lens it should be stopped down as much as possible. Unfortunately this cuts down the light falling on the sensor and thus reduces the sensitivity of the system. Typical industrial scenes feature a chronic shortage of ambient light, making the control of scene illumination even more vital.

Another, more fundamental phenomenon, known as the 'diffraction limit' comes into play at small apertures. Therefore, general purpose lenses usually offer their optimum resolving power, typically about 100 lppm, at a moderate aperture of around f8. However, it has to be noted that the lppm specification for lenses intended for general imaging is quoted at the limit of contrast which is acceptable to *human vision*. The contrast required for machine vision systems is much higher, and this therefore considerably downgrades the resolving power of the lens.

The message for the machine vision system designer, therefore, is that the quoted resolution of an image sensor in terms of pixels can only be realised if the spacing of the pixels in the focal plane is significantly greater than the line spacing implied by the lppm figure for the optics. Great care must be taken when matching lenses to modern ultra-high density CCD image sensors (e.g. '1/3 inch' format), which feature very small photosites.

## 2.4.5 *Practical lenses*

When selecting lenses for use in a machine vision system it is unlikely that a simple, 'thin' lens of the type described above will provide an adequate solution. Typical commercially available lenses are more complex, consisting of a number of lens elements in different glasses and with different curvatures. They are designed so that the combination of elements corrects for aberrations as much as possible for a given size and cost. Such lenses have considerable thickness (see figure 2.16).

Their focal lengths are measured from two planes called the 'principal planes', and labelled $H_1$ and $H_2$ in the diagram. The lens acts as if it were a thin lens placed at the position of $H_1$ when considered from the object side, and placed at $H_2$ when considered from the image side. The lens formulae can then be applied as though the space between the planes did not exist.

All practical lenses feature a minimum focusing distance. This is the minimum distance between the front of the lens and the object ($u$) at which a

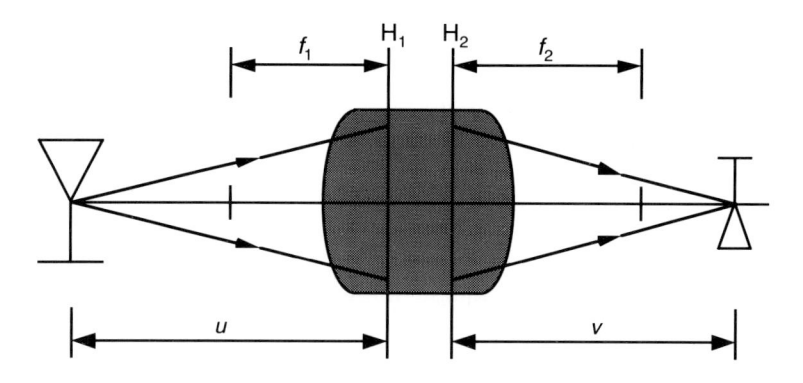

Figure 2.16  Equivalent ray diagram of a practical lens

sharply focused image can be produced. The distance is indicated by the lower limit on the lens focusing ring. If extra magnification is required from a lens and this is prevented by the minimum focusing distance then an extension tube may be inserted between the lens and the sensor. This increases $v$, which allows $u_{min}$ to be reduced according to the lens formula (2.1).

Many types of high quality commercial camera lenses can be utilised in industrial machine vision systems. Lenses intended for 35 mm, cine and TV cameras, photographic enlargers, and microscope objectives can all prove useful. Ideally one should choose a lens which is working at or near its optimum magnification. As noted above, lenses need to be carefully matched to particular sensors. Therefore care should be taken to select one which will allow a high quality image to be formed in the image plane which is sufficiently large to cover the whole area of the sensor being used.

The use of commercial 'zoom' lenses, with variable focal length as well as aperture, often helps to speed up the design and development phase of machine vision systems. However, their complex construction makes them bulky, costly and more prone to aberrations than their fixed focal length counterparts. Consequently, they are seldom used in the final dedicated application.

It has been noted that some types of lamp emit considerable quantities of infra-red radiation. Normal lenses will be unable to bring infra-red radiation to a focus in the same plane as the visible light for which they are intended. If the image sensor is insensitive to infra-red this may not be a problem, but all silicon sensors are strongly sensitive. Thus aberrations may be unavoidable, or even worse, the sensors photo-sites may become saturated. The simple solution is to remove the unwanted infra-red with a high-pass filter in front of the lens. Note that there are some applications where infra-red illumination is deliberately exploited, for example in electronic PCB inspection, or security, but then *visible* wavelengths are explicitly filtered out.

### 2.4.6 Specialised optics

Most optical systems are carefully designed to form a simple (de)magnified image of a scene with the minimum of distortion. However there are some cases when specialised optical systems must be used which deliberately distort the image in some way.

For inspection of blind holes etc., the use of an ultra-wide angle, or 'fisheye', lens with an angle of view of 150° or more may be appropriate – see figure 2.17. Although the images are severely distorted, circular arcs which are concentric with the lens remain circular and radial lines remain radial. Cheap examples of these lenses may be obtained by dismantling the domestic security 'peep-holes' often fitted through doors.

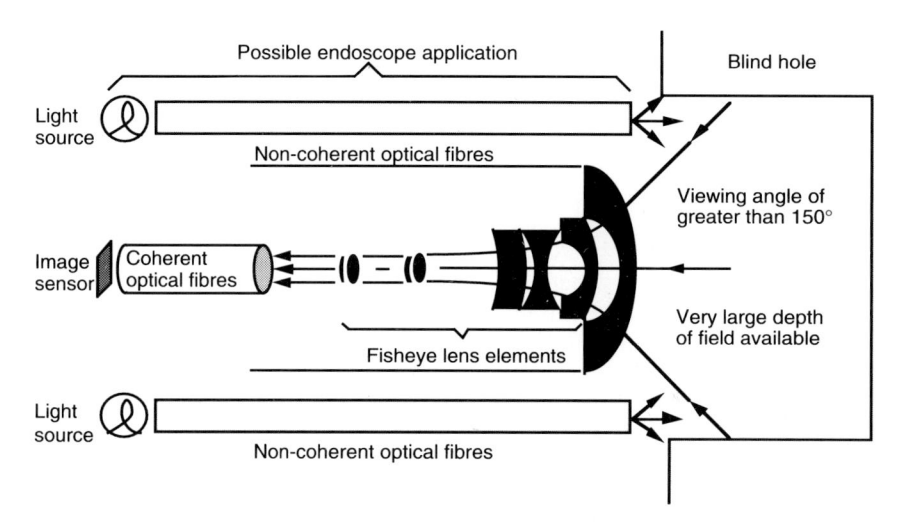

Figure 2.17   Fisheye lens combination in blind hole inspection

Fisheye lenses are often combined with a 'ring-light' to ensure that light is available over the wide area being observed, especially when working in deep holes. The lens/light combination may be built into an 'endoscope' where the image is delivered by a coherent bundle of optical fibres to a remotely located camera. A set of non-coherent fibres may also be used to deliver light from a remote source, and this ensures cool lighting for delicate subjects (see figure 2.17). The fibre optic bundle provides a flexible light medium although there is a limit to the amount of mechanical stress that such a bundle can withstand. Therefore small radius bends and continuous flexing should be avoided.

Where the application demands light collection over a large area the use of a Fresnel lens is appropriate (see figure 2.18). Here the bulk of the lens material is removed but the surface shape retained. These lenses can only be used for the limited range of object–image positions for which they were designed and for positions near the axis. For other positions the rays in the lens will not be nearly parallel to the axis and some will be affected by the discontinuities in the surface resulting in a deterioration of the image. The lenses are produced by moulding either glass or plastic. Common examples are those found in overhead projector light diffusers, reversing aids for cars and commercial vehicles and large area magnifying lenses for map reading etc.

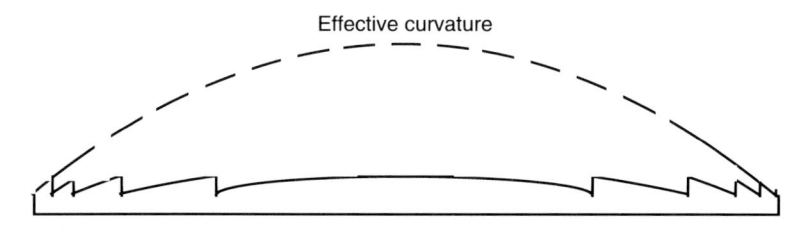

Figure 2.18   Schematic diagram of a Fresnel lens

There are many cases where several parts of the scene to be viewed are important, but are well separated, or even in different planes. In the former case, best use of the sensor resolution may be made by using plane mirrors to split the image so that only the regions of interest are visible to the sensor – see figure 2.19a. In the latter case, mirrors may also be used to provide simultaneous views of an object from two or more viewpoints – see figure 2.19b.

Note that mirrors should preferably be the more expensive front-silvered types to avoid double images. Also in complex applications it may be necessary to ensure that the lenses used to form the final image have good depth of field because of the unequal path lengths of images from different parts of the scene.

## 2.5   Achieving robust solutions

Even after every effort has been made to generate the best quality images, encapsulating the maximum amount of relevant information, there are a number of basic, practical issues which system designers should consider.

Never assume that it will be possible to achieve as good a result with scene constraints on production systems as those achieved on the prototype,

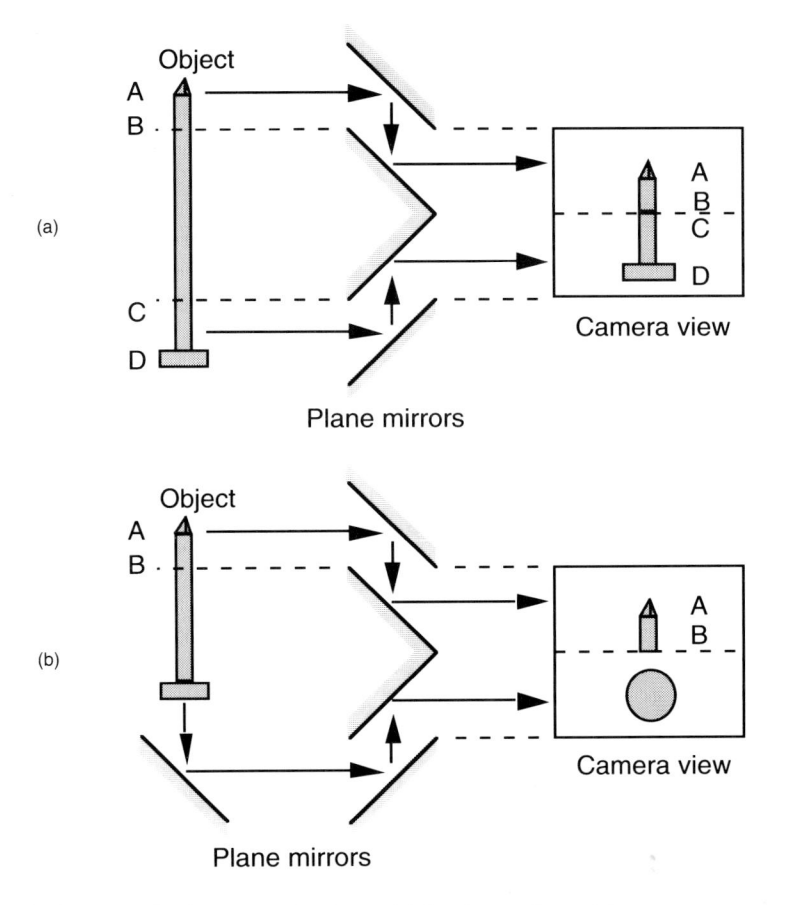

Figure 2.19  Use of mirrors to obtain multiple views of an object

because the factory floor is a very different environment from the laboratory. Therefore it is necessary to make all stages in the machine vision system sufficiently 'robust' to cope with less than ideal images.

Always take steps to ensure the long term *accuracy*, *reliability* and *survivability* of the machine vision system front-end. Accuracy is maintained by incorporating a *calibration* system. Such a system will permit regular accuracy checks, eliminate drift problems, and should be capable of operation without taking the system out of commission.

Reliability and survivability are assisted by avoiding dirt on the camera lens, by controlling the ambient temperature and by considering appropriate radiation hazards. The lens can be kept clean by means of a curtain of air, or occasional air blasts, and in extreme temperatures the camera can be protected by a water-cooled jacket and/or dichroic mirrors, which reflect long-wave infra-red radiation but pass visible light. Nuclear radiation quickly causes conventional lens coatings to become 'browned', therefore specialist lenses must be used in such circumstances.

Reliable and cost-effective operation can be further enhanced by regular monitoring of failure and reject rates etc., during the normal operation of the system. It is also necessary to be aware of the possibility of damage to equipment wreaked by the human labour-force. This can either be due to ignorance of its sensitive nature or through more deliberate Luddite tendencies! This can only be avoided by proper and careful consultation with the workforce and appropriate training programmes.

## References

1. P. Kitchin and A. Pugh, 'Processing of binary images', in A. Pugh (Ed.), *Robot Vision*, pp. 21–42, IFS (Publications), Bedford, 1983.
2. H. Geisselmann *et al.*, 'Vision systems in industry: application examples', *Robotics*, **2**, pp. 19–30, Elsevier Science Publications, (North-Holland), Amsterdam, 1986.
3. C. Loughlin, 'Automatix provides the seal of success for Austin Rover', *The Industrial Robot*, **14**(3), pp. 145–148, IFS (Publications), Bedford, 1987.
4. R. M. Pluta, *Dot Code*, Philips Industrial Automation, Eindhoven, Netherlands.
5. J. Edwards, 'Machine vision and its integration with CIM systems in the electronics manufacturing industry', *Computer-Aided Engineering Journal*, Feb. 1990, pp. 12–18.
6. J. Hollingum, *Machine Vision: The Eyes of Automation*, pp. 94–98, IFS (Publications), Bedford, 1984.
7. B. Batchelor, D. Hill and D. Hodgson, *Automated Visual Inspection*, IFS (Publications), Bedford, 1985.
8. J. D. Todd, 'Advanced vision systems for computer-integrated manufacture: Part II', *Computer-Integrated Manufacturing Systems*, **1**(4), pp. 235–246, Butterworth, London, 1988.
9. J. Henry and C. Preston, 'Implementing machine vision: an IBM case study', *Sensor Review*, **8**(2), pp. 73–78, IFS (Publications), Bedford, 1988.
10. J. Hermann, 'Pattern recognition in the factory: an example', in *Proc. 12th ISIR*, IFS, Bedford, UK, 1982.

# 3 Image Acquisition

The general aim of the image acquisition sub-system can be summarised as:

*'The transformation of optical image data into an array of numerical data which may be manipulated by a computer, so the overall aim of machine vision may be achieved'.*

In order to achieve this aim three major issues must be tackled – these are representation, transduction (or sensing) and digitisation.

## 3.1 Representation of the image data

The method of representation that is chosen should fulfil two requirements:

(a) it should facilitate convenient and efficient processing by means of a computer;
(b) it should encapsulate all the information that defines the relevant characteristics of the image.

It is only after due consideration of the method of image representation that techniques for transduction and digitisation can be specified.

### 3.1.1 *What are the issues in image representation?*

The conventional optical sub-system will deliver a continuous two-dimensional function, $f(x, y)$, where the value of the function at any pair of spatial coordinates is the intensity of the light at that point.

In order to fulfil requirement (a) above, it is necessary that the continuous function should be 'quantised' so that it can be represented as an array of numbers. This of course requires that the representation is only an approximation to the original image, but that is the price that always has to be paid for the convenience of using numerical computers to process analogue data.

Two forms of quantisation are necessary, as follows.

## Spatial quantisation

Here the image is 'sampled' at $m \times n$ discrete points in the image. Each sample is called a 'picture cell' ('pixel' or 'pel' for short). $m$ and $n$ are integers, and it is quite common for $m = n$ and for $m$ and $n$ to be integral powers of 2.

Like all forms of sampling, image sampling is subject to the Nyquist criterion. In this case the sampling frequency determines the maximum spatial frequency which can be accurately resolved in the digitised image. As the number of samples per unit distance is reduced, the high spatial frequency content in the image is lost. For example, in an image of printed cloth the weave of the cloth represents the highest spatial frequency while the pattern printed on the cloth will be of lower spatial frequency. Therefore a reduction in sampling rate may well result in the texture of the cloth becoming invisible while the printed pattern is still clearly visible.

This type of degradation will only be properly seen as such if the signal is low-pass filtered prior to sampling such that its maximum frequency component does not exceed twice that of the sampling frequency, as specified by the Nyquist criterion. In practice this is seldom done and reduction of the spatial resolution below a critical limit causes the image to break up into obvious blocks owing to quantisation (also known as 'pixellation'). This quantisation is the counterpart of 'aliasing' in the spatial frequency domain, which is why the reduction of this effect in applications such as computer drawn fonts is known as 'anti-aliasing'.

Figure 3.1 shows an image sampled at $512 \times 256$, $256 \times 128$, $128 \times 64$ and $64 \times 32$ pixel spatial resolutions. To a certain extent the perception of

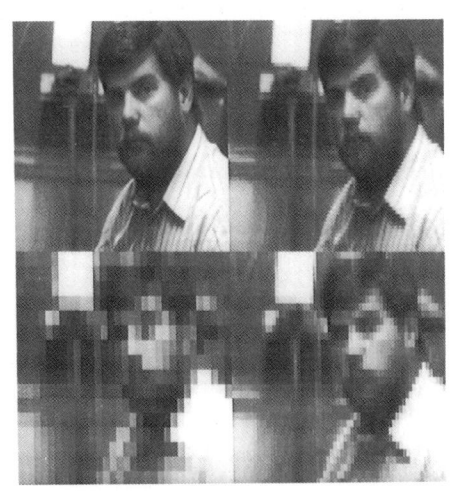

Figure 3.1 A grey scale image seen at various spatial resolutions. Starting from the top left and going clockwise: (a) $512 \times 256$, (b) $256 \times 128$, (c) $128 \times 64$, (d) $64 \times 32$

aliasing varies according to the reproduced size of the images but it is certainly obvious in the lowest resolution image.

*Amplitude (intensity) quantisation*

Here each pixel must be assigned a numerical code which represents the intensity of the image function at that point, as closely as the 'resolution' of the code allows.

The resolution of the code is determined by the number of quantisation levels ('grey levels') that are available between the extremes of intensity, namely black and white. The set of grey levels ranging from black to white is called the 'grey scale' of the system.

In order to satisfy requirement (a) above, the number of grey levels is usually an integral power of 2, such that:

black = 0

white = $2^l - 1$

where $l$ is an integer and there are $2^l$ grey levels in the grey scale.

The specification for grey scale resolution is often not as clear cut as that for spatial resolution. The former can be determined by the smallest feature(s) that must be detected, but the latter depends on the design of the whole machine vision system. If the details that need to be observed by the machine are concerned simply with outline shape then it is usual to perform all processing from segmentation onwards with binary images, where $l = 1$.

However, if surface detail must be checked, or if the pose of the objects which contain information in three dimensions cannot be predetermined, then the grey scale must be expanded to allow the subtleties of shading etc. to be observed. In demanding measurement applications the use of grey scale images allows the positions of edges, etc. to be determined to sub-pixel accuracy by means of interpolation. The value of $l$ is then application specific but it is seldom necessary, or efficient, for values of $l$ to exceed eight.

Figure 3.2 shows an image sampled at values of $l$ of 6, 4, 2 and 1. Six bits per pixel is generally considered to be equivalent to the intensity resolution of the human visual system. Notice how reducing the value of $l$ causes intensity 'contours' to become readily visible in the image. Although these are objectionable to humans they may not prevent useful results being obtained by a machine.

*Systemic impact of the choice of quantisation parameters*

Generally, in order to satisfy requirement (b) above, the values of $l$, $m$ and $n$ should be as high as is necessary to encapsulate relevant information in the

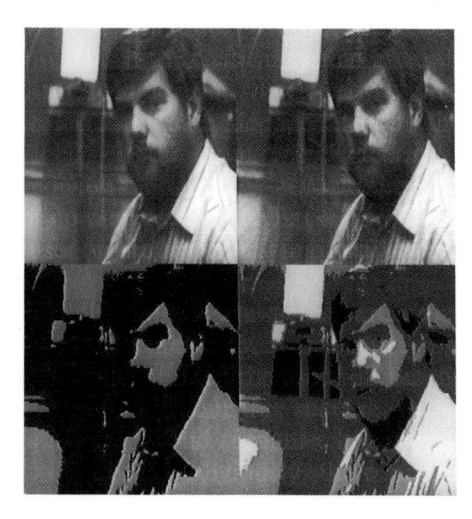

Figure 3.2  A grey scale image seen at various grey scale resolutions. Starting from the top left and going clockwise: (a) 6 bits/pixel; (b) 4 bits/pixel: (c) 2 bits/pixel; (d) 1 bit/pixel

image. However, this brings requirement (b) into conflict with requirement (a) on the grounds of computational efficiency. This is because a single image contains $l \times m \times n$ bits of data and so the processing efficiency is greatly aided by keeping $l$, $m$ and $n$ low (particularly $m$ and $n$).

Thus it is necessary to be aware of these conflicting requirements when selecting values for the quantisation parameters $l$, $m$ and $n$. In this way an appropriate compromise can be reached.

In summary:

(a) $l$ determines the grey scale resolution of the system;
(b) $m$ and $n$ determine the spatial resolution.

### 3.1.2  *Binary images*

There is a special case of intensity quantisation called 'binarisation' where an image is generated with only two grey levels, black and white (i.e. $l = 1$). Binary images are highly compliant with requirement (a) above, since they are very simple to store and manipulate as each pixel is represented by a single bit. Fortunately they are also very easy to generate from a more general grey scale image.

A threshold value, $T$, is used to partition the image into pixels with just two values, such that:

IF $f(x,y) \geq T$ THEN $g(x,y) = 1$
IF $f(x,y) < T$ THEN $g(x,y) = 0$

where $g(x,y)$ denotes the binarised version of $f(x,y)$.

In many practical applications it is possible to ensure that binary images are also compliant with requirement (b) above. This is typically achieved by the proper manipulation of scene constraints to ensure that the relevant information in the scene is encapsulated in simple high contrast images which ensure that a value of $T$ can be easily chosen that will generate helpful and meaningful binary images. If thresholding is applied to normal contrast grey scale images then the features which result in the corresponding binary image will be highly dependent upon the value of $T$ that is chosen – see figures 1.3c and 1.3d.

Thus this is a particularly important class of images in industrial machine vision because such images can in turn make segmentation and feature extraction relatively trivial tasks.

### 3.1.3 Array tessellation

When deciding how an image is going to be sampled, the physical arrangement, or tessellation, of the sampled pixels in the digitised image needs to be considered. By far the most dominant method is the 'orthogonal' tessellation shown in figure 3.3a. However, it has significant limitations which cause other tessellations to be of interest.

Consider the problem of connectivity. In figure 3.3a if pixels 1, 2, 3 and 4 are connected then are pixels 5 and 6 connected? This represents an ambiguity concerning connectedness which will complicate any processing task which relates to connectedness – e.g. 'boundary following' and 'thinning' operations.

Now consider that pixels 1 and 2 are separated by one unit, but pixels 2 and 3 are separated by $\sqrt{2}$ units. Thus there is also a distance inconsistency which needs to be carefully considered. This reveals that the orthogonal tessellation is not actually optimal with regard to representation requirement (a) above, despite its prevalence.

As an alternative, consider the hexagonal tessellation depicted in figure 3.3b. In this case it is clear that connectedness is unambiguous and that all pixels are separated by a distance of one unit. While the hexagonal tessellation is actually very common in nature (consider the compound eye of a fly, for example) it is seldom encountered in man-made vision systems. There are several reasons for this, including:

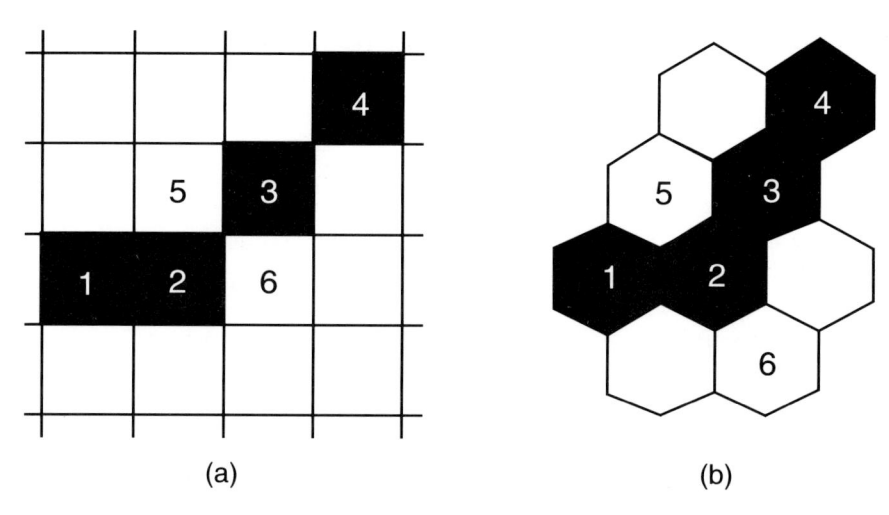

Figure 3.3  Array tessellations: (a) orthogonal; (b) hexagonal

(a) A preponderance of horizontal and vertical features in the real world which cannot be portrayed completely satisfactorily in a hexagonal system. This has led to:
(b) A lack of proprietary image sensors and display technology featuring hexagonal tessellations.
(c) Engineers and scientists who are generally trained in Cartesian geometry and experience some difficulty (unnecessarily) coming to terms with hexagonal geometry.

However, it has to be remembered that in machine vision the most efficient extraction of the minimum necessary data is the goal, and therefore hexagonal tessellations could have a valid role to play.

*Non-uniform sampling*

Orthogonal and hexagonal tessellations are uniform sampling schemes. Young [1] notes that uniform sampling is very useful for image processing systems:

'... where the aim is to produce enhanced pictures for a person to look at ...'

since

'... one part of the original image is as important as any other part; uniform sampling is needed, because it is not possible to predict where in the output viewers will direct their gaze or attention'.

When the goal of a system is the *interpretation* of images it may perhaps be achieved more easily if non-uniform sampling schemes are used appropriately. In particular, log-polar mapping of samples has interesting and useful properties.

This arrangement is shown in figure 3.4a. Each 'ring' contains the same number of samples and the radius of the ring is proportional to the logarithm of its 'ring index number'. The radial line of samples is known as a 'wedge' and the wedge index is used together with the ring index to address a sample uniquely. Thus the ring and the wedge indices are equivalent to $r$ and $\theta$, respectively, in a conventional polar coordinate system.

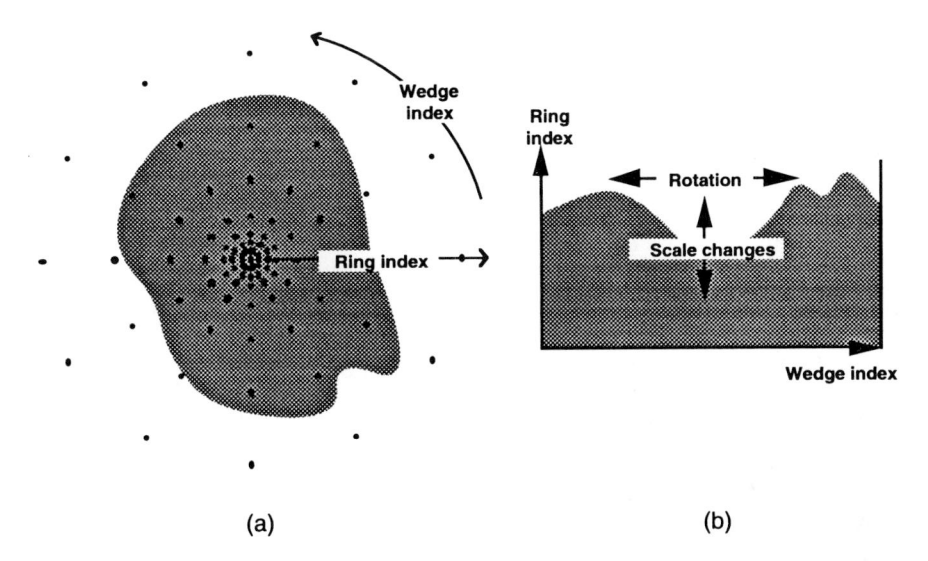

Figure 3.4  The properties of a log-polar sensor array: (a) photosite disposition; (b) simplified handling of scaling and rotation

The net result is an array which has very dense sampling at the centre and a rapid fall-off in density towards the edge. Thus it is a way of achieving reasonable resolution *and* reasonable field of view within a single sensor, without generating excessive amounts of data. However, this means that the sensor must always be centred on the region of interest in the scene. As such, its utility is dependent upon the adoption of an 'attentive vision' strategy in which the gaze of the vision system is fixed upon an object of interest. The human retina features a 'pixel' distribution of this general type and attentive vision is supported by eye, head and even body movements. This effect can be partially simulated in machines by the use of 'eye-in-hand' cameras used in a 'visual servoing' role.

Once the log-polar sensor is centred on the object of interest, its peculiar properties come into their own. When the image samples are mapped by their ring–wedge indices into Cartesian space it is noted that changes of scale become simple translation towards or away from the wedge axis – see figure 3.4b. Furthermore, rotation becomes a simple cyclic shift parallel to that axis. Therefore scale- and rotationally-invariant matching of objects, object recognition and orientation measurement become relatively trivial matters [2].

## 3.2 Image transduction (or sensing)

### 3.2.1 What is an image?

The *Collins Concise English Dictionary* defines an image (in this context) as:

*'the visual impression of something produced by a mirror, lens etc.'*

This is a sensibly loose definition because it is now common for images to be formed from radiation of all sorts. Figure 3.5 shows the electromagnetic spectrum. Within it can be identified an 'optical spectrum' with loosely defined wavelength limits of 10 nm to 1 mm [3].

Despite being invisible to the unaided human eye, radiation such as radar, X-rays, gamma rays, ultrasound etc. have become important sources of coherent information about our world. Until recently all such data have ultimately had to be converted into 2-D images formed in the visible portion of the electromagnetic spectrum in order to be intelligible to humans.

This is no longer always necessary, thanks to the development of machine intelligence which allows sensed information to be interpreted directly from

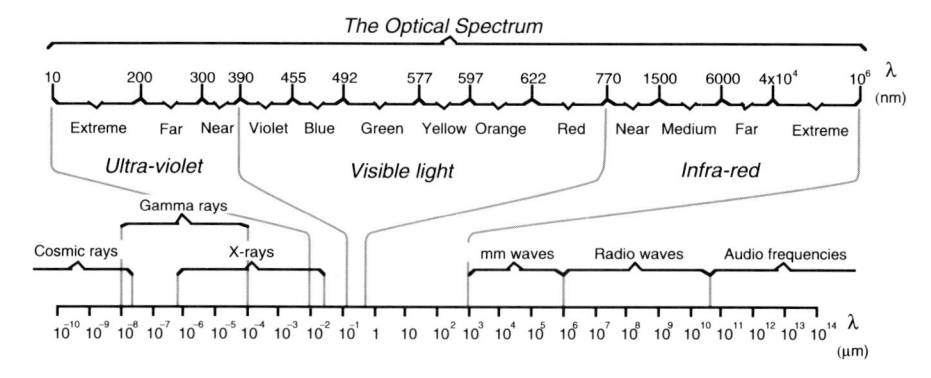

Figure 3.5 The optical spectrum in perspective

numerical representations. However, the wavelengths from about 300 nm (near ultra-violet, or UV-B) to 1.5 µm (near infra-red, or IR-A) currently dominate this field and seem likely to do so for the foreseeable future.

For convenience, this range of wavelengths is conventionally referred to as 'light' [4]. There are a number of reasons why these wavelengths will remain prevalent in industrial machine vision applications:

(a) Light is familiar and inherently safe. In applications where machines must work in close proximity with humans this is an important factor. By definition, humans can detect visible light and are familiar with its properties and the relatively mild health risks of IR and UV.
(b) Light is easy to generate. There are many different types of light sources which are well characterised and reliable.
(c) Light is easy to control and process. There is a wealth of knowledge about optics and the limitations of common materials and configurations [5]. Many different types of optical sub-system are available in proprietary form and may be adapted with a little thought and care. Materials such as germanium which pass wavelengths substantially beyond the visible band are expensive and difficult to work with.
(d) Light is easy to detect. There is a wide range of photo-sensitive mechanisms that can be successfully exploited using well understood materials. In particular, silicon has a useful response to most wavelengths in the range identified, having high quantum efficiency in the visible and near-IR. Furthermore, the use of silicon allows photo-detectors to be routinely integrated with signal processing electronics.

### 3.2.2 Principal photosensitive mechanisms

There are two fundamentally different mechanisms for the transduction of optical radiation into electrical signals, as follows.

*Thermal detectors*

In thermal detectors the absorption of photons causes a temperature rise which in turn produces the desired output. The transduction methods employed include thermistors, bolometers, thermocouples, golay cells (pneumatic detectors) and pyroelectric materials [4]. Detectors of this type are used to sense infra-red in astronomical applications because the attenuation due to interstellar dust is dramatically reduced at wavelengths greater than 2.2 µm [6]. A single detector is 'raster-scanned' over an area of sky, either by moving the whole telescope, or a secondary mirror.

However, these methods are either too slow in response, fragile or bulky to be useful in machine vision applications. The only exception is perhaps the pyroelectric effect which has found application in the tactile sensing arrays. Even this is mostly by default because the material concerned, polyvinylidene fluoride (PVF$_2$ or PVDF), is selected mainly for its piezoelectric properties [7].

*Quantum detectors*

Quantum detection is by far the most important mechanism in image sensing for automation. It relies upon the energy of absorbed photons being used to promote electrons from their stable state to a higher state above an energy threshold. When this occurs the properties of the material are altered in some measurable way.

The wavelength of the incident photon is related to the energy that it carries by the familiar Planck/Einstein relationship:

$$e = \frac{h \times c}{\lambda} \tag{3.1}$$

where $e$ = energy carried by a photon,
$h$ = Planck's constant, $6.626 \times 10^{-34}$ J s,
$c$ = speed of light *in vacuo*, $2.998 \times 10^8$ m s$^{-1}$,
$\lambda$ = wavelength of incident radiation.

On collision with an electron either all or none of this quantum of energy is transferred to the electron. Therefore, eqn (3.1) reveals that the *maximum* wavelength to which the quantum detector will respond is determined by the energy threshold, and therefore by material selection. The other extreme of the response curve is limited by the ability of the material to absorb high energy radiation quanta, i.e. by the 'opacity' of the material at short wavelengths.

The fundamental sensitivity of a quantum photodetector is measured by the 'quantum efficiency' ($\eta$). This is the average number of electrons promoted by the arrival of an incident photon. In primary photosensitive mechanisms it is always a number less than unity.

*The external photoeffect*

In the external photoeffect the energy threshold of interest is the 'surface work function' of the material. When the incident photons have an energy greater than this threshold, electrons are emitted.

This gives rise to 'photoemissive' detectors where the electrons are emitted into a vacuum and then collected and measured in a variety of ways. The work function of common photoemissive materials generally restricts their use to visible light or shorter wavelengths only.

This mechanism finds application mainly in non-imaging photomultipliers or qualitative image intensifiers. It is also used in very high performance quantitative imaging sensors called 'image dissectors'. However these are too slow and bulky for use in the majority of industrial automation applications.

*The internal photoeffect*

The internal photoeffect requires the incident photon to promote an electron across an energy gap which exists within the bulk of the material. Many intrinsic semiconductor materials feature energy gaps between their valence and conduction bands which are of similar order to the energy carried by photons of light. Also, the established technique of doping to achieve well-defined extrinsic energy levels widens the choice of photosensitive materials even further.

There are several modes of operation for internal quantum detectors, but the most important ones are:

(a) photoconductive,
(b) photovoltaic.

The resistance of photoconductive materials drops in the presence of light owing to the generation of free charge carriers. An external bias must be applied across the material to measure this change. The transfer characteristic is generally logarithmic and they exhibit relatively slow response times and high hysteresis. The major use of materials such as cadmium sulphide is as non-imaging, discrete sensors, but compounds of selenium or antimony (e.g. $Sb_2O_3$) are used to coat the targets of 'Vidicon' imaging tubes.

Photovoltaic devices consist of semiconductor junctions. They may respond to photons by generating a logarithmic voltage in open-circuit mode, or a linear current in short-circuit mode. Under these circumstances no external bias is required and the devices are generally called 'photocells'. Their chief application is in 'solar panels' for the generation of power and in non-imaging, discrete form.

Alternatively such junction devices may be operated in photoconductive mode, under conditions of reverse bias. This widens the 'depletion region' which is established when the recombination of free (majority) carriers across the junction reaches equilibrium. The minority carriers that flow under these conditions form the 'dark current' which is effectively constant for a given device, but which limits the lower extent of the dynamic range. When the junction is illuminated the reverse photocurrent rises in proportion to the incident radiation, rapidly swamping the dark current. The reverse bias produces a smaller junction capacitance than the unbiased photocell which improves response times.

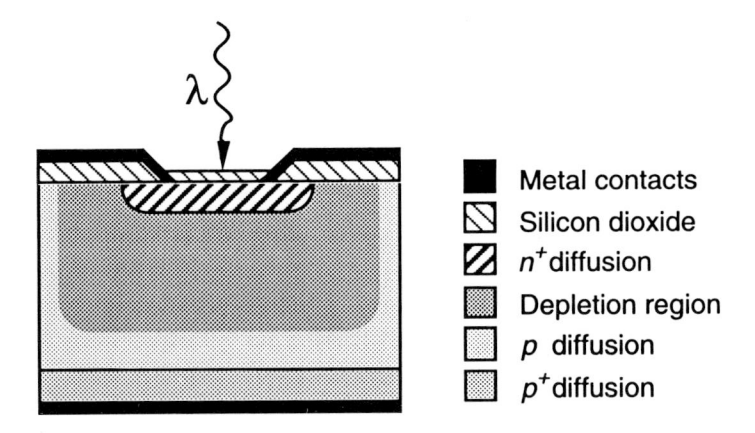

Figure 3.6  A schematic representation of a silicon photodiode

In practical devices the top layer of the junction is made thin and is heavily doped compared with the bottom layer (see figure 3.6). This ensures that the depletion region comes close to the surface and therefore captures carriers generated by short-wavelength radiation. Conversely, the light doping of the lower layer ensures that the depletion region extends deep enough into the silicon to capture carriers generated by the longest wavelengths of interest.

The extent of the depletion layer may be limited by the reverse breakdown voltage of the diode in which case $p$–$i$–$n$ diodes structures may be used. The layer of intrinsic material has high resistivity and can be kept depleted by relatively small reverse biases. Therefore the inclusion of a relatively thick intrinsic layer allows a large sensitive volume to be achieved. This structure also improves the response time of device.

The simple architecture of photodiodes ensures a minimum of light attenuation, yielding high quantum efficiencies and wide spectral responses. Such photodiodes also feature good stability and dynamic range. Efficiency can be further improved by coating the active surface with an anti-reflection layer of silicon dioxide, $\lambda/2$ thick.

Photodiodes used in photoconductive mode therefore offer attractive properties, but are not used in this way in large arrays. This is because the periodic sampling implied by an efficient scanning mechanism would only allow each diode to be active for a short time, giving poor overall responsivity. To overcome this problem and make the diode responsive for virtually the whole scanning cycle, 'photon flux integration' mode may be employed [8].

In this case the 'space-charge' of the reverse biased depletion region is treated as a charged capacitor which is left open-circuit for most of the scanning cycle. Under these conditions, the space-charge capacitance would normally discharge over a period of tens of seconds owing to the action of the thermally driven dark current. However, when the junction is exposed to light the effects of the photocurrent predominate and the capacitance is discharged at a virtually constant rate in a few milliseconds. Furthermore, the rate of decay of charge is linearly dependent upon the intensity of the illumination. Therefore the change in voltage across the space-charge capacitor is proportional to the time integral of the illumination incident during the period when the diode was left open-circuit. Hence this may be called the photon flux integration period (integration period).

This is the basis of a photosite which is useful in imaging arrays. It retains the advantages of the photoconductive photodiode and offers good responsivity. The only disadvantage is in applications where the light incident on the photosite fluctuates during the integration period. Naturally, this will result in a blurred response. However, if the light intensity is sufficient it may be possible for the integration period to be kept short even if the system operation demands a fixed period scanning cycle. The 'fast shutters' available in many modern video cameras are implemented in this way.

If the junction depletion region of the photodiode is replaced by an 'inverted' region induced by an insulated electrode or 'photogate', then another photosensitive element is created. This is called the 'photoMOS' element (see figure 3.7), which depicts a simple 'surface channel' device. This device operates exclusively in photon flux integration mode.

When an appropriate potential is applied to the photogate a 'potential well', depleted of majority carriers, is created in the substrate material. This behaves as a capacitor which can store minority carriers. When photons are incident upon this region hole–electron pairs are generated and these are

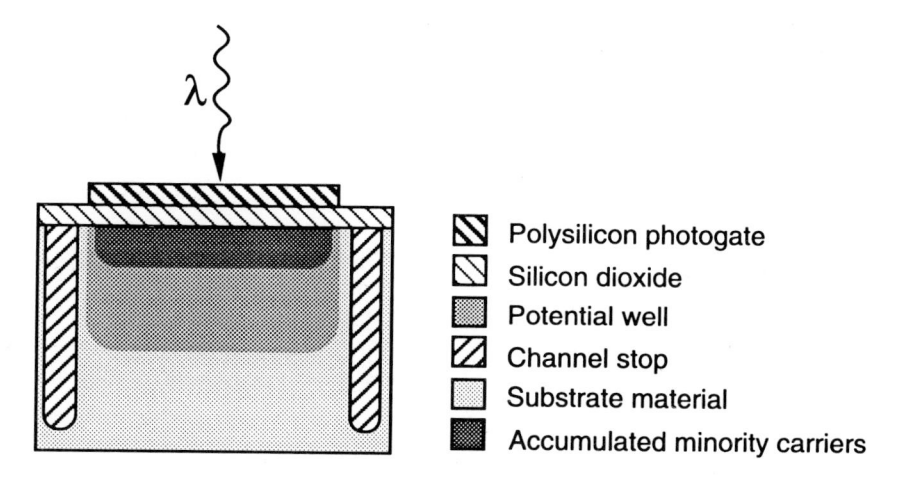

Figure 3.7 A schematic representation of a single photoMOS element

separated by the electric field in the well. The majority carriers are repelled while the minority carriers are contained within the well. Thus the well accumulates a charge of minority carriers which is proportional to the photon integration time and the intensity of the incident illumination.

The size and shape of the potential well is controlled by the photogate dimensions and applied voltage. In practical devices it is also constrained by a 'channel stop' diffusion. The capacity of the well is finite, being of the order of one million carriers for practical devices [9], and this represents the upper limit of the dynamic range of photoMOS elements. If the number of minority carriers generated by light is excessive, then the potential well will ultimately overflow. This is a problem when photoMOS devices are made into arrays because localised saturation in a photosite can corrupt its neighbours. This effect is known as 'blooming'. In practical devices it is controlled by the use of an 'anti-blooming sink electrode' which causes the excess charge to leak harmlessly into the substrate.

The other extreme of the dynamic range is determined by the thermal generation of minority carriers within the well. These are indistinguishable from genuine photon-induced carriers and give rise to the 'dark current' in the element. The dark current doubles for every 8–10°C temperature rise above −25°C. These effects can be minimised by cooling the device, although the dynamic range of practical devices at room temperature is typically of the order of 2000:1.

The necessity for the photogate in standard photoMOS devices means that the incident light must pass through a layer of polycrystalline silicon ('polysilicon'). This absorbs some light, especially at short wavelengths. Consequently, such elements feature a blue-deficient spectral response and lower overall quantum efficiency than photodiodes. However, for specialist applications such as astronomy the blue response can be improved by etching the substrate thickness down to around 10 μm and arranging for the light to enter the element from the rear (substrate side).

### 3.2.3 *Principal imaging architectures*

Industrial automation has a number of uses for non-imaging light sensors – such as optical safety curtains and object-in-position detectors. However, the major impact of light sensing on its future has been identified as being through image sensing for the purpose of machine vision. This requires that arrays of light sensitive elements can be efficiently implemented.

There are two important array configurations, as follows.

*Linear arrays*

Linear arrays, or linear imaging devices (LIDs) consist of a one-dimensional array of photosites. They are made using diode or MOS photosite technologies and practical devices have anything from a few elements up to 6000 or more (see figure 3.8).

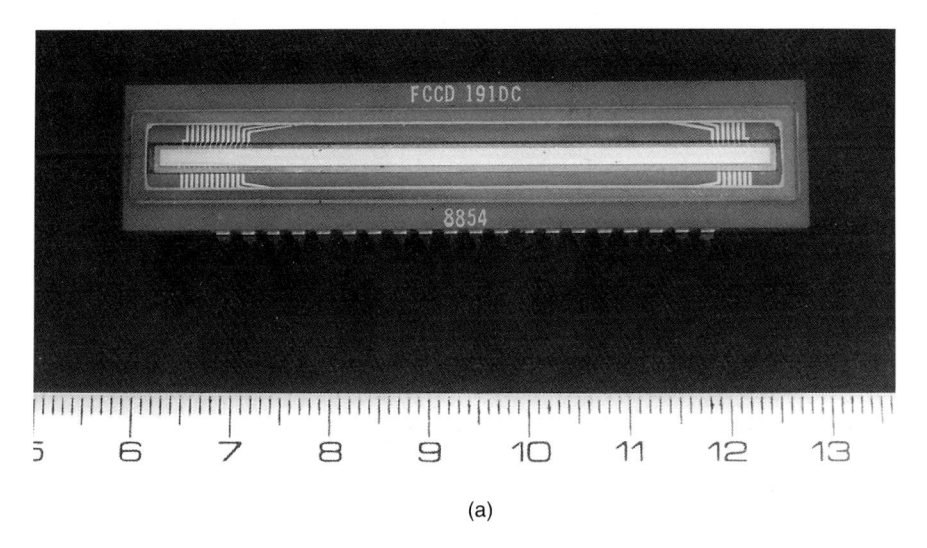

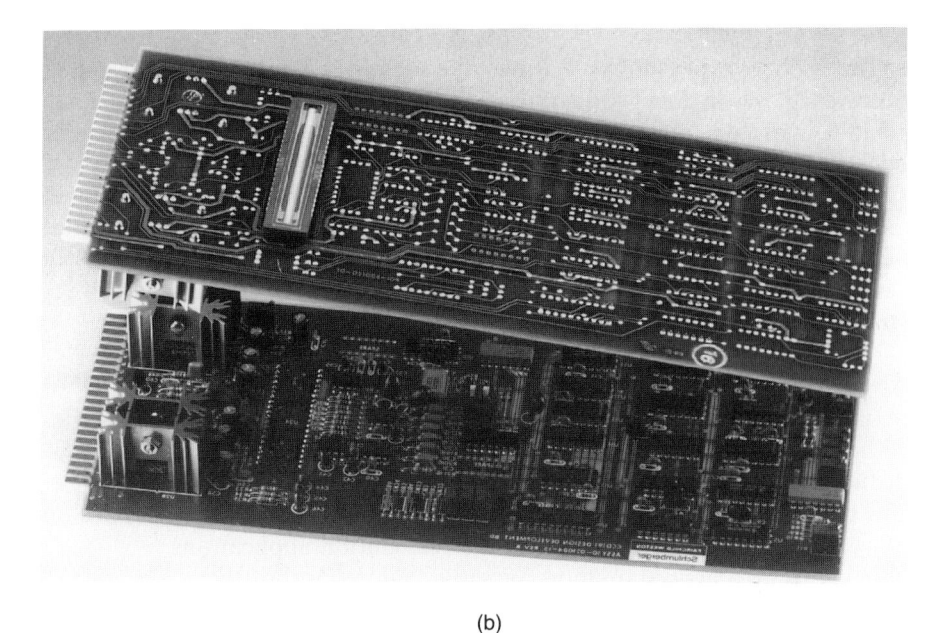

(a)

(b)

Figure 3.8  The Loral Fairchild CCD 191 6000 pixel line-scan image sensor: (a) close-up of the sensor chip; (b) the interface electronics development board *(Photographs courtesy of Loral Fairchild Image Sensors (USA) and Optimum Vision Limited (UK))*

The wide range of available resolutions means that it is relatively easy to select an image sensor which is well matched to a task. Hollingum [10] identifies many typical applications within industrial automation. These fall into two major groups:

(a) one-dimensional gauging;
(b) two-dimensional imaging where subject motion or mechanical scanning provides the second dimension.

Examples of the first type are the alignment and placement of car windscreens at Austin–Rover Motor Company (Cowley, UK) and the inspection of bolts in a cylinder block assembly at the Ford Motor Company (Bridgend, UK). The second type are more numerous, including:

(a) identification of car body types – Ford Motor Company (UK);
(b) inspection and measurement of corn cobs – Green Giant (USA);
(c) measurement and inspection of steel bar – British Steel Corporation (UK);
(d) inspection of glass containers – United Glass Containers (UK);
(e) inspection of endless woven belts – Scandura (UK).

The second type of application is common because of the frequent occurrence of flow-line production methods in industry. These often involve smooth linear or rotary movement which is ideally suited to this mode of operation.

It should only be necessary to resort to mechanical scanning when area sensing is required but the resolution requirements of the application cannot be met by an available area array.

*Area arrays*

Area arrays, or matrix arrays, consist of a two-dimensional array of photosites. They make it possible to investigate static real world scenes without any mechanical scanning. Thus much more information can be deduced from a single 'real-time' glance than would be possible with LIDs. The use of an area array also makes it easier to accommodate tolerance on object positioning in measurement and recognition applications.

In short, area arrays are more versatile than LIDs, but there is a price to be paid for this – cost, complexity and a limited range of available resolution. Area arrays really fall into two classes: those which are intended for broadcast television compatible applications, and those which are aimed at a more limited scientific market.

The first class is characterised by relatively affordable cost but virtually fixed resolution in a band from around 490 × 380 to 780 × 575 pixels, with a

*Image Acquisition*

4 : 3 pixel aspect ratio. These are factors optimised for the RS170 (525-line) and CCIR (625-line) video standard formats. These video standards also impose a pixel-serial, 'raster-scan' data output format with 2 : 1 interlacing onto the sensor architecture. However, complete monochrome cameras compatible with these standards can be bought for as little as £200 ($300) at present.

Figure 3.9   Loral Fairchild's range of ultra-high resolution full-frame CCD image sensors. 1048 × 1048, 2048 × 2048 and 4096 × 4096 pixels (left to right) *(Photographs courtesy of Loral Fairchild Image Sensors (USA) and Optimum Vision Limited (UK))*

The second class are characterised by wider choice of resolution, square pixels and freedom from some of the restrictions imposed by the broadcast video formats. However, they are available at prices which reflect their limited markets. Perhaps the ultimate in available resolution is a 4096 × 4096 full-frame (parallel/serial) CCD device from Loral Fairchild Imaging Sensors (USA) (see figure 3.9). However, this device costs around £90 000 ($135 000) for a perfect sample, falling to around £3000 ($4500) with 2% blemishes (specified as photosites having a response which differs from the mean by $< \pm 10\%$)!

*Vacuum-tube technologies*

Unlike LIDs, the range of available area arrays is not totally dominated by solid-state devices. Vacuum tube technologies were the first to be developed to meet the needs of area arrays, dating back to the iconoscope in 1923. There has been a steady programme of development from that time to the present, and there is no sign of obsolescence yet. Indeed it is recognised that vacuum tubes represent the state of the art for maximum resolution and sensitivity, and they meet many specialised needs.

The majority of current applications use a tube architecture known generically as a 'Vidicon'. It is shown schematically in figure 3.10. The light image is focused onto a transparent signal plate coated on the inside with a photoconductive target material. The front face of the signal plate is coated

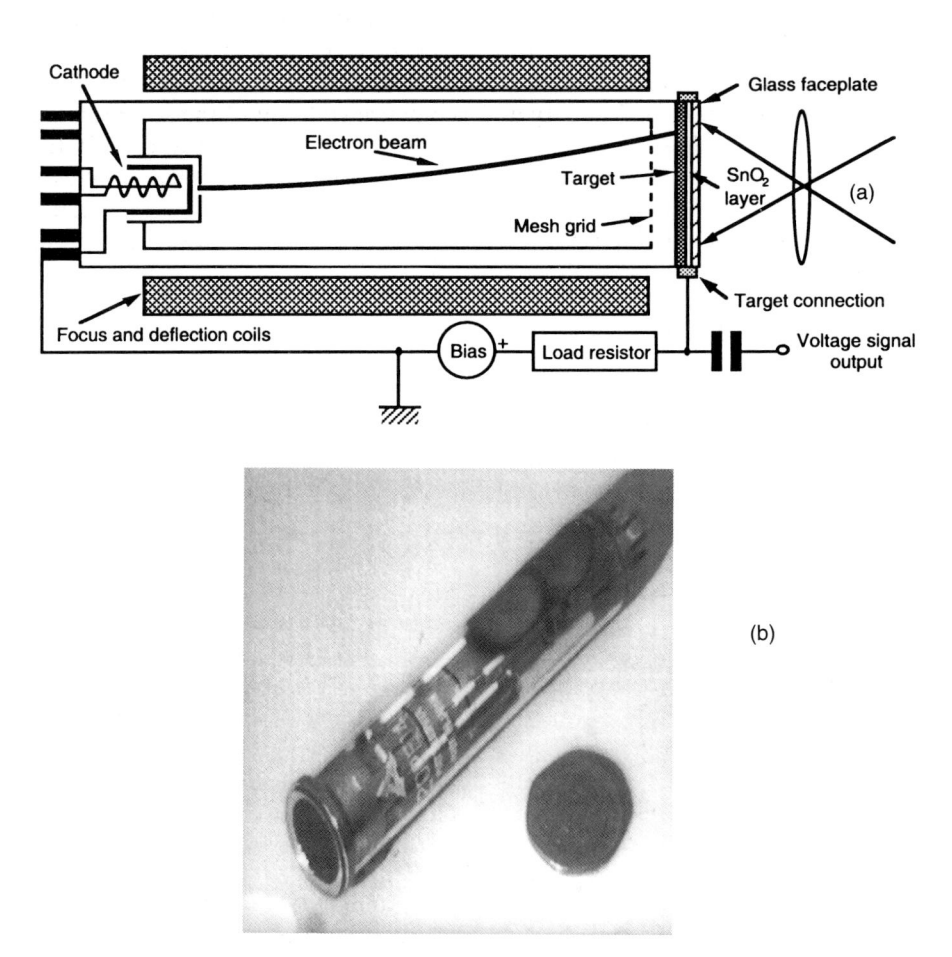

Figure 3.10  Vidicon tube image sensors: (a) schematic representation of construction; (b) a ½-inch format, general purpose Vidicon tube

with a transparent conducting layer (e.g. $SnO_2$) which is biased to a positive potential of about 50 V. The target coating is maintained at approximately 0 V by a mesh grid just behind it. So the signal plate and target behave as a continuous array of capacitors containing a photoconductive dielectric.

When no light is incident on the target, its resistance is very high and the charge on the 'capacitor' element is maintained. However, when light strikes the target the dielectric material becomes leaky and the target side voltage rises locally. The target is scanned by a low velocity electron beam (decelerated by the mesh grid) and, when this strikes a discharged 'capacitor' beam, current flows to restore the target side to 0 V. This current also flows in the external bias circuit, producing a volt drop in a load resistor which is capacitively coupled as the voltage signal output.

In order to produce the raster output necessary for broadcast video applications, the electron beam is scanned across the surface of the target by electromagnetic deflection coils. This is the great weakness of vacuum tubes, principally because the linearity and geometric stability of the scanning are relatively poor. The coils make the sensor bulky and are responsible for major power drain. The tube is obviously also sensitive to external fields.

The original Vidicons used antimony trisulphide as their target coating. This is a photoconductive material which exhibits a non-linear, but panchromatic, response to light of the form [11]:

$$V_{out} = R_L [K(E_{in})^\gamma + l_{dc}] \qquad (3.2)$$

where $V_{out}$ = output signal,
$R_L$ = load resistance,
$E_{in}$ = incident irradiation,
$\lambda$ = linearity coefficient (typically 0.65–0.75),
$l_{dc}$ = dark current.

Thus such devices exhibit what is known as 'gamma distortion' which must be corrected in many applications. This target material also suffers from image 'lag', where the coating takes time to recover from bright illumination. These problems were largely overcome in professional applications by the use of lead oxide as a coating material in tubes often known as 'Plumbicons'. The inside of the lead oxide coating is doped to make it act as a $p$-type semiconductor so that the target behaves as a continuum of $p$–$i$–$n$ photodiodes, operated in photon flux integration mode. This bestows a linear characteristic and reduced lag, but at the expense of more costly manufacturing procedures.

Other forms of Vidicon are made using more complex target materials and construction, for specialised applications. 'Newvicons' and 'Chalnicons' use zinc and cadmium tellurides for much improved sensitivity, but with a spectral response biased towards the near-IR. The silicon-target Vidicon, or

'Sivicon', achieves improved sensitivity and a spectral response similar to that of the human eye. The pyroelectric Vidicon ('Pevicon' or 'Pyricon') extends its spectral response into the medium-far infra-red.

While the standard Vidicon is the least sensitive of all of the family, it has the greatest dynamic range and resistance to blooming because the low gamma value resists saturation. The Plumbicon has the lowest dark current, which is also independent of temperature, and the shortest lag.

The standard Vidicon is susceptible to 'image retention', where highlights of static scenes form a semi-permanent image on the target. In severe cases this will lead to permanent damage known as 'burn'. Newvicons and Plumbicons are relatively insensitive to this problem and Sivicons are virtually immune.

Other types of vacuum-tube cameras are still made for even more specialised applications. The silicon-intensifier target ('SIT') and 'image isocon' tubes give incredible sensitivity, allowing images to be recorded under overcast moonlight conditions corresponding to around $10^{-6}$ lux.

*Solid-state technologies*

LIDs were the first devices to take advantage of the small size, robustness, low power consumption and geometrical stability of solid-state imaging technologies. This was because of the relatively small number of elements involved. However, rapid progress in VLSI technology has made it possible to manufacture solid-state broadcast-compatible area arrays more cost-effectively than vacuum-tube types.

The considerable strengths of the vacuum-tube area array technologies should not be ignored, but they now find only niche applications in industrial machine vision applications. Their bulk, fragility, short working life, lack of geometric stability and linearity, and their susceptibility to magnetic fields mean that the majority of new applications will exploit solid-state technologies.

The two most important solid-state transducer technologies for image sensing arrays have already been introduced: photodiodes and photoMOS. Array sensors require hundreds, thousands or even millions of such photosites to be efficiently connected together. This is currently achieved in two ways: digitally controlled analogue multiplexing and charge-coupled analogue shifting. Coupling these array scanning methodologies to the two major types of photosite yields four fundamental array imaging architectures. These are summarised in figure 3.11.

Digitally controlled analogue multiplexing requires that a MOS pass transistor should be fabricated adjacent to each photosite. When the transistor is energised it connects the photosite output directly to a common video line. This is terminated in a charge-to-voltage buffer amplifier stage to

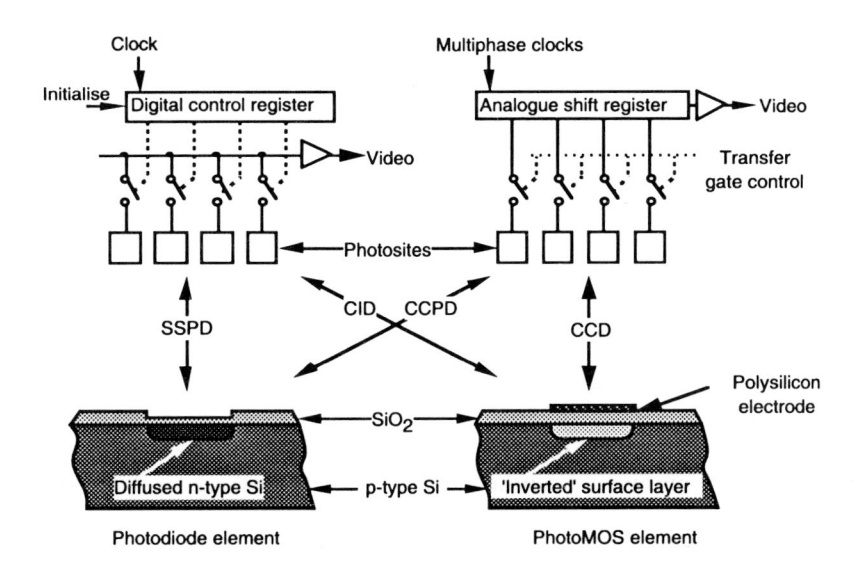

Figure 3.11  A summary of fundamental imaging array architectures – after Batchelor *et al.* [5]

provide the output video signal. Each photosite is read out in turn under control of the multiplexing logic.

Charge-coupled analogue shifting exploits the same 'potential well' phenomenon as the photoMOS element. Sets of electrodes deposited on the surface of the silicon can be energised in sequence to transfer analogue charge 'packets' towards the output. Each electrode in a set is controlled by one 'phase' of the overlapping transfer clock waveforms. The earliest devices used a three-phase process but this is generally being superseded by a two-phase process which is more complicated to fabricate but easier to drive [12]. The charge packets ultimate destination is a charge-to-voltage buffer/amplifier which supplies the video output signal.

The self-scanned photodiode (SSPD) architecture features digitally controlled multiplexing of diode photosites. When the pass transistor is energised, current flows into the depletion region capacitance to 'top-up' the charge lost during the photon integration time. The video signal is generally derived by sensing this current as each photosite is selected in turn. The major problem with this approach is the large capacitance of the lengthy common video line. This increases the noise level of SSPD arrays in comparison with competing architectures.

The need to fabricate a junction photodiode and a pass transistor in each unit cell means that it is larger than that achievable with photoMOS elements. However, the isolation of each photosite means that SSPD arrays are highly resistant to blooming and smearing. As already noted, photodiodes exhibit high quantum efficiency and broad spectral response.

In area array applications SSPDs are generally known as MOS $X$–$Y$ sensors. The digital scanning architecture potentially confers random access to the photosites, although this is seldom exploited. The pass transistor in each unit cell may reduce the overall density of placement, but devices are offered commercially with 577 × 388 resolution. Since only a part of the unit cell is photosensitive, its signal-to-noise ratio tends to be lower than other types. The potential for aliasing is also greater, for the same reason.

In basic charge-coupled device (CCD) architectures, photon integration may be performed within the shift register itself by stopping the transport clocks. However, this approach is prone to 'smearing' unless the readout time is very short compared with the photon integration time. Generally it is preferable to separate the functions of sensing and scanning by connecting the photosite to the shift register by means of a transfer gate. The transfer gate electrode is pulsed at the end of the photon integration time to load the photocharge into the shift register. From there it can be transported to the output. The shift register is shielded by metallisation to eliminate smearing.

During every transfer operation some charge may be lost; this effect is measured by a parameter called 'charge transfer efficiency' (CTE). The type of photoMOS device described earlier was a surface-channel structure where charge packets move along the interface between the silicon and its oxide. Unoccupied energy states exist here, trapping some signal carriers and reducing CTE to around 99.5%. In view of this, most good quality CCDs employ a more complicated 'buried-channel' structure. This uses an $n$-type diffusion into the surface of the substrate to move the channel away from the interface. Thus the number of traps is reduced and CTE is increased to around 99.999%, ensuring that packets can be shifted from remote parts of the array with negligible loss.

Even after the effects of CTE have been considered the CCD architecture achieves low noise figures because the charge packets only see the small input capacitance of the buffer amplifier. It also achieves much higher densities because the fundamentally simple structure requires no metallic contacts to be made throughout the majority of the array. However, the loss of sensitivity and blue response due to the photogate, and susceptibility to blooming and crosstalk, are noteworthy disadvantages.

However, an interesting compromise is the charge-coupled photodiode (CCPD) architecture which is becoming quite commonplace for LIDs [13]. This uses diode photosites to gain their advantages, interconnected by CCD shift registers for compactness and low noise.

The final architectural combination is the charge-injection device (CID). This consists of a MOS photosite which is scanned digitally by removing bias voltages that maintain the potential well. When this happens the integrated charge is injected into the substrate, causing a photocurrent to flow. This is sensed and converted to a video output voltage.

Since CIDs do not shift charge like CCDs, they benefit from reduced blooming, smearing and crosstalk. It is also possible to read out the photosites in random sequence although many implementations disallow this in favour of producing standard video format outputs. The readout process cannot be accomplished as quickly as for CCDs although some designs allow non-destructive readout. This permits multi-frame integration times compatible with standard video formats which is useful for some noisy applications.

CIDs retain much of the compactness of CCD photosites and are mostly used in area array applications which exploit their special characteristics. However, the unit cost is considerably higher owing to their fabrication on non-standard epitaxial substrates.

Although CIDs and MOS $X-Y$ sensors do hold a percentage of the market for solid-state broadcast-type area arrays, it is really dominated by CCDs. This is because they offer the greatest density and cost-effectiveness. There are three principal CCD architectures, and these are summarised in figure 3.12.

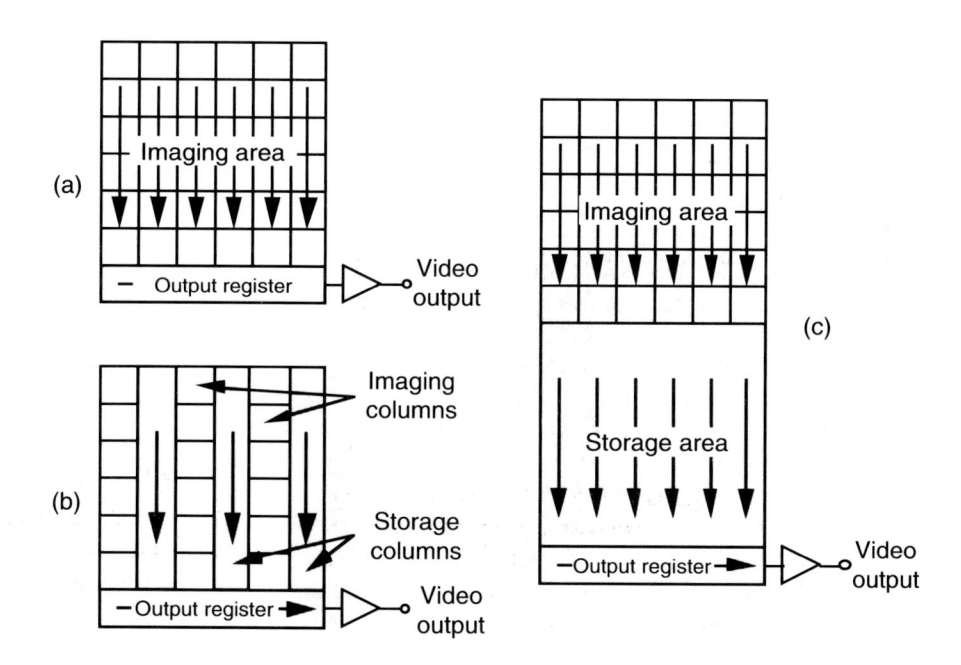

Figure 3.12 Schematic representations of CCD area array architectures: (a) parallel/serial; (b) interline transfer; (c) frame transfer. After Batchelor et al. [5]

The parallel/serial (P/S) architecture is the most basic form, where photon flux is integrated within the CCD shift registers. Therefore its whole surface is active, maximising responsivity and minimising aliasing. Stopping the transport clocks initiates the photon integration period. When they are restarted the data are shifted vertically down one row and the horizontal output register is rapidly clocked to remove one row of data. This process is repeated until the whole array is read out. A mechanical shutter is the only way of stopping light from smearing the image during this period.

Therefore, if a shutter is to be avoided, this approach can only used when the shifting time is much less than the integration time. Thus it is not used in broadcast-type formats, but it is especially well suited to astronomical applications owing to the long exposure times given.

The interline transfer (ILT) architecture is similar to an array of LIDs. The rows of photosites are interspersed with shielded vertical shift registers which are loaded during field blanking time. Then the vertical registers are clocked during the next photon integration period at line rate. The horizontal output register is clocked at pixel rate to provide the video output.

The vertical shift registers in the ILT architecture reduce sensitivity and increase aliasing effects somewhat. However it is a cost-effective design which is common in very low-cost commercial broadcast-type cameras.

The frame transfer architecture is basically a P/S type which is shielded from light, supplemented by another full size array which is exposed to light. During the field blanking period the whole imaging area is shifted vertically into the shielded storage area in typically 470 µs, minimising smearing. While the imaging area integrates the next field, the storage area is read out like the P/S design.

The imaging area is completely active, conferring the same advantages as the P/S design discussed above, but standard video format readout rates are achieved. The silicon area for this design is double that of the ILT architecture and the yield must be lower, but at present it is probably the most popular commercial type of all.

### 3.2.4 *Comparison of vacuum-tube and solid-state area arrays*

The solid-state sensor largely dominates the use of area arrays in machine vision. However, the vacuum-tube technology has a significant number of attributes which make it clearly superior for niche applications. In particular, specialised tubes can offer the best possible spatial resolution, sensitivity and range of spectral response, but of course not in any single design. On the other hand, solid-state sensors can always offer the best performance in terms of life, robustness, lag, power consumption and geometric stability.

## 3.3 Digitisation

This section is concerned with the conversion of signals derived from a camera into an array of numbers so that a computer may process the data efficiently. If a user is required to interact effectively with the system, some form of image display mechanism must also be provided. The elements necessary to perform these operations are illustrated in the block-diagram of figure 3.13, and discussed below.

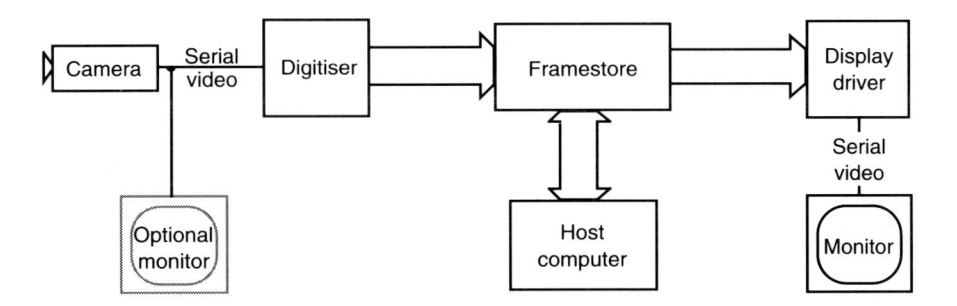

Figure 3.13 Block diagram of a basic system for image capture and display

Although there are many possible types of camera, area arrays producing broadcast-compatible raster scan video format are by far the most common in current general image processing applications. There is also no established standard for interfacing line-scan cameras, etc. Therefore discussion of digitising hardware will concentrate on that which is suitable for use with the video format outlined in the Appendix. The discussion will also be limited to the digitising of monochrome signals because these are most common. However, the principles presented can be readily extended to the digitisation of RGB-encoded colour video.

### 3.3.1 Image capture

The first step in digitisation is sampling by means of an analogue-to-digital converter (ADC). This must occur at the correct rate to achieve the desired spatial resolution ($m \times n$) and must be performed in synchronism with the arrival of the serial video data. The synchronisation is achieved with the aid of 'sync separator' circuits which extract the horizontal and vertical synchronisation pulses embedded within the video signal.

In addition to synchronisation pulses, the horizontal retrace period of the video waveform also maintains the video signal at 'black level' for a short period. This is used by a 'black-level clamp' circuit in the digitiser to establish a reference with respect to which the rest of the active video will be compared. After stripping of the synchronisation signals and suitable buffering the active video is fed to an ADC. This samples the video at regular intervals and converts the instantaneous difference between it and the reference black level into a digital number. The ADC must convert at a very high rate (up to 15 MHz) and so 'flash', or more recently 'semi-flash', devices are used. Such devices exploit considerable hardware parallelism to achieve the necessary conversion rates.

The sampling frequency, and hence the value of $m$, are determined by the sample clock. This is may be derived by a free-running crystal oscillator or a phase-locked loop (PLL) frequency multiplier system. In the first case, the oscillator often runs at a multiple of the required sample frequency and is divided down by a counter which is enabled by the trailing edge of the line synchronisation pulse. In the second case, a high frequency voltage controlled oscillator (VCO) is divided down to match the frequency of the line synchronisation pulses. The incoming synchronisation pulses and the generated frequency are compared by a phase comparator and the error signal is used to control the VCO. The net result is a sample clock whose frequency is determined by the incoming line synchronisation frequency and a constant equal to the division ratio applied to the VCO output. Although this method is considerably more complex, it has the advantage that it automatically accommodates any variation in the line frequency from the video source. In practice, this variation can be quite significant, especially if a video tape recorder is used to supply the video.

The sample clock is divided by counter chains to provide appropriate addresses for writing the data into the 'framestore' memory. This is an area of memory which must be capable of storing one complete frame of video information. In practice it is much simpler, and often quite sufficient, to capture video fields rather than frames (so the memory may more accurately be called a fieldstore), but hereafter the term framestore will be used regardless. In either case, memory is normally organised as an array of $m \times n$ locations which can each store $l$ bits. For typical values of $m = 512$, $n = 256$ and $l = 8$, the storage requirement is 128 kbyte.

The vertical spatial resolution is fundamentally limited by the line structure which is inherent in the raster scanned format of broadcast standard video (see Appendix). Therefore there is no choice over the sampling frequency in the vertical direction, only the number of samples ($n$) in use can be selected, up to a limit of approximately 580 in the CCIR system ($\approx 480$ in the RS170 system). However, when fields are captured, rather than frames, the maximum value of $n$ is limited to half of these figures.

The value of $l$ is determined by the resolution of the ADC used. Values in excess of eight are unusual because of a sharp increase in cost of conversion and memory hardware in relation to the rapidly diminishing benefit derived from greater grey scale resolution for most applications.

The framestore must be accessible to the host computer for image processing operations. The method of achieving this is highly dependent upon the bus-structure of the host, but it will either be I/O-mapped or memory-mapped. In the former case, a simple interface with the framestore is achieved at the cost of speed of data transfer. In the latter case, much higher transfer rates may be achieved, especially if DMA techniques are employed to move the data. This is much more in keeping with the volume of data and the typical speeds of processing required, but naturally requires that the framestore has a dedicated interface to the host computer in question.

If the machine vision system is to capture and process each and every frame generated by the camera, a severe constraint is placed upon the bandwidth of the host-framestore interface. This is because the only time that it can use to transfer the data is the frame blanking period of approximately 1.5 ms, implying a bandwidth requirement of 85 Mbytes s$^{-1}$ for a 128 kbyte image array! Fortunately this is a relatively rare requirement and usually the digitiser can simply stop acquiring new data while transfers to and from the host take place.

It is unusual for the host to perform image processing operations directly in the framestore, for a number of reasons. Firstly, the host may be relatively slow in fetching data and writing it back again – this is particularly true if the framestore is I/O-mapped. Secondly, if just one framestore exists for the image capture and display system then updating that framestore must either be performed only in horizontal and vertical blanking periods, with obvious performance implications, or serious disturbances on the displayed image must be tolerated. Thirdly, most image processing operations cannot be performed 'in-place'. This means that processed data cannot be written back into the framestore without modifying data which have yet to be processed. This problem will affect all neighbourhood processes, for example.

Therefore, the well specified host computer must usually supply an area of its local memory for use as a 'frame buffer', which can be used to store processed results temporarily. Ideally, several frame buffers should be provided so that the effects of various processing techniques can be directly compared with one another, as well as with the original image. 'Dyadic' operations, such as the subtraction of one image from another, also demand the simultaneous availability of two input images.

*Pixel aspect ratio and the need for calibration*

In digitisation of conventional video signals, the horizontal spatial resolution of the system is determined by the number of times that the active video line is sampled along its length. This may be varied over a wide range by simply adjusting the frequency of the sampling pixel clock. However, it has already been noted that the vertical spatial resolution is fundamentally limited by the raster scan video line structure.

This limitation means that it is very often the case that the aspect ratio of the resulting pixels is not 1 : 1. This means that an apparently square pixel in the video signal does not represent a square area in the real-world. This issue is complicated by the fact that the display monitor may not show a square pixel in response to video data that *does* have a 1 : 1 aspect ratio (consider that all monitors have height and width controls which may even be under user control).

Finally, there is a further complication in the case of solid-state sensors because they produce video which is inherently sampled in the horizontal direction. For example, a CCD sensor may have 580 effective pixels per horizontal line. When these data are read out it is converted to an analogue video waveform which will be re-sampled by a digitiser in an asynchronous manner, with quite possibly a different number of pixels per line, say 512. This becomes a particular problem in some measurement applications where the dimensions of real-world objects are determined by relation to the known size of pixels in the sensor.

In addition to all these problems the magnification in the optical system is often not precisely known. The only sure answer to any of these issues is carefully to calibrate the whole system in its final form using a scene containing objects of known dimensions and/or at known distances, etc.

## 3.3.2 Image display

The basic hardware for video image digitising can be extended to incorporate display of the processed data by feeding the output of the framestore to a digital-to-analogue converter (DAC). In such a simple system the display circuitry would usually operate in synchronism with the incoming video signal. Therefore no extra synchronisation generation circuitry would be necessary, but the problems of disturbances on the display during processing by the host would be virtually inevitable.

A more complex system would have a separate framestore to hold the processed data ready for output to the display monitor. This would be scanned out in the standard video format, but usually asynchronously with the incoming video. This approach has the benefit that images of processed data can be displayed even when there is no video source connected to supply synchronisation pulses. Also the host is less constrained about when it can transfer processed data for display purposes while minimising visible disturbances. The ultimate way to overcome this problem is to employ two display framestores so that one can be on display while the other one is being updated by the host. When the update is complete the two frame stores can be swapped invisibly during frame blanking. The only problem is the cost of the additional memory and the increased complexity of the control circuitry.

In the case of these more complicated systems a separate 'dot-clock' must be used to control the synchronisation generation and read-out process. However, its frequency should be the same as that of the sample clock used by the digitiser if the maximum information content of the digital representation is to be displayed to the user. Each 'dot' will be able to exist in one of $2^l$ intensity levels, where $l$ is determined by the resolution of the DAC. Therefore, the active video will consist of an analogue waveform reconstructed from the contents of the framestore memory by the DAC. Special DACs are available for video signal generation which can automatically mix synchronisation with active video and produce a CCIR standard output waveform with the correct 75 ohms output impedance.

A common requirement for the display of images that were originally monochrome is that of 'pseudo-colouring'. This may allow subtle detail in some images to be enhanced by using clearly distinguishable colours to represent closely related grey levels in the original image. This kind of enhancement can be performed in real time by the image display system if extra circuitry is provided between the framestore and the monitor.

One DAC must be provided to drive each of the red, blue and green (RGB) channels of a colour monitor system. Between the framestore memory output and the three DACs a special bank of memory called a colour look-up table (CLUT) is inserted. This is RAM which can be written by the host computer (off-line) with data in the form of intensity values for each of the three colour channels. Thus when the framestore supplies a grey level value it is not output directly but is used instead as an address for the CLUT. This address recalls a set of colour intensity values such that a particular colour is displayed instead of the original grey level.

If the framestore uses $l$ bits/pixel the CLUT would typically be organised as $2^l \times 24$ bits so that each of the three colour channels is supplied with 8 bits of intensity data per pixel. In this case the colour output will be one of a 'palette' of $2^{24}$ (i.e. $\approx 16.7$ million) different colours, but of course only $2^l$ of these are available simultaneously. However, the host computer can reprogram the CLUT to ensure that an appropriate sub-palette is in use for any purpose.

In particular, if each colour channel is programmed to give the same intensity level for a given input address then a grey scale, rather than coloured, image will be displayed. If the intensity value 'looked-up' is the same as the input address then the original grey scale image can be viewed unaltered. More usefully, however, the CLUT can be written so that although it only outputs shades of grey, they are non-linearly related to the input address. This allows certain forms of contrast manipulation (e.g.

contrast stretching) to be performed in 'real-time' (defined as within one TV frame time of 40 ms, or field time of 20 ms, as appropriate). The only constraint is that the contrast modification function must be predictable in advance – i.e. it cannot vary from image to image without reprogramming the CLUT. That is a task which cannot generally be achieved in real-time.

### 3.3.3 General system performance issues

Dedicated hardware in the form of arithmetic logic units and pipeline processors may be added to allow functions such as convolution and morphological operations to be performed in real-time. For example, a '3 × 3 mask convolution', which is a common image processing requirement (see chapter 4), could be performed over the whole image area within one frame time of 40 ms. The use of such hardware adds considerably to the complexity and cost of machine vision systems.

While the use of specialised hardware architectures undoubtedly speeds up many simple operations, it brings with it the penalty of limited flexibility. The ideal arrangement for the development of any machine vision application is to investigate the problem using a software image processing package and then install software-compatible dedicated boards to speed up time-critical operations. Such hardware is often available from the manufacturers of commercial digitiser/framestore boards [14, 15]. Unfortunately such boards are usually proprietary to their own particular 'family' and so care must be taken in selecting a family that has all of the required capabilities. Obviously the best time to do this is *after* the requirements of the system have been clearly identified.

Note, too, that great care must be taken over the definition of real-time processing if unnecessary expense and complexity is to be avoided in machine vision systems. As noted above, for processing of video-sourced imagery, real-time means completing an operation within one frame time and this is very expensive to achieve. However, for the more general machine vision case, real-time means that the processing task does not interfere with the throughput of the vision controlled system. In some cases, many minutes may be available for processing, in others just a few milliseconds.

One way of easing the complexity of real-time processing is to recognise that a lengthy delay in making a decision can sometimes be tolerated, even when throughput rates are very high indeed. Under these circumstances it is possible to exploit 'pipelined parallelism' by executing a sequence of simple processes one after another. Therefore the result is not known until some time after the processing started, but the throughput of the processing can perhaps be made to match the real-time requirement.

This kind of system organisation lends itself well to flow-line production techniques such as those using a conveyer-belt to transport objects. An image of the object can be acquired upstream of the point where a decision such as accept/reject needs to be made. Provided that the information processing pipeline delay is less than the transport delay between the two points, real-time operation of the manufacturing process is possible.

**References**

1. D. Young, 'Representing images for computer vision', *AISB Quarterly Newsletter*, **63**, pp. 8–11, University of Sussex, 1987.
2. J. Wilson and R. Hodgson, 'A pattern recognition system based on models of the human visual system', In *Proc. IPA '92*, pp. 258–261, Maastricht, Netherlands, IEE, 1992.
3. L. J. Pinson, 'Robot vision: an evaluation of imaging sensors', in A. Pugh (Ed.), *Robot Sensors: Vol. 1 – Vision*, pp. 15–63, IFS (Publications), Bedford, 1986.
4. J. Wilson and J. Hawkes, *Optoelectronics: An Introduction*, Prentice-Hall, Englewood Cliffs, New Jersey, 1983.
5. B. Batchelor, D. Hill and D. Hodgson, *Automated Visual Inspection*, IFS (Publications), Bedford, 1985.
6. I. McLean, *Electronic and Computer-Aided Astronomy*, Ellis Horwood, Chichester, 1989.
7. H. R. Nicholls and M. H. Lee, 'A survey of robot tactile sensing technology', *Int. J. Rob. Res.*, **8**(3), pp. 3–30, 1989.
8. G. P. Weckler, 'Operation of p–n junction photodetectors in a photon flux integrating mode', *IEEE Journal of Solid-State Circuits*, **SC-2** (3), pp. 65–73, Sept. 1967.
9. Technical brief: 'Electrons vs. voltages', *CCD Solid State Imaging Technology Data Book*, p. 354, Loral Fairchild, Milpitas, California, 1991.
10. J. Hollingum, *Machine Vision: The Eyes of Automation*, pp. 94–98, IFS (Publications), Bedford, 1984.
11. S. R. Ruocco, *Robot Sensors and Transducers*, Chapter 3, Open University Press, Milton Keynes, 1987.
12. G. F. Amelio, 'Charge-coupled devices', *Scientific American*, **230**, pp. 22–31, Feb. 1974.
13. K. Miwada *et al.*, 'Third generation CCD linear sensor', *ESSCIRC '85*, pp. 346–358, Toulouse, France, IEEE, New York, 1985.
14. D. Koenig, 'The boards are back in town: board survey', *Image Processing*, **3**(3), pp. 18–36, July/August 1991.
15. P. Atkin, 'Better boards by design: boards survey', *Image Processing*, **4**(2), pp. 31–45, May/June 1992

# 4 Image Preprocessing

## 4.1 Introduction and theoretical background

Image preprocessing seeks to modify and prepare the pixel values of a digitised image to produce a form that is more suitable for subsequent operations within the generic model. There are two major branches of image preprocessing, namely image enhancement and image restoration.

Image enhancement attempts to improve the quality of the image or to emphasise particular aspects within the image. Such an objective usually implies a degree of subjective judgement about the resulting quality and will depend on the operation and the application in question. The results may produce an image which is quite different from the original and some aspects may have to be deliberately sacrificed in order to improve others.

The aim of image restoration is to recover the original image after it has been degraded by 'known' effects such as geometric distortion within a camera system or blur caused by poor optics or movement. In all cases a mathematical or statistical model of the degradation is required so that restorative action can be taken.

Both types of operation take the acquired image array as input and produce a modified image array as output, and they are thus representative of *pure* 'image processing'. Many of the common image processing operations are essentially concerned with the application of linear filtering to the original image 'signal'. An obvious example is the elimination of noise from an image, such as might be acquired in very low light conditions. However, while it may not be as intuitively obvious, removal of motion blur is also a linear filtering operation. Therefore, in order to assist with the basic understanding of these digital signal preprocessing operations, some initial reference to linear systems theory is appropriate [1, 2, 3].

Consider the system depicted in figure 4.1a, where a continuous input signal $f(t)$ is applied to a system, S, which produces an output signal $g(t)$. If the system may be defined as being linear and shift-invariant, such that it obeys the principle of superposition and its properties do not change with time, then the output can be defined mathematically as

$$g(t) = \int_{-\infty}^{\infty} f(\tau) \, h(t-\tau) \, d\tau$$

where $h(t)$ is a function which completely characterises the system response and is known as the 'impulse response' because it represents the output of that filter if it was stimulated by a perfect impulse (an infinitely narrow pulse of infinite amplitude), and the term $h(t-\tau)$ identifies a version of $h(t)$ time-shifted by the amount $\tau$. This integral is called the 'convolution integral' and can be written in short form as

$$g = f * h \quad \text{where } * \text{ implies the 'convolution' operation}$$

Engineers have traditionally been trained to analyse and design such system characteristics in the frequency domain, while observing and measuring the effect of their implementations in the time domain. The mathematical definition of convolution, as given above, shows that filtering cannot only be implemented in the frequency domain, but also in the time domain, if it is more appropriate to do so.

Consider an audio frequency amplifier. Its input signal, whether speech or music, can be depicted as a time-dependent voltage waveform $f(t)$, as in figure 4.1b. Such a time domain waveform can be readily viewed on an ordinary oscilloscope, and this is where the engineer will measure amplitude levels, rise times etc. However, the same input signal can also be depicted in the frequency domain, albeit only with the aid of a sophisticated spectrum

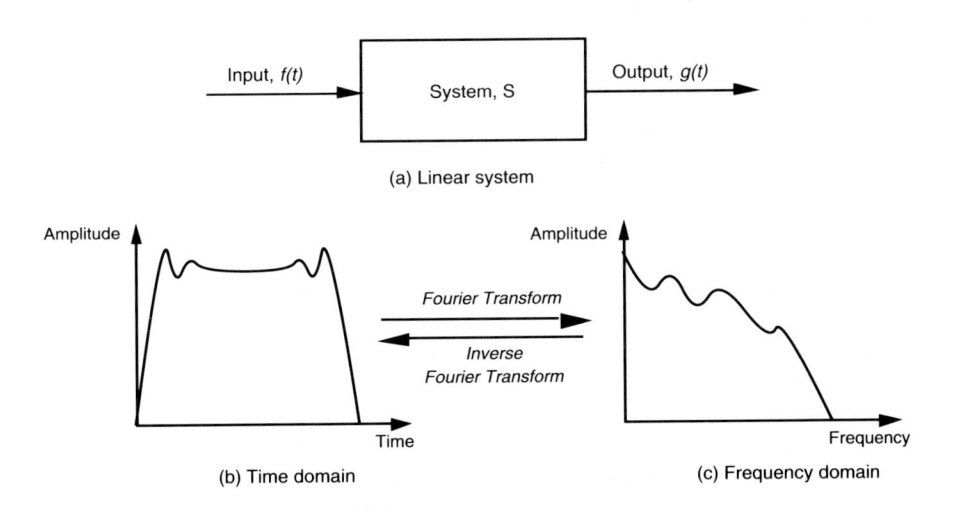

Figure 4.1 Linear systems and signal representations

analyser. Here it will be viewed as a distribution of different voltage amplitudes occurring at particular frequencies in its audio range, or spectrum $F(\omega)$, as illustrated in figure 4.1c. This is the kind of representation that engineers will use to conceive of the amplifier frequency response etc. Therefore, it is useful to be able to study signals in either the time domain or the frequency domain.

Mathematically one can switch between the two domains or representations using the Fourier ($\Im$) and Inverse Fourier ($\Im^{-1}$) Transforms.

The Fourier Transform $F(\omega)$ of $f(t)$ is defined as

$$F(\omega) = \Im f(t) = \int_{-\infty}^{+\infty} f(t) e^{-j2\pi\omega t} dt$$

The Inverse Fourier Transform, transforming $F(\omega)$ back into $f(t)$, is defined as

$$f(t) = \Im^{-1} F(\omega) = \int_{-\infty}^{+\infty} F(\omega) e^{j2\pi\omega t} d\omega$$

Although the above Fourier Transform relations, from an engineering viewpoint, were related to the frequency and time domains associated with electrical signals, the same relation holds for spatially distributed signals where spatial frequency ($u$) and spatial dimension ($x$) are the parameters of interest. Hence we have

$$F(u) = \Im f(x) = \int_{-\infty}^{+\infty} f(x) e^{-j2\pi u x} dx$$
$$f(x) = \Im^{-1} F(u) = \int_{-\infty}^{+\infty} F(u) e^{j2\pi u x} du$$

which allows all of the foregoing mathematical relationships to be applied more widely, and to one-dimensional 'image' data in particular.

Now that the digital processing of signals is becoming increasingly common, the ability to transform between the two domains has acquired a new significance because it is feasible to implement signal processes in either domain. The choice of which domain to work in is merely a question of conceptual and computational convenience.

In practice, some processes like low pass filtering which are *conceptually* most straightforward in terms of the frequency domain may now actually also be realised in that domain thanks to hardware implementations of the Discrete (Fast) form of the Fourier Transform that allow real-time transformation of input signals [4, 5]. Equally, some operations such as 'correlation', are more easily understood in terms of the time domain and are being implemented by digital hardware in that domain.

It is appropriate to note here that correlation, which is the process of quantifying the similarities between two signals, can be expressed mathematically as

$$g(t) = \int_{-\infty}^{+\infty} f(\tau) \, h(t + \tau) \, d\tau$$

where $g(t)$ is known as the correlation coefficient.

Comparison of this equation with the convolution integral given above reveals that the two operations become identical if $h$ is a symmetrical function.

Returning to the above example of a continuous audio signal, we have seen that when such a signal passes through a linear system whose response is characterised as $h(t)$, the impulse response, then the output signal $g(t)$ may be deduced mathematically from the convolution integral. However the system characteristic is normally *expressed* in terms of the frequency domain, e.g. as in 'low pass', 'high pass' filters etc., and denoted by the term $H(\omega)$. $H(\omega)$ and $h(t)$ are also related by the Fourier Transform thus

$$H(\omega) = \Im h(t) \quad \text{and} \quad h(t) = \Im^{-1} H(\omega)$$

Since the time and frequency domains are related by the Fourier Transform, and since the system (or filter) input and output are related by the system frequency response, it is thus possible to determine the output function $g(t)$ in one of two ways – see figure 4.2.

The first option is the direct application of the convolution integral (shown by the left-hand vertical anticlockwise arrows in the figure). The second option (shown by the clockwise arrows in the figure) is the indirect, but conceptually familiar, determination where the input function $f(t)$ is first transformed into the frequency domain $F(\omega)$, multiplied by the filter frequency response $H(\omega)$, and then converted back into the time domain to produce $g(t)$. Thus

$$g(t) = \Im^{-1}\{H(\omega).F(\omega)\} = \Im^{-1}\{H(\omega).\Im\{f(t)\}\}$$

### 4.1.1 Digital convolution

For a digital representation of the continuous signals referred to in the previous section, the integrals can be replaced by summations of a discrete number of points used to represent the signal. Remember that we should always determine how often to sample in the time (or spatial) domain in order to represent the signal correctly. Shannon's sampling theorem states that a band-limited signal (one that comprises a finite set of (spatial)

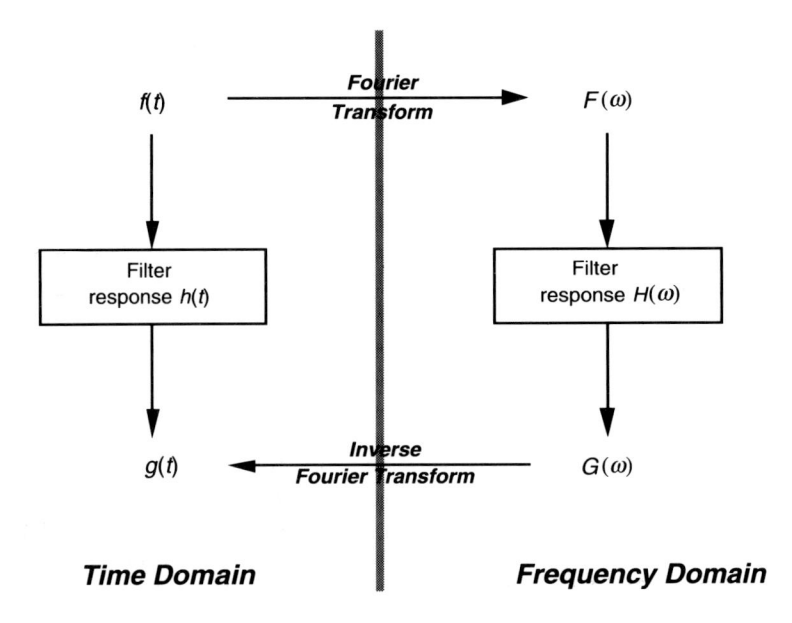

```
                    Fourier
     f(t)  ─────Transform─────▶   F(ω)
      │                             │
      ▼                             ▼
  ┌─────────┐                  ┌─────────┐
  │ Filter  │                  │ Filter  │
  │response │                  │response │
  │  h(t)   │                  │  H(ω)   │
  └─────────┘                  └─────────┘
      │                             │
      ▼                             ▼
               Inverse
     g(t) ◀──Fourier Transform──  G(ω)

        Time Domain              Frequency Domain
```

Figure 4.2   Relation between filtering in the time and frequency domains

frequencies) can be represented or reconstructed if the image is sampled at a frequency which is twice that of the highest (spatial) frequency present in the signal. This sampling frequency is referred to as the 'Nyquist frequency' [6]. Bearing this limitation in mind, the convolution integral for a one-dimensional digitised signal becomes a summation expressed as

$$g(i) = f(i) * h(i) = \sum_{k=-\infty}^{+\infty} f(k)h(i-k)$$

where $i$ is the sample number and $k$ is a shift expressed in terms of an integer number of samples

Although the limits on the summation are infinite, the function $h$ usually takes on finite values only over a limited set of values (say from $-w$ to $+w$) and hence the equation becomes

$$g(i) = \sum_{k=i-w}^{i+w} f(k)h(i-k)$$

Thus the output $g(i)$ at point $i$ is given by the *weighted* sum of the $i$th input sample and its neighbouring samples, where the weights are given by $h(k)$, the sampled representation of the impulse response. The full output is thus deduced by a series of shift, multiply and accumulate operations and the process is termed digital convolution.

Images are two-dimensional signals but the extension of a one-dimensional transformation into two dimensions is straightforward. Consider a picture of a scene containing a picket fence: this represents a regular, periodic pattern in the horizontal dimension of space. When transformed this will be represented by a strong response at a single point in the horizontal axis of the 'spatial frequency domain' corresponding to the fence posts. Similarly, a ladder, resting vertically up against a wall, would produce a substantially strong response in the vertical axis of the spatial frequency domain corresponding to the rungs.

Each pixel (image sample) position can be defined by its $x$ and $y$ coordinate position within the $m \times n$ image array. Thus the earlier transform equations can be adapted to incorporate this two-dimensional aspect by writing $f(x, y)$ instead of $f(i)$. The Fourier Transform of this two-dimensional signal results in $F(u, v)$ which is a function in the spatial frequency domain, where $u$ and $v$ are spatial frequency variables.

The two-dimensional transform definitions are summarised in figure 4.3.

Mathematically, the two-dimensional convolution operation for a continuous signal can be written as

$$g(x, y) = f(x, y) * h(x, y) = \int_{-\infty}^{\infty} \int_{-\infty}^{\infty} f(\alpha, \beta).h(x - \alpha, y - \beta) d\alpha d\beta$$

where $\alpha$ and $\beta$ are dummy variables of integration (cf. $\tau$ earlier).

The two-dimensional Fourier Transforms are

$$F(u, v) = \int_{-\infty}^{\infty} \int_{-\infty}^{\infty} f(x, y).e^{-j2\pi(ux+vy)} dx dy$$

and

$$f(x, y) = \int_{-\infty}^{\infty} \int_{-\infty}^{\infty} F(u, v).e^{j2\pi(ux+vy)} du dv$$

The two-dimensional impulse response of the filter, $h(x, y)$, is also known as the 'point spread function' because it describes the way that the filter in question would spread the energy of a central impulse into its surrounding area.

The discrete form of the two-dimensional convolution can now be written as

$$g(x, y) = f(x, y) * h(x, y) = \sum_i \sum_j f(i, j).h(x - i, y - j)$$

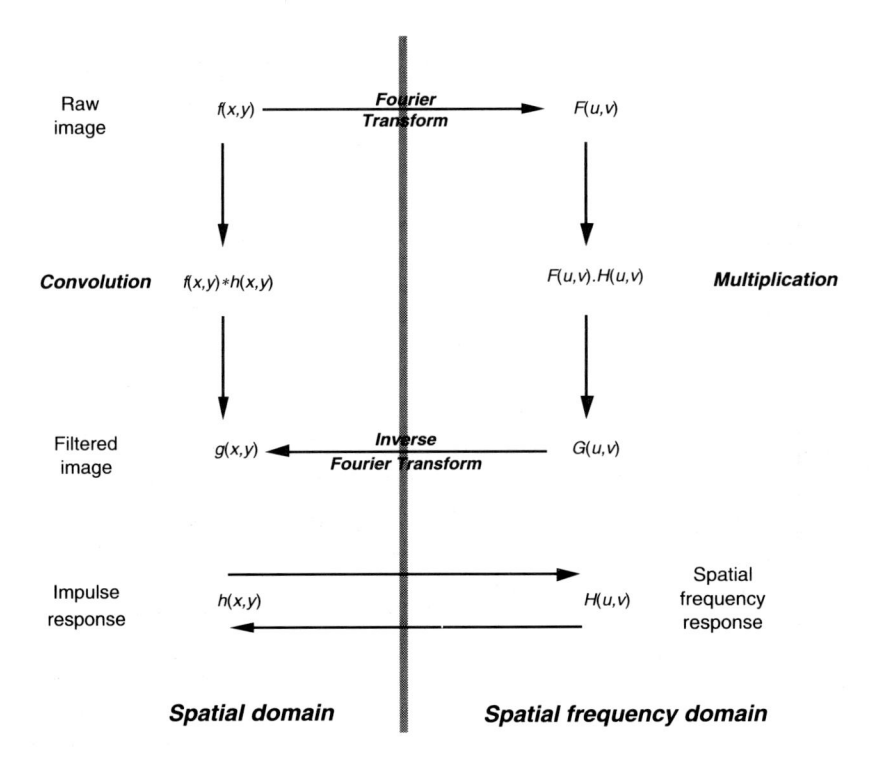

```
                    Fourier
Raw      f(x,y) ─── Transform ──▶  F(u,v)
image
  │                                  │
  │                                  │
  ▼                                  ▼
Convolution  f(x,y)*h(x,y)      F(u,v).H(u,v)    Multiplication
  │                                  │
  │                                  │
  ▼                                  ▼
Filtered              Inverse
image   g(x,y) ◀── Fourier Transform ── G(u,v)


                                                Spatial
Impulse    h(x,y) ──────────────▶  H(u,v)      frequency
response          ◀──────────────               response


       Spatial domain            Spatial frequency domain
```

Figure 4.3  Two-dimensional transform definition summarised.

In this equation, integer values $i$ and $j$ have replaced the linear variables $\alpha$ and $\beta$. The point spread function, $h$, takes the form of a two-dimensionally sampled representation of the continuous impulse response of the filter. The summation is limited to the area where $f$, the image, and $h$ overlap. Thus $h(x, y)$ can be regarded as a window (mask, kernel or template) which is moved over the complete digital image $f(x, y)$.

Figure 4.4 illustrates the convolution operation for a $3 \times 3$ point spread function window. It is superimposed upon the input image, commencing at the origin, and each input pixel is multiplied by the corresponding window value. These nine results are summed and the final value returned to the output image at a position corresponding to the centre element of the window. The window is then moved by one pixel (not the window size) to its next position and the operation is repeated. Alert readers may spot that this implementation actually corresponds mathematically to *correlation*, rather than convolution, because the order of the samples in the window is not

*Image Preprocessing* 97

**Input image**

|    |    |    |    |    |    |    |    |
|----|----|----|----|----|----|----|----|
| 29 | 03 | 06 | 12 | 78 | 79 | 93 | 80 |
| 14 | 25 | 68 | 60 | 66 | 69 | 97 | 83 |
| 16 | 67 | 20 | 16 | 95 | 84 | 68 | 89 |
| 04 | 29 | 08 | 21 | 99 | 85 | 63 | 75 |
| 68 | 62 | 66 | 127| 113| 63 | 48 | 57 |
| 120| 33 | 37 | 121| 67 | 112| 52 | 49 |
| 62 | 65 | 61 | 109| 60 | 64 | 66 | 59 |
| 121| 106| 111| 100| 41 | 56 | 57 | 31 |

**Output image** — output cell value: **-86**

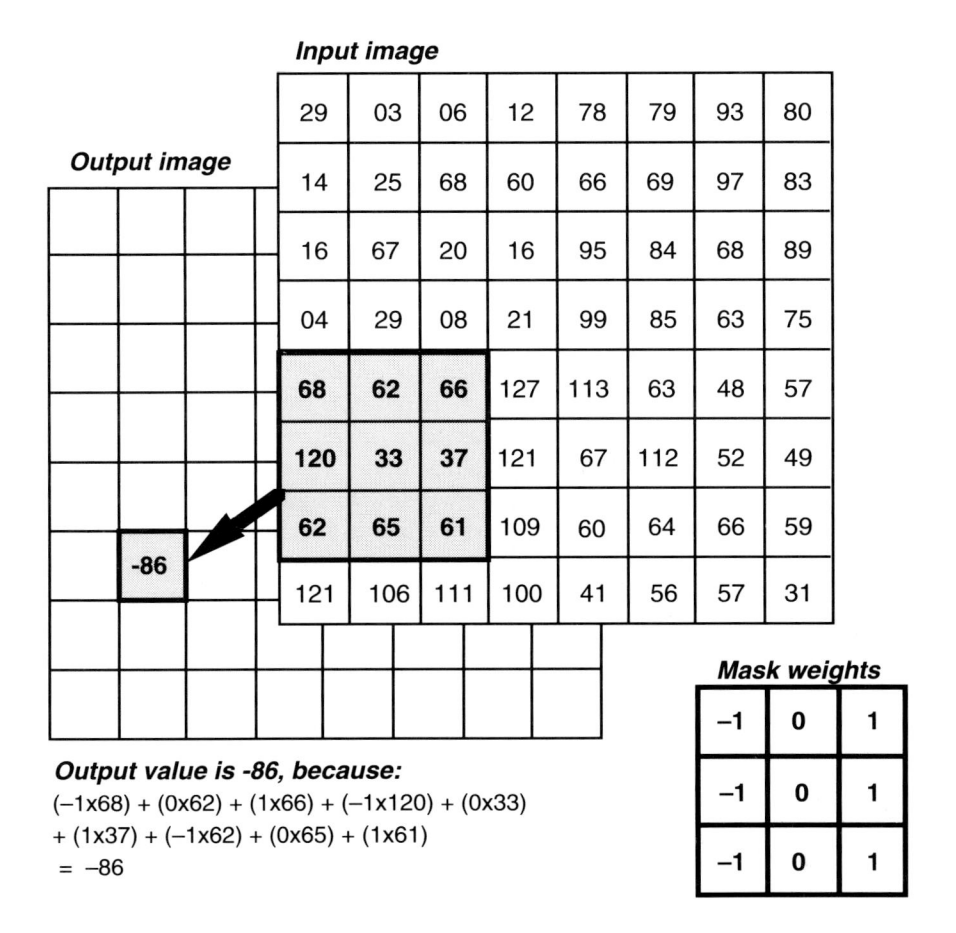

**Mask weights**

| -1 | 0 | 1 |
|----|---|---|
| -1 | 0 | 1 |
| -1 | 0 | 1 |

**Output value is -86, because:**
(−1x68) + (0x62) + (1x66) + (−1x120) + (0x33)
+ (1x37) + (−1x62) + (0x65) + (1x61)
= −86

Figure 4.4   Digital convolution (neighbourhood window operation)

reversed. However, most common image processing operations use *symmetrical* windows, whereupon the two mathematical definitions become identical.

The point spread function may be defined in many ways and this allows many system response functions or filter types to be implemented. Furthermore the basic idea of convolution can be extended by varying the weights of $h$ over the image, and by varying the size and shape of this point spread function window. These operations are no longer linear, and are no longer convolutions in the strictest sense (i.e. they do not follow the summation equations), but they become 'general moving window' operations. These give rise to a very useful family of 'filtering' operations which may be classified as linear (e.g. 'high pass' and 'low pass'), non-linear (e.g. 'mode', '$k$-nearest neighbour' and 'median') and adaptive (e.g. 'sigma') [7].

In order to exploit the versatility of the moving window operation fully, it is necessary to conceptualise the values, or weights, in the window in different ways, according to the task in hand. For example, when performing convolution, the window must be adjusted to approximate to the point spread function of the desired filter characteristic. Unfortunately this is not intuitively trivial for all but the most simple functions, such as a Gaussian (whose Fourier transform is another Gaussian!). Alternatively, when performing a correlation, the window is best thought of as representing a template – i.e. a numerical 'model' of an image feature which is required to be found. This interpretation can assist the intuitive design of window weights, especially for edge operators (which are actually high pass filters), where the window values model a grey level step function (e.g. consider the weights shown in the window in figure 4.4 in this light).

It is common for complex processing operations to be performed by passing a series of windows over the image and combining their outputs to form the final result. The Sobel edge operator is a good example of this. It uses one mask to estimate edge strength in the vertical direction, and a second one to estimate it in the horizontal direction (see section 4.4.2).

The size of the window can be as large as necessary but naturally the computational expense rises according to the square of the window size (each output pixel requires $i \times j$ multiplications and accumulations). Therefore, it is normal to use small square windows, typically $3 \times 3$ and exceptionally $5 \times 5$, and accept the errors which result from the crude approximations to the filter point spread functions. Such small convolution windows only allow for the simplest filtering operations and are much cruder than those normally found in digital signal processing.

A practical issue associated with the digital convolution operation is what to do with the pixels around the border of the image. If the centre of the window is placed upon a pixel at the very edge of the input image, then some of the window weights will operate upon invalid 'pixels' beyond the boundary of the real image. Clearly, any output data derived from this operation will be meaningless. The practical result of this is that the valid output image must be smaller than the input image by $(b - 1)/2$ pixels, for a window size $i \times j$, where $i = j = b$, and $b$ is an odd integer. Some decision about what to do with this border in the output image must be made – it is normally set to black.

If the point spread function, $h$, has the property of separability, whereby $h$ may be represented by a vector output product of a vertical component and a horizontal component, then the system or filter response may be applied to the input image by first convolving it with the horizontal component and then convolving the result with the vertical component. This replaces a two-dimensional convolution with two one-dimensional convolutions and hence reduces processing time from the order $n^2$ to $2n$.

Clearly, digital image processing has its roots in linear systems theory and many operations are implemented by means of a digital form of convolution derived directly from that mathematical basis. However, we have seen how this implementation of convolution has been adapted to a more general moving window class of operations, which includes non-linear ones. For this reason it is appropriate to focus future discussion on a classification of image array 'mapping functions' that are commonly identified by the applied image processing community [8, 9]. Five main operational approaches, for input array to output array mappings, are identified:

(a) Point operations: essentially each pixel is processed in isolation, independent of all other pixels in the image array.
(b) Global operations: the global characteristics (statistics) of the image array are used to modify the pixel values.
(c) Neighbourhood operations: data from the immediate neighbours is used to modify a pixel value.
(d) Geometric operations: the pixel values are modified according to the structural content of the image.
(e) Temporal (frame-based) operations: the resulting image is a combination of more than one unprocessed image.

Each approach is illustrated in more detail in the following sections.

## 4.2  Point operations

A pixel, or point, operation is one in which the output image is a function of the grey-scale values of the pixel at the corresponding position in the input image, and only of that pixel. They do not alter the spatial relationships between pixels within the image at all.

When considering the application of point operations it is often helpful to interpret the image via its grey-level histogram. This histogram is a graphical representation of the number of occurrences of each grey-level intensity (or frequency count) in an image (see figure 4.5a). The abscissa, or $x$-axis, refers to the quantised grey-level value and the ordinate, or $y$-axis, refers to the number of pixels having that grey level.

### 4.2.1  Image brightness modification

Perhaps the simplest pixel operation is a brightness adjustment across the whole image. The need for this can be easily confirmed by looking at the

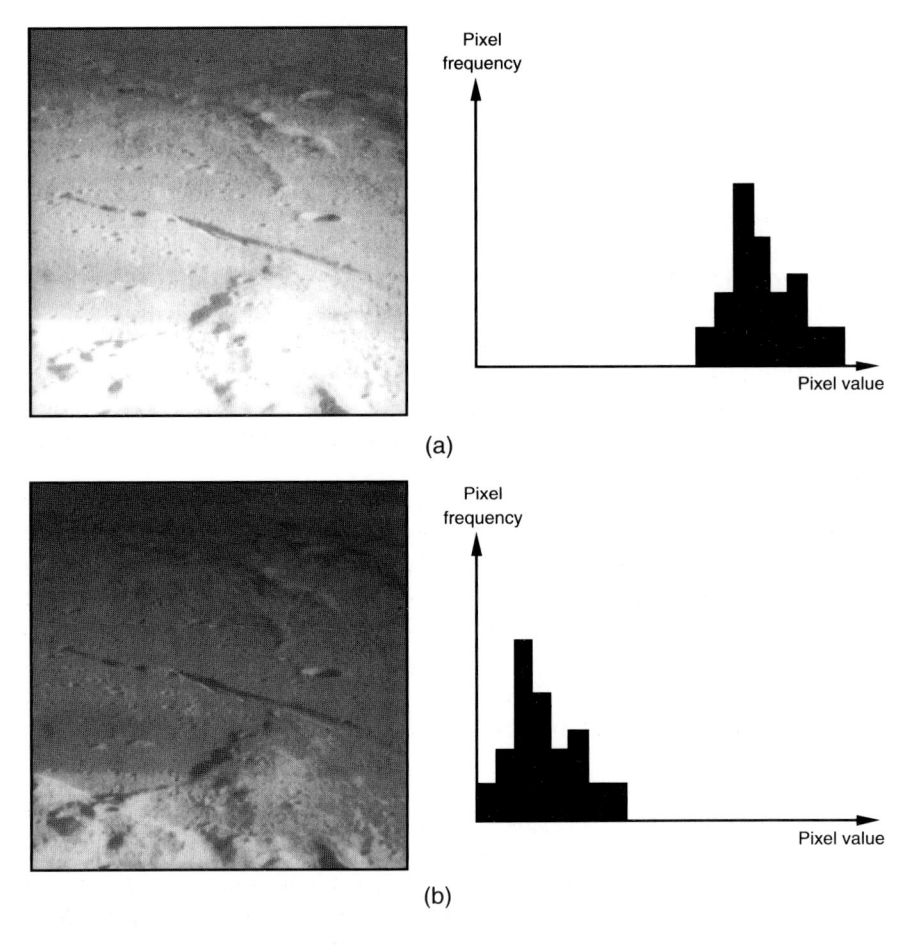

Figure 4.5 Brightness adjustment: (a) raw image and its histogram; (b) brightness reduction and corresponding histogram

histogram; all of the pixels will be concentrated at one end of the range of grey levels and the levels at the other end will be sparsely populated (see figure 4.5a).

Decreasing brightness can be thought of as the simple subtraction of a constant from all pixel intensity values stored in the image array. Such an operation will move the histogram to the left along the abscissa as shown in figure 4.5b. Clearly the image brightness can also be increased by addition of a constant. In general the brightness modification operation can be expressed as

$$P' = A + P$$

where $P'$ is the pixel value after enhancement
$P$ is the pixel value before enhancement
$A$ is the enhancement factor (constant).

### 4.2.2 Contrast enhancement

The brightening operation does not alter the distribution of the pixel intensity values in the histogram in any way, so it does not adjust the image contrast. However, the histogram in figure 4.5b shows that the image does not make full use of the available grey-scale range. In other words, it lacks *contrast*. This can be improved by grey-level scaling where a multiplication operation is used to stretch the histogram to cover the complete range of grey-level values.

Such scaling factors are generally constructed in a piecewise-linear fashion. This allows a compressed portion of the histogram to be spread out more than a sparsely populated portion of the same histogram (e.g. figure 4.6a, function (i)). Figure 4.6b illustrates the effect of function (ii) on the image shown in figure 4.5b. In this example, each pixel intensity value is simply multiplied by $2^l/\{P_{max}\}$, where $\{P_{max}\}$ is the maximum grey-level value of significance in the original image. This ensures that the brightest pixels in the original image are scaled to 'peak white' intensity in the output image.

In this case contrast modification (and brightness adjustment) has been used for image restoration to make up for inappropriate camera work, and has been used to achieve the optimal usage of the available grey levels. However, it can equally well be used for image enhancement, for example of radiograms, where one part of the grey scale is emphasised at the expense of other parts, to reveal specific application-dependent information. Figures 4.6d and e show respectively an original dental radiogram and a version contrast enhanced to give more detail in the spongy bone area, gum and root of the second tooth from the left. Overall, figure 4.6e represents the middle third of the grey scales of the original image re-mapped over the full black–white range.

### 4.2.3 Negation

It is sometimes helpful to be able to work with negative images, where black is mapped as white and vice versa. This can be particularly useful when imaging photographic negatives.

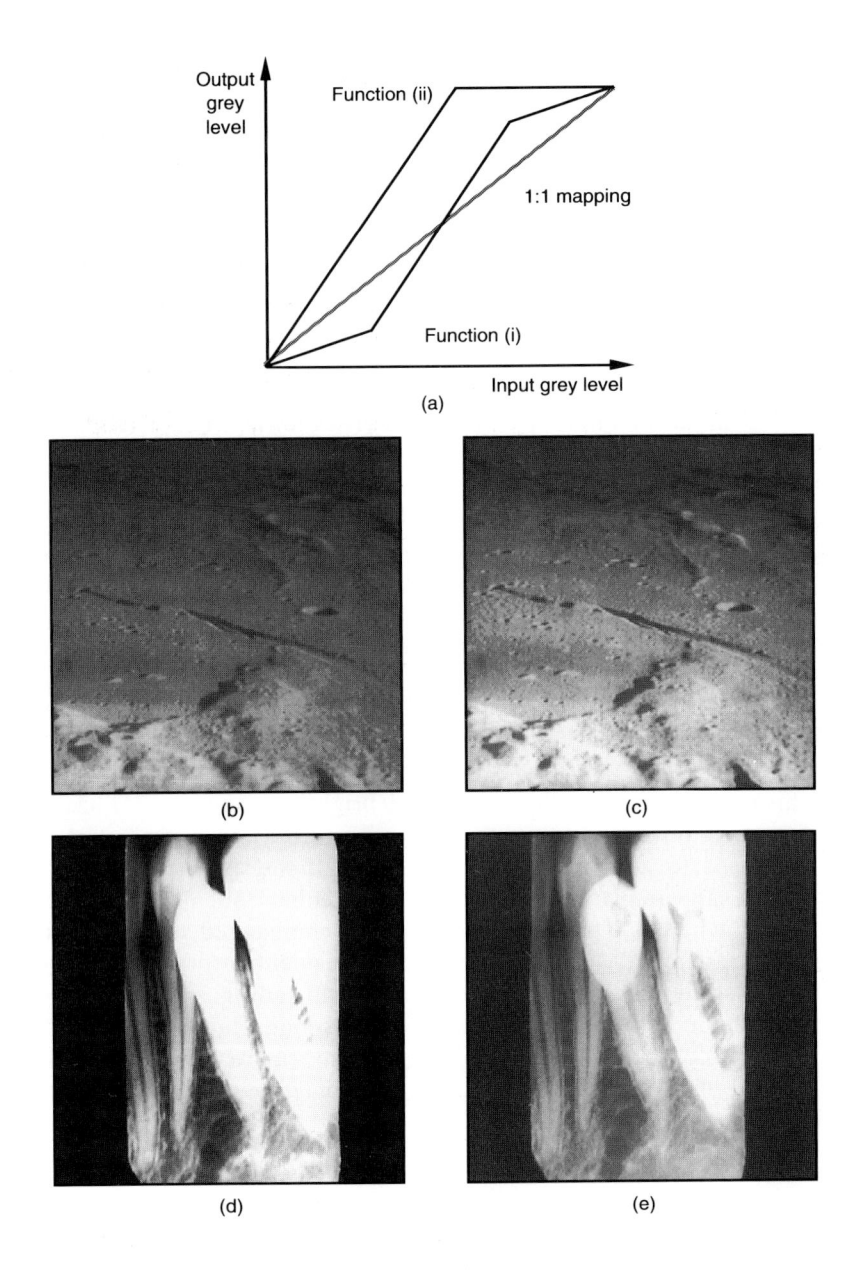

Figure 4.6 Contrast adjustment: (a) examples of piecewise-linear input to output mappings; (b) lunar image (same as figure 4.5b); (c) contrast stretched version; (d) original dental radiogram; (e) enhanced version

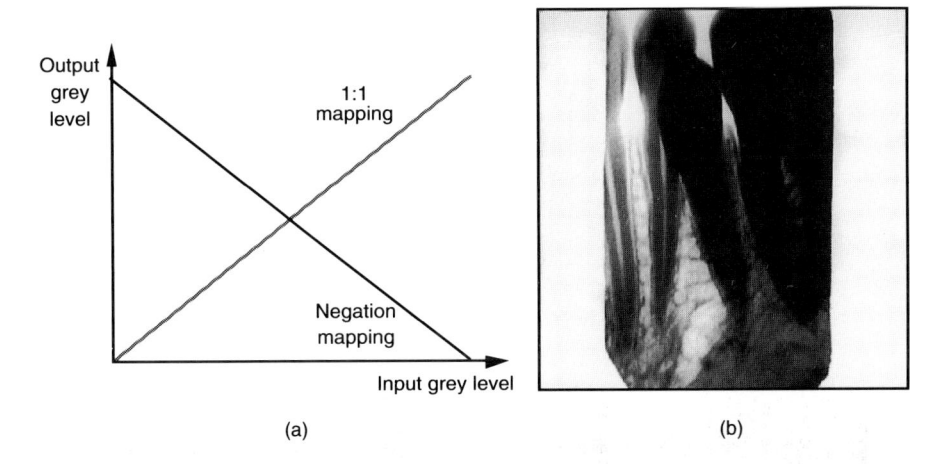

Figure 4.7 (a) Negation mapping; (b) effect of negation on image of figure 4.6(e)

This can be achieved quite simply by subtracting the stored pixel value from the maximum grey-level value being used. This is illustrated in figure 4.7 and defined by the expression

$$P' = 2^l - P$$

where $l$ is the number of grey level bits used
$P'$ and $P$ are as defined previously.

## *4.2.4 Thresholding*

Binary images are much simpler to analyse than grey-scale images, but raw images often cannot be converted directly to binary without some preprocessing. Therefore, there is often a need to threshold a grey-scale image to obtain a binarised version so that the image can be segmented into foreground and background regions. The mathematical definition of the standard binary threshold was discussed in section 3.1.2.

Selection of the value of the threshold, T, is a critical issue. It is common to study the image histogram in order to do this. An image which is well suited to binarisation will feature two or more very clear peaks, separated by well defined troughs. The classic example is the 'bimodal histogram' produced by a high contrast scene and illustrated in figure 4.8a. In such

104                    *Applied Image Processing*

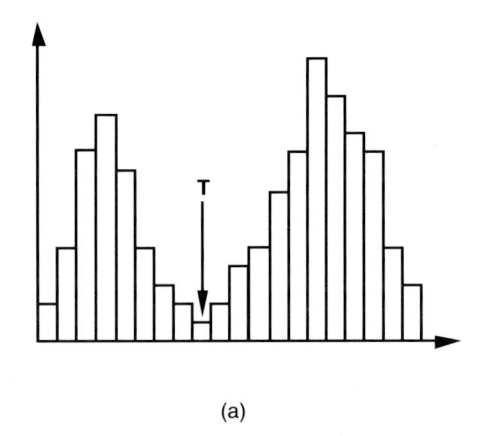

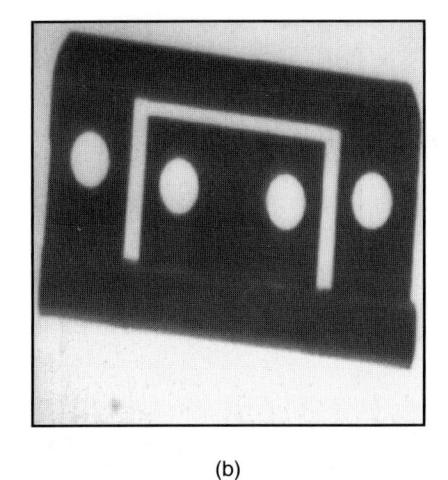

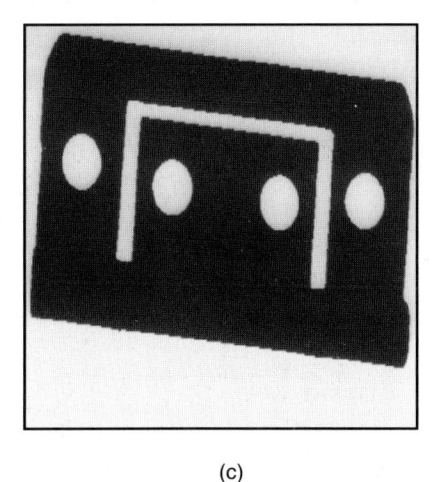

(a)

(b)                                  (c)

Figure 4.8   Selecting a suitable threshold value for binarisation: (a) an ideal, 'bimodal' grey-level histogram showing the ideal threshold value, T; (b) a typical grey-scale image; (c) the result of thresholding at T

cases the threshold can simply be selected either manually or automatically by finding the lowest point of the trough.

One variation on the simple threshold is the dual, or interval, threshold operation (also known as 'window comparison'). Here a binary output image is produced where all grey-level values falling *between* two threshold values $T_1$ and $T_2$ are converted to logic 1 (white) and all grey level values falling *outside* this interval (or window) are converted to logic 0 (black).

Once again, threshold selection is a major issue. The keyboard photograph given in figure 4.9a exhibits a *trimodal* grey-level histogram (see figure 4.9b). Selecting dual thresholds at values $T_1$ and $T_2$ will give the

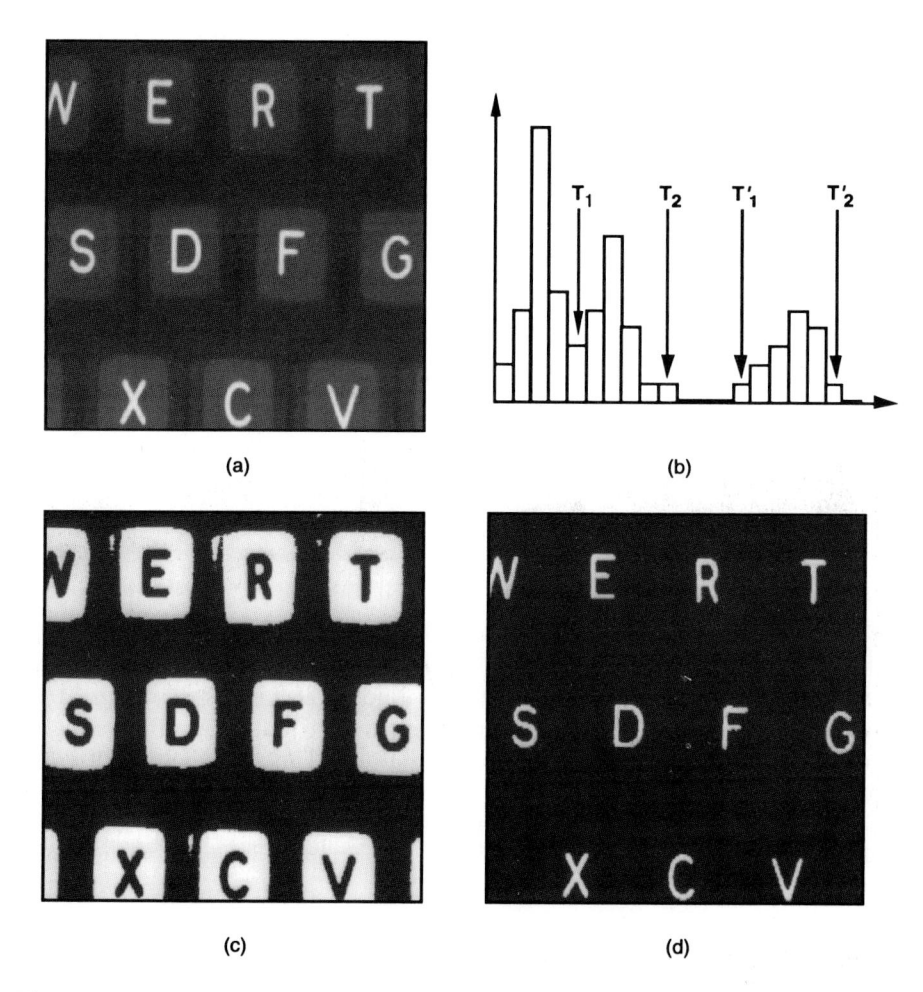

Figure 4.9 The dual threshold operation with examples: (a) the original grey image; (b) its trimodal histogram; (c) the result of thresholds at $T_1$ and $T_2$; (d) thresholds at $T'_1$ and $T'_2$

binarised images shown in figure 4.9c – notice how the top surface of each key is picked out, but also contrasted with the lettering. However, selecting a dual threshold of $T'_1$ and $T'_2$ respectively will produce the output image shown in figure 4.9d – in this case only the key legend is highlighted. Note that figure 4.9d could also have been produced by selecting a single threshold at $T'_1$.

Other variations of thresholding include a 'grey-scale threshold' in which all values between $T_1$ and $T_2$ retain their input grey-scale values but all input values outside this interval are set to zero. Sometimes grey levels that fall within an interval can be highlighted in colour, while the rest are left as is, and so on. Multiple thresholds can be used to reduce the number of grey-level values in an image, i.e. to lower the quantisation level.

### 4.3 Global operations

Global operations rely upon some decision being made about the whole image, based upon its statistics, before being applied locally. It can be argued that thresholding is a global operation because the threshold, T, is determined by inspection of the globally derived grey-level histogram. However, the actual process is *locally* a point operation, without doubt.

However, the following section describes a much more clear-cut global operation.

#### *4.3.1 Histogram equalisation*

The contrast enhancement techniques described above all seek to move or change the grey-level histogram of the image in order to improve contrast and observed image quality. One well known technique is designed to achieve an optimal contrast improvement by redistribution of pixel values in order to produce a *uniform* (flat) histogram. This is called 'histogram equalisation'.

The output image histogram should ideally contain an equal number of pixels at every discrete grey-level value. Given an input image of $m$ rows by $n$ columns using $l$-bit grey-scale resolution then an 'ideal' histogram would be flat with $(m \times n/2^l)$ pixels at each grey level. To produce such a result the distribution of pixels in the input image has to be spread more evenly over the whole image.

Intuitively, the lowest grey-level value in the original image should become zero; pixel values near to the lowest value should also become zero until the new level has its allocation of pixels; this allocation process continues until all grey levels have equal allocations.

Mathematically, this is a mapping function of old grey values to new values and can be expressed as

$$N(g) = \text{Max} \left\{ 0, \text{Round} \left( \frac{2^l \cdot c(g)}{m.n} \right) - 1 \right\} \qquad (4.1)$$

where $N(g)$ is the new grey value
    $c(g)$ is the cumulative pixel count up to old grey level $g$
and   Round implies a rounding to the nearest integer value.

*Image Preprocessing* 107

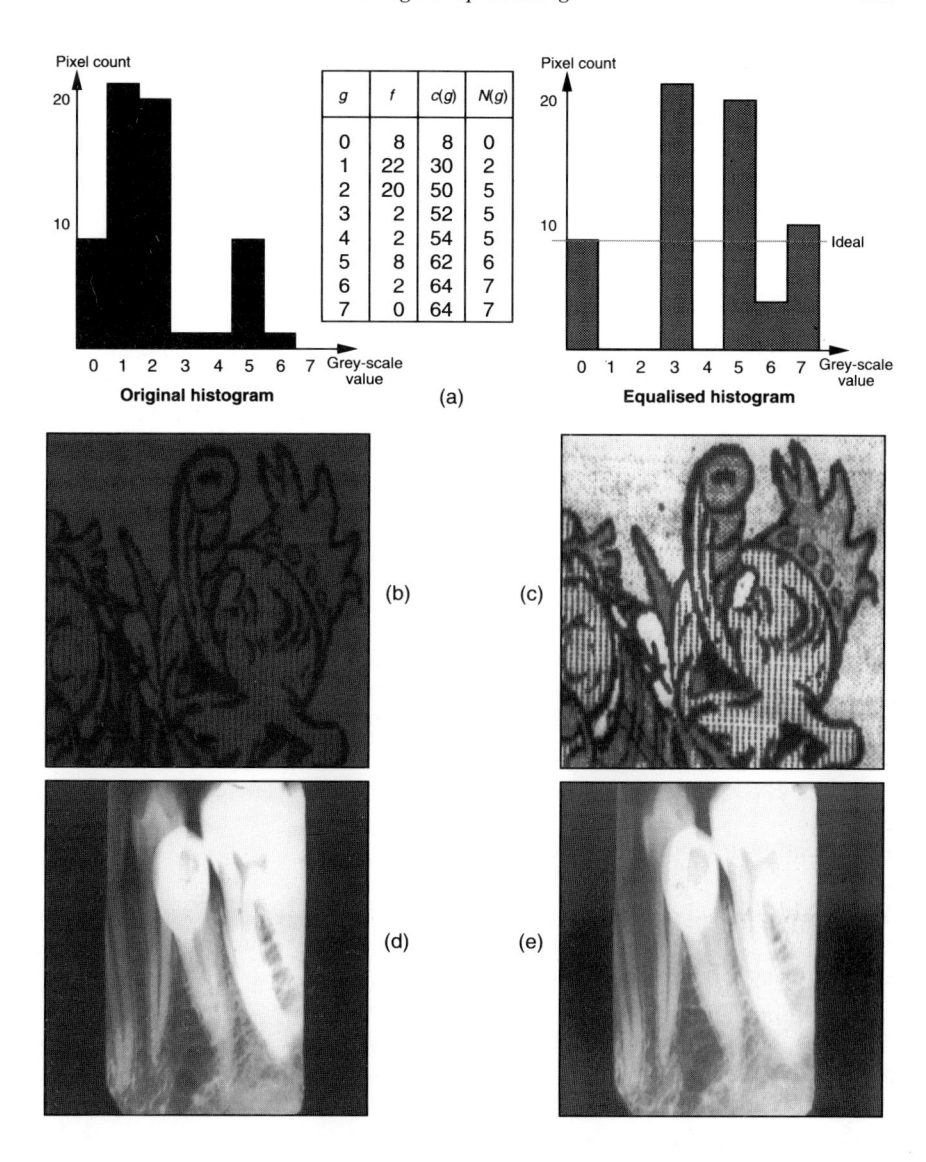

| g | f | c(g) | N(g) |
|---|---|------|------|
| 0 | 8 | 8 | 0 |
| 1 | 22 | 30 | 2 |
| 2 | 20 | 50 | 5 |
| 3 | 2 | 52 | 5 |
| 4 | 2 | 54 | 5 |
| 5 | 8 | 62 | 6 |
| 6 | 2 | 64 | 7 |
| 7 | 0 | 64 | 7 |

(a)

Figure 4.10  Histogram equalisation worked example and images: (a) worked example (see main text for definitions, $f$ = pixel frequency count); (b) raw image well suited to histogram equalisation; (c) the result; (d) raw image unsuited to histogram equalisation; (e) the result

Figure 4.10a illustrates the above mapping using a histogram of a simple image where $m = n = 8$ and $l = 3$ (i.e. 64 total pixels with 8 available grey levels). In the original image histogram, grey levels 0 to 2 contain a total of

50 pixels; the pixel count per grey level is given in the table together with the cumulative total $c(g)$. The new grey-level values $N(g)$ are determined by application of eqn (4.1) to each row of the table.

Since the pixel count per grey level can take on any value between 1 and $(m \times n)$ the equation contains a rounding function because only discrete grey-level values can be used in the output histogram. This rounding process will give rise to some zero pixel allocations at certain grey-level values in the new image.

Although the equalised histogram may still be uneven, the technique is nevertheless very powerful and often used. However, the aim of producing a flat histogram does not suit all images. Like most forms of image processing it is essentially a process of trial-and-error to determine whether a particular image will benefit or suffer from this technique, but the success rate improves markedly with experience. Figure 4.10b shows an image well suited to histogram equalisation, while 4.10d shows one which is not – result images are shown in 4.10c and e respectively.

'Histogram specification' is a variation of the technique where something other than a flat histogram is identified as the goal. This may be used to compensate for the non-linear response of the human visual system. Adaptive histogram equalisation is a technique which divides the image into sub-images and performs separate equalisation operations on each; in some cases it yields better results, but it can be very slow.

## 4.4 Neighbourhood operations

A neighbourhood operation is one in which the output pixel value is determined not only by its input value but by the influence of its neighbouring pixels. Such an operation is thus defined as being 'spatially dependent' since it depends on the pixel values at positions other than the pixel under immediate consideration. Global image statistics do not influence the operation at all.

The neighbourhood is often taken to mean the four nearest neighbours (N, S, E & W; '4-C' connectedness) or the eight nearest neighbours (as 4-C plus NE, NW, SE & SW; '8-C' connectedness), since a neighbourhood of $3 \times 3$ pixels is very commonly considered. However, any neighbourhood size is theoretically allowable up to the maximum spatial resolution of the system but the 'computational expense' of processing increases dramatically as the size of the neighbourhood increases.

In neighbourhood operations the shape of the input histogram may be fundamentally changed and a wide range of complex operations or *image transforms* can be implemented in this way. Therefore many, but not all, neighbourhood processes use the digital convolution technique discussed in section 4.1.1. The following sections discuss some typical neighbourhood processes. However, the list is not exhaustive.

## 4.4.1 Image smoothing

The process of image smoothing seeks to remove unwanted noise from an image while at the same time preserving all of the essential details that an observer would wish to see in the original image. Image noise generally manifests itself as random fluctuations in grey-level values superimposed upon the 'ideal' grey-level pixel value, and it usually has a high spatial frequency.

In these cases it is often acceptable to apply a low pass filter to the input image in order to allow the low spatial frequencies in the 'ideal' image to pass through while attenuating the high spatial frequencies of the noise components. Unfortunately it is impossible to retain all the 'ideal' image detail in such a smoothed image and some degradation will occur. This degradation is usually in the form of blurring where edges in the original image become less well defined as a result of the low-pass operation. Thus all smoothing filters will seek a compromise in removing as much noise as possible while still preserving the detailed edge information.

The filtering operation can be implemented by convolving the entire image with a simple $3 \times 3$ or $5 \times 5$ window or mask as defined in section 4.1.1.

### Spatial averaging (neighbourhood averaging)

This is one of the simplest approaches to smoothing and operates by replacing each pixel value by the average or mean of its immediate neighbours. The averaging process has the effect of ironing out significant grey-level differences between pixels in a neighbourhood. This technique reduces image noise but at the expense of significant image blur as it tends to smear the edges of objects in the image. Spatial averaging corresponds to the convolution of a rectangular 'impulse response' with the image and is therefore equivalent to multiplication by a $\sin x/x$ (or sinc) function in the frequency domain. It is thus possible to see how smoothing is achieved but note also that the large side-lobes which are characteristic of the sinc function tend to cause a degree of rippling.

The smearing effect increases as the mask size increases. The window used in spatial averaging has all of its weights set to $1/(i \times j)$ – where $i$ and $j$ are the dimensions of the window. When this is 'convolved' with the original image, local averaging will occur. The blurring of edges can be partially overcome by incorporating a threshold T – such that the $3 \times 3$ window smoothed value for each pixel may be defined as

$$\text{Smoothed value} = 1/9 \sum (\text{Neighbourhood values})$$

provided that: | Smoothed value − Original value | > T
otherwise: use original pixel value.

Such a scheme will leave unchanged those pixels whose grey-level value is close to the average value of the neighbourhood.

*Gaussian filter*

The Gaussian filter is widely used in image analysis as it enables the system designer to select the degree of smoothing to be achieved. This can be done by changing the convolution mask coefficients so that an appropriate degree of detail is retained within the smoothed image. The one dimensional Gaussian function can be expressed as

$$G(x) = \frac{1}{\sqrt{2\pi}\sigma} \exp\{-x^2/2\sigma^2\}$$

where $\sigma$ is the standard deviation for this distribution.

The two-dimensional Gaussian function, which has circular symmetry, is written

$$G(x,y) = \frac{1}{2\pi\sigma^2} \exp\{-(x^2+y^2)/2\sigma^2\} \qquad (4.2)$$

Two-dimensional filtering can be achieved by convolving the image with a one-dimensional Gaussian mask in both the vertical and then the horizontal directions. The degree of smoothing is controlled by the shape of the Gaussian curve used, that is, it depends on the selected value of standard deviation, $\sigma$ – see figure 4.11. The Gaussian curve is also characterised by a completely smooth decay to zero, unlike the ripple induced by the sinc function, as implemented by spatial averaging.

The digital version of the function uses a sufficient number of mask coefficients to effect the degree of accuracy required. A small value of $\sigma$ will effect a limited degree of smoothing while retaining fine edge detail; a large value of $\sigma$ will effect greater smoothing (hence excluding fine detail) while retaining only the major edges and lines. Mask sizes are typically chosen to accommodate spreads of $\pm 3\sigma$ since 99% of the area under the Gaussian curve lies within this range. Even so, with the large mask sizes generally used, the speed of processing is often too slow for industrial machine vision systems, though the technique is generally popular with the computer vision community and those seeking to manipulate and enhance images 'off-line'.

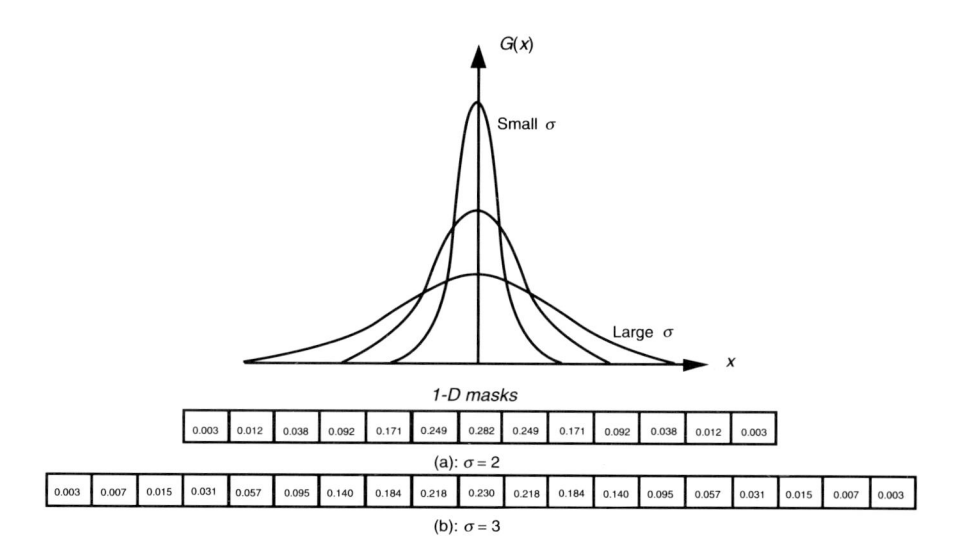

Figure 4.11  One-dimensional Gaussian curves and discrete convolution masks

*Other smoothing filters*

These include the '*weighted mean*' filter where the mask weight for a pixel is related to its distance away from the centre of the window.

The '*mode*' filter replaces a pixel by the value of its most common neighbour. This is useful when trying to label large similar areas within an image, as for fields and crops in remotely sensed images.

The '*k-nearest neighbour*' filter sorts all the pixels in the window and then averages the $k$ pixels closest to the value of the pixel at the centre of the window; this sets the new pixel value. $k$ is selected as a constant value less than the window area. Similarly the '*sigma*' filter sets the new pixel value by averaging all the neighbourhood pixels whose value is within some parameter, $p$, of the original pixel value. This parameter, $p$, is derived from the standard deviation of the neighbourhood pixel value distribution. All of these filters seek to preserve as much as possible of the edge detail in the original image.

The 'mode', '$k$-nearest neighbour' and 'sigma' filters involve non-linear operations such as sorting and thus *do not* represent convolution of a mask with an image, yet they are examples of neighbourhood moving window operations. Another simple non-linear moving window operation which has a particular application niche is the '*median*' filter. This requires that each pixel value in the original image is replaced by the median of the pixels in a window (see figure 4.12).

In order to find the median value for a $3 \times 3$ neighbourhood it is necessary to sort the nine pixel values into either ascending or descending order; the middle or median value is then selected as the result to be placed in the corresponding pixel in the output image. In the example shown in figure 4.12a the filtered value is 17. Intuitively, this is representative of the nine neighbourhood values, however the neighbourhood *average* would give 38.5 which is not representative of the neighbourhood.

The median filter disregards extreme values (high or low) and does not allow them to influence the selection of a pixel value which is truly representative of the neighbourhood. It is therefore good at removing isolated extreme noise pixels (often known as 'salt and pepper' noise), while substantially retaining spatial detail (see figure 4.12b, c). However, its performance deteriorates when the number of noise pixels is more than half the number of pixels in the window.

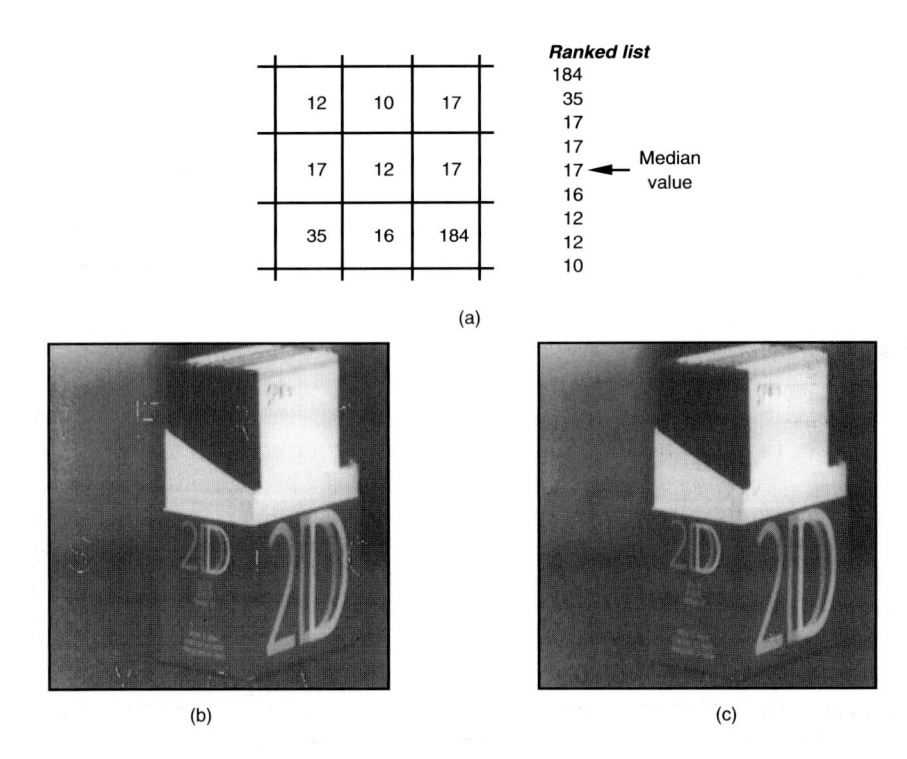

Figure 4.12 Worked examples of median filter and images: (a) worked example; (b) original image contaminated with simulated 'salt and pepper' type 'noise'; (c) the result of median filtering (b)

## 4.4.2 Image sharpening

Another class of filters can be defined by their effect of emphasising or strengthening the edges within an image. In general a high pass filter has the inverse characteristic of a low pass filter; it will not change the high frequency component of the signal but will attenuate the low frequencies and eliminate any constant background intensity.

Therefore the high pass filter has the property that pixel data in the region of an object edge is modified and the edge effect is enhanced. This property of edge detection is a prerequisite for many image analysis operations (see Chapter 5), and hence filters which locate and enhance edge information are of major importance.

### Edges and lines

In a digitised image an edge is defined as a sequence of connected edge points (see figure 4.13).

An *edge* is characterised by an abrupt change in intensity indicating the boundary between two regions in an image, and is often referred to as a discontinuity. In practical images an edge is often seen as a slow change in grey-level values between connected pixels. If noise is present in the image then the profile of the edge will contain random fluctuations. Ideally an edge is a marked change in grey-level value. A *line* is a region of constant

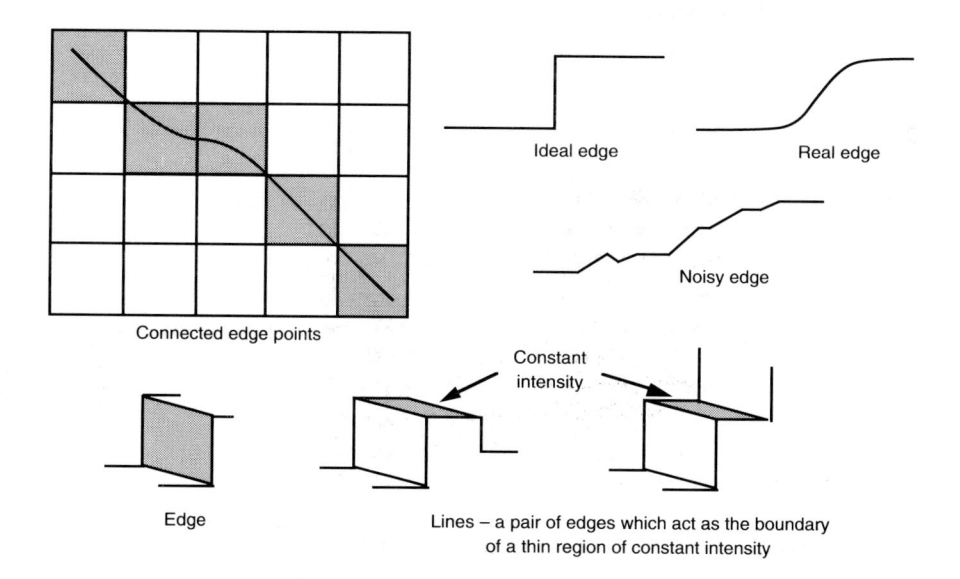

Figure 4.13 Edges and lines

intensity found between two edges which act as a boundary for the line. Two examples of such constant intensity lines are illustrated in figure 4.13.

*Gradient operators and edge enhancement*

Methods of identifying the intensity discontinuities that mark object edges within an image are usually based on the calculation of intensity gradients across the image. The occurrence of a high local intensity gradient, indicating a sudden intensity transition, is evidence for the existence of an edge discontinuity. The gradient calculations may utilise first or second order operators.

A first-order gradient, defined for a continuous function $f(x,y)$, is given by

$$\frac{df}{dx}.\bar{i} + \frac{df}{dy}.\bar{j}$$

where $\bar{i}$ and $\bar{j}$ are unit vectors in the $x$ and $y$ directions.

A second-order gradient (Laplacian) can be formally defined as

$$\frac{d^2f}{dx^2} + \frac{d^2f}{dy^2}$$

The first-order gradient operator when applied to a continuous function produces a vector at each point whose angle gives the direction of maximum change of the function at that point, and whose amplitude gives the magnitude of this maximum change. The separability criterion allows a digital gradient to be found by convolving the image with two windows, one giving the $x$ component, $g_x$, of the gradient, the other giving the $y$ component, $g_y$. Thus

$$G_x(i,j) = (x\text{-window}) * n(i,j) \quad \text{and} \quad G_y(i,j) = (y\text{-window}) * n(i,j)$$

where $n(i,j)$ is some neighbourhood of $(i,j)$, and
    $*$ represents the usual digital convolution calculation.
An appropriate set of window weights for a first-order gradient operator might be

$$x\text{-window} = [-1\ 0\ 1] \quad \text{and} \quad y\text{-window} = \begin{bmatrix} 1 \\ 0 \\ -1 \end{bmatrix}$$

since these replace the pixel at the centre of the window with the finite-difference of its neighbours on either side (x-window), or above and below (y-window). This pair of masks responds to edge intensity changes in both the horizontal and vertical directions and the combined result yields a gradient vector. The magnitude, $G_m$, and angle, $G_\phi$, of the gradient vector are calculated as follows

$$G_m(i,j) = \sqrt{G_x^2 + G_y^2}$$
$$G_\phi(i,j) = \tan^{-1}(G_y/G_x)$$

Note that the direction of this vector is normal to the edge in the scene.

In a uniform region of an image the resultant gradients will be low value, whereas in a region containing edges the resulting gradient calculations will produce high values. One disadvantage of the proposed operator is its susceptibility to noise since it simply computes the finite-difference between every pixel and its immediate vertical or horizontal neighbours, and this can lead to spurious responses. Other common masks attempt to introduce a degree of 'smoothing' into the estimation of the gradient by applying weights to the pixels *along* the direction of the edge component. Common examples are Roberts, Prewitt and Sobel operators [7, 10, 11].

The Roberts edge detector is one of the simplest practical gradient operators. It uses two $2 \times 2$ masks convolved with the image. Each mask determines the intensity change in one of two diagonal, but mutually orthogonal, directions and the resultant overall gradient vector is evaluated by appropriately combining these two orthogonal results (see figure 4.14).

The Prewitt gradient operator is clearly related to the simple conceptual operator proposed above, but incorporates a degree of smoothing by virtue of giving equal weightings to pixel finite-differences which are horizontally or vertically adjacent to the origin – see figure 4.15a.

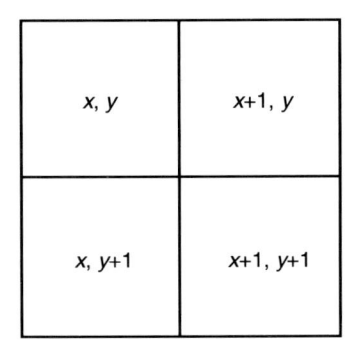

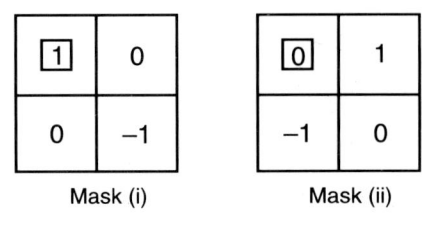

| x, y | x+1, y |
|---|---|
| x, y+1 | x+1, y+1 |

| 1 | 0 |
|---|---|
| 0 | −1 |

Mask (i)

| 0 | 1 |
|---|---|
| −1 | 0 |

Mask (ii)

|Gradient| $= \sqrt{\{f(x, y) - f(x+1, y+1)\}^2 + \{f(x+1, y) - f(x, y+1)\}^2}$
$\approx |f(x, y) - f(x+1, y+1)| + |f(x+1, y) - f(x, y+1)|$

Figure 4.14 The Roberts gradient operator. Note that the boxed element indicates the 'origin' of the mask

(a) $x$-window $= \begin{bmatrix} -1 & 0 & 1 \\ -1 & \boxed{0} & 1 \\ -1 & 0 & 1 \end{bmatrix}$ and $y$-window $= \begin{bmatrix} 1 & 1 & 1 \\ 0 & \boxed{0} & 0 \\ -1 & -1 & -1 \end{bmatrix}$

(b) $x$-window $= \begin{bmatrix} -1 & 0 & 1 \\ -2 & \boxed{0} & 2 \\ -1 & 0 & 1 \end{bmatrix}$ and $y$-window $= \begin{bmatrix} 1 & 2 & 1 \\ 0 & \boxed{0} & 0 \\ -1 & -2 & -1 \end{bmatrix}$

Figure 4.15  Simple 3 × 3 gradient operators: (a) the Prewitt operator; (b) the Sobel operator

The Sobel operator is recognised as one of the best 'simple' edge operators. It utilises two 3 × 3 masks as defined in figure 4.15b. The origin of the weights in the masks is obscure, and they were probably derived empirically in the first place, however it is clear that they apply more weight to the central pixel finite-differences than the horizontally or vertically adjacent ones. A convincing justification for the particular effectiveness of the masks, presented in terms of their close approximation to a circular weighted neighbourhood, is provided by Davies [11].

In well constrained applications of machine vision, images are likely to feature high contrast between regions of interest and their backgrounds. This may mean that it is not important to get an accurate estimate of the gradient vector, since it will be subsequently thresholded to indicate merely the presence or absence of an edge point. In such circumstances it may be considered unnecessary to incur the computational expense of calculating the direction of the gradient vector, and its magnitude may be estimated by one of the approximations which are in common use, as follows [12]:

$$G_m(i,j) \approx |G_x| + |G_y| \quad \text{or} \quad G_m(i,j) \approx \max\{|G_x|, |G_y|\}$$

Alternatively, in other edge detection applications where *accuracy* of estimation of the gradient vector is not of paramount importance, 'compass operators' may be used instead of gradient operators; – see section 5.2.1. This involves the correlation of a set of 'templates' (usually eight), with the image, and is thus an example of a general moving window operation rather than strict convolution. Each template models an edge at a discrete orientation. The template which gives the maximal response indicates the presence of a candidate edge point and the orientation of the edge is indicated by the (crudely quantised) orientation of that template.

Furthermore, an approach to the identification of *specific* structural features in an image, such as edges and corners, is presented in Chapter 6 under mathematical morphology; the hit–miss transform (section 6.5.2).

*Unsharp masking*

This technique derives its name from the idea that subtracting from an image an amount which is proportional to the 'unsharp', or low-pass filtered, version of the same image will 'crispen', or sharpen, the appearance of edges. This is common practice in the printing industry to improve the appearance of printed photographs. However, it is normally easier to achieve this effect by *adding* a fraction of the gradient, or high-pass-filtered, version of the image to the original.

This process is illustrated in figure 4.16 for a one-dimensional continuous signal. It is clear that subtracting 4.16b from 4.16a results in 4.16c which has the form of the second derivative of the original signal (known as the '*Laplacian*'). It is clear that the enhanced edge contour in figure 4.16d can be produced by adding 4.16c to 4.16a.

For digital images it is easier to derive the Laplacian directly by means of finite difference operators. In one dimension, the 3-element mask is $[-1 \ 2 \ -1]$ because the first difference subtracts pixel(1) from pixel(2) and

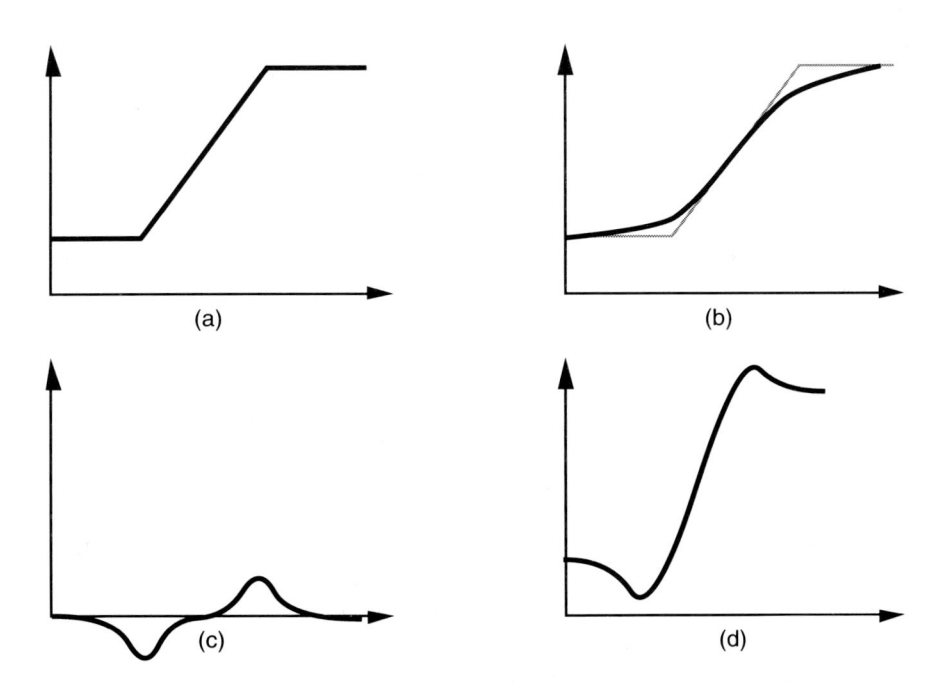

Figure 4.16 Laplacian unsharp masking (in one dimension): (a) a typical discontinuity; (b) the result of low-pass filtering; (c) $[(a) - (b)] \equiv$ | Laplacian | of the original signal; (d) the 'sharpened' result

pixel(2) from pixel(3) to produce two intermediate finite-difference results. The second difference subtracts the second of these intermediate results from the first and therefore implements the equation:

$$-\text{pixel}(1) + 2.\text{pixel}(2) - \text{pixel}(3)$$

When extended to two dimensions, the 3 × 3 Laplacian mask becomes

$$\begin{bmatrix} 0 & -1 & 0 \\ -1 & \boxed{4} & -1 \\ 0 & -1 & 0 \end{bmatrix}$$

This is a common form of the two-dimensional Laplacian mask; it can be used alone to locate discontinuities but unfortunately it amplifies noise and produces double responses to edges. However, when used in unsharp masking a proportion of this mask is added to the original image according to the following equation:

$$g(x, y) = f(x, y) + \alpha.G(x, y)$$

where $G(x, y)$ is the Laplacian gradient operator, and
$\alpha$ controls the degree of sharpening achieved.

Since integer arithmetic is very often used for computational efficiency in the implementation of moving window operations, the parameter $\alpha$ is usually set to unity, whereupon the mask operations involved in sharpening can be simplified to

$$\begin{bmatrix} 0 & 0 & 0 \\ 0 & \boxed{1} & 0 \\ 0 & 0 & 0 \end{bmatrix} + \begin{bmatrix} 0 & -1 & 0 \\ -1 & \boxed{4} & -1 \\ 0 & -1 & 0 \end{bmatrix} = \begin{bmatrix} 0 & -1 & 0 \\ -1 & \boxed{5} & -1 \\ 0 & -1 & 0 \end{bmatrix}$$

This ensures that only one mask need be convolved with the image to achieve the desired effect (note that the use of floating-point convolution allows $\alpha$ to assume any value, giving a much finer control of the degree of sharpening achieved).

Figure 4.17 shows an original image and then results of the Roberts, Sobel and (integer) sharpening operators.

## 4.5 Geometric operations

In geometric operations the spatial distribution of pixels is deliberately changed to achieve the desired effect. This might be because of a known distortion in an image, perhaps because of adopting a viewpoint which is

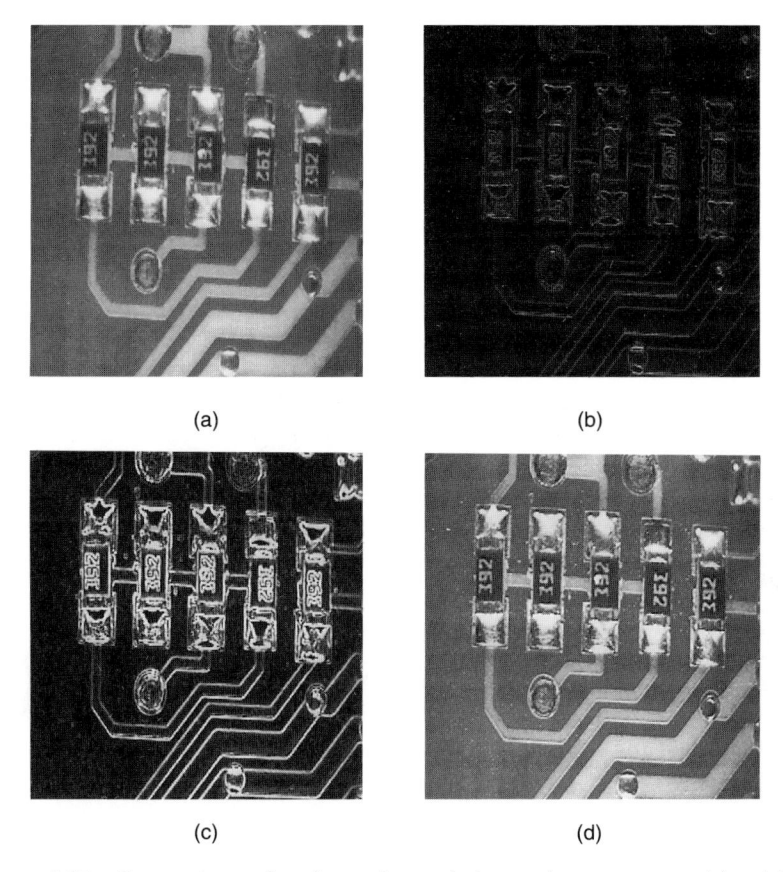

Figure 4.17 Comparison of various edge and sharpening operators: (a) original image; (b) result of the Roberts operator; (c) result of the Sobel operator; (d) result of the sharpening operator

not normal to the surface under investigation. Alternatively geometric corrections may be required simply for image-enhancement purposes or even just for fun! The 'hall of mirrors' at a funfair is an excellent low-technology example of geometric image manipulation, and the computer graphic effect known as 'morphing', where one object transforms its shape into another one, is currently very popular in feature films and TV advertising.

## 4.5.1 Display adjustment

The simplest form of geometric process is to read the data out of the frame buffer in a different order from that in which it was stored. This can achieve inversion and lateral inversion, either of which may be necessary because of inversions encountered in the optics of machine vision systems.

Figure 4.18a identifies one standard convention used to identify the $x$ and $y$ axes for an image. Note that the origin is taken at the top left of the image and increasing $y$ means movement *down* the image. Normally the stored image array values are read out in row order 0 to $(n-1)$, taking on each row the column order 0 to $(m-1)$. Changing the order in which these values are fed to the monitor display circuitry will produce the displays depicted in figure 4.18b.

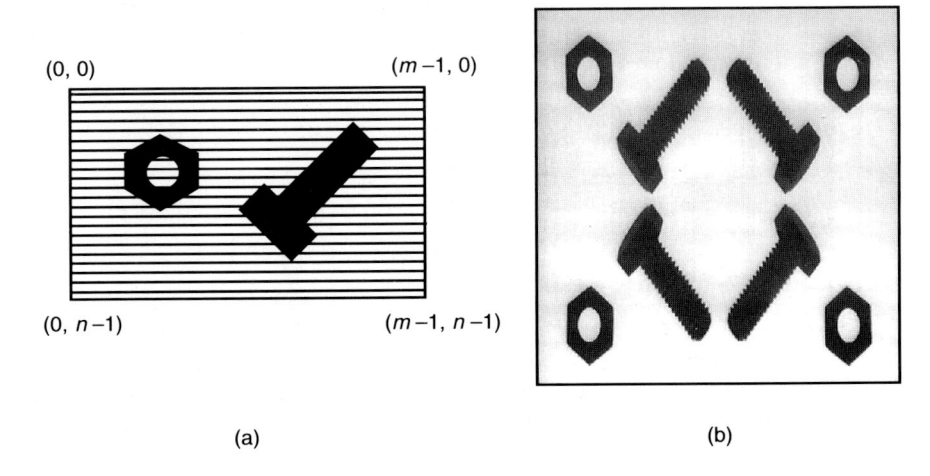

Figure 4.18 Display manipulation: (a) pixel identification standard; (b) illustration of inversion and lateral inversion

### 4.5.2 Image warping, magnification and rotation

Image warping is typically used to correct for distortion introduced by the image acquisition system or to distort an image deliberately to produce more meaningful results. One such example is in the processing of satellite imagery where the image may be warped to produce an apparent view of one area of land as if an imaging satellite was positioned directly overhead and not, for example, in orbit over the equator. Such a warping would produce lines of latitude and longitude which would appear as straight lines on a rectangular grid and not as curved lines intersecting at varying angles [13].

An image warping operation consists of two basic steps. The first is a spatial transformation in which pixels in the input image are mapped onto the new output image plane. The second is a grey-level interpolation to estimate the grey-level values for the warped image.

The spatial transformation can be expressed as a mapping function for an input image pixel $(x,y)$ to become an output image pixel $(x',y')$ as

$$x' = W_x(x,y) \quad \text{and} \quad y' = W_y(x,y)$$

In general the mapping functions $W_x(x,y)$ and $W_y(x,y)$ will be very complex and thus will not be derived by rigorous mathematical analysis. However the spatial warping can be modelled by selecting a number of pixels in the input image (control points) for which the output positions are known. The input image is overlaid with a grid of control points which divide the image into rectangles. These control points can then be moved so that the rectangles become arbitrary, but tessellated, polygons. This is a four-point warp which for the polygon under consideration is defined by the following generalised mapping expressions

$$x' = W_x(x,y) = c_1 x + c_2 y + c_3 xy + c_4$$
$$y' = W_y(x,y) = c_5 x + c_6 y + c_7 xy + c_8$$

where $c_1$, $c_2$, $c_3$, $c_4$, $c_5$, $c_6$, $c_7$ and $c_8$ are coefficients determined from the four corners of the new polygon.

These equations map every point in the original rectangle to a new point in the output polygon, but the values of $x'$ and $y'$ will be fractional. Therefore it is not possible simply to copy the original pixel intensity value to a new location in the output image. Instead it is necessary to 'interpolate' the grey-level intensity values of the output image which must still lie upon a regular orthogonal grid. Interpolation may be performed on a nearest-neighbour basis, where the pixel in the output image is assigned the unmodified grey level of the re-mapped input pixel which falls closest to its location.

A more accurate form is bilinear interpolation where the output pixel value is accurately calculated by assuming that straight lines link the four re-mapped grey levels that surround the desired output pixel location, making up a planar surface. The coordinates of the desired output pixel determine where its grey level lies on that surface. As may be reasonably expected, bilinear interpolation gives much better results than nearest-neighbour, but executes much more slowly.

A special case of warping is image magnification. While this is a trivial thing to perform in an optical system, it is far from trivial as an image processing operation. This is because interpolation must be performed unless the desired magnification factor is an integer power of 2.

Rotation of an image or part of it brings similar problems if the angle of rotation is not an integer multiple of 90°, and even then there can be

Figure 4.19  Various geometric operations on a digital image: (a) the original image; (b) image with 36 polygon warping grid superimposed; (c) the image after warping; (d) 4× linear magnification using nearest-neighbour interpolation; (e) similar magnification using bilinear interpolation; (f) translation and rotation (through 274°) of a portion of the image

problems if the image aspect ratio is not 1:1. Once again these are indications that image processing should NOT be used to compensate for poorly controlled images!

Figure 4.19 depicts the various operations discussed in this section on a digital image.

## 4.6 Temporal (frame-based) operations

The preceding discussions have been entirely concerned with carrying out processes upon a single static image in isolation. Clearly there are many modern uses of images which rely upon a sequence of images being produced by, say, a video camera. Correspondingly, there are many image processing routines that operate upon a sequence of two or more images in order to achieve the desired effect.

Perhaps the most fundamental technique is the 'frame difference' operator where the output image is the product of a point-by-point subtraction of one image from another. This versatile technique is used as the basis of some automated visual inspection processes where a prototype scene is compared with a scene which is known to contain a reference image. For example, in PCB inspection, a newly constructed board can be compared with a 'known good board' in such a way and any missing or misoriented components will be clearly highlighted. Since the basic operating principle of the frame difference operator is so simple, it is possible easily to build hardware systems that will allow real-time comparison of a reference frame with standard video camera outputs. Figure 4.20 shows two images of keyboards, one of which has had faults introduced into the key legends. The subsequent images show the results of the frame difference operator and the subsequent application of a median filter to remove spurious noise spots.

Another application of the frame difference operator is motion detection. Objects which do not move between the two frames will be cancelled in the output image, but anything that moves will be clearly highlighted and the amount and direction of motion can be easily estimated. This has clear potential in security applications and is the basis of many compression algorithms for moving image sequences (see Chapter 9).

Generally, the technique is useful for eliminating unwanted systematic effects that are known to be constantly present in a vision system – as such it represents a form of calibration. For example, a lighting gradient can be removed by storing its effect on a plain white background as a reference image and subtracting that reference from all subsequent images. However, from a systemic machine vision viewpoint it would be a much better idea to rearrange the lighting to eliminate the gradient, and hence the unnecessary processing step! A better example is taken from astronomy where CCD

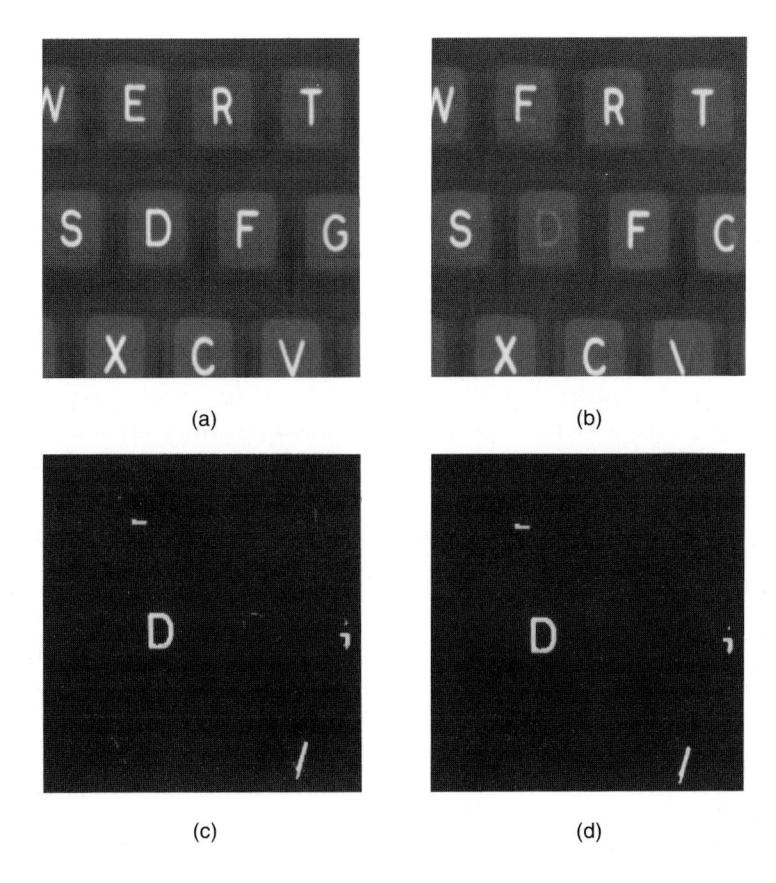

Figure 4.20 An example of the frame difference operator in inspection: (a) 'Known good' image of a keyboard; (b) image of a 'faulty' keyboard; (c) result of difference operator acting on binary versions of (a) and (b); (d) result of median filter applied to (c)

image sensors are widely used for their sensitivity and geometric stability. However, video signals from CCDs do exhibit a low level of fixed patterning as a result of mis-match of individual photosites etc. which cannot be avoided and which interfere with the work of the astronomers. This patterning can be eliminated from data frames by subtracting a dark reference image taken with the telescope's shutters closed.

Point-by-point addition of images can also be useful in cases where random noise is a problem and there is not much movement between images. If the noise can be regarded as purely random variations in pixel intensity values, averaging the pixel values from a number of 'identical images' will improve the signal-to-noise ratio. The number of images combined in this way must be a compromise between the desired signal-to-noise improvement and the risk of blurring due to motion. This technique is referred to as signal averaging in digital signal processing literature, and the improvement in signal-to-noise ratio is equal to the square root of the number of additions performed.

## References

1. M. L. Mead and C. Dillon, *Signals and Systems*, Van Nostrand, Wokingham, 1986.
2. P. A. Lynn, *Electronic Signals and Systems*, Macmillan, London, 1986.
3. H. Baher, *Analog and Digital Signal Processing*, Wiley, Chichester, 1990.
4. C. S. Burrus and T. W. Parks, *DFT/FFT and Convolution Algorithms*, Wiley, Chichester, 1985.
5. R. C. Gonzalez and R. E. Woods, *Digital Image Processing*, 2nd edn, Addison-Wesley, London, 1992.
6. N. Storey, *Electronics – A Systems Approach*, Addison-Wesley, London, 1992.
7. W. Niblack, *Digital Image Processing*, Prentice-Hall, Englewood Cliffs, New Jersey, 1986.
8. B. Batchelor and F. Waltz, *Interactive Image Processing for Machine Vision*, Springer-Verlag, London, 1993.
9. T. Fountain, *Processor Arrays – Architectures and Applications*, Academic Press, New York, 1987.
10. R. J. Schalkoff, *Digital Image Processing and Computer Vision*, Wiley, Chichester, 1989.
11. E. R. Davies, *Machine Vision: Theory, Algorithms, Practicalities*, Academic Press, New York, 1990.
12. I. E. Abdou and W. K. Pratt, 'Quantitative design and evaluation of enhancement/thresholding edge detectors', *Proc. IEEE*, **67**, pp. 753–763, 1979.
13. R. A. Schowengerdt, *Techniques for Image Processing and Classification in Remote Sensing*, Academic Press, New York, 1983.

# 5 Segmentation

## 5.1 Introduction

The principal objective of the segmentation process is to partition an image into meaningful regions which correspond to part of, or the whole of, objects within the scene. This is done by systematically dividing the whole image up into its constituent areas or regions. If the regions do not correspond directly to a physical object, or object surface, then they should correspond to some area of uniformity as defined by some predetermined assertion, or predicate.

The subdivision process should cease when all regions of interest have been identified and no further subdivision should occur. This may be easy to achieve in certain controlled applications where the outcome of the segmentation is well defined (visual inspection) but it is very difficult in applications where the outcome is not known (robotic guidance). In all cases the extent to which the segmentation process is carried out will depend on the particular problem to be solved.

Segmentation has a unique position within the generic model of a machine vision system (see figure 1.2) as it forms the bridge between the low-level and the high-level processing operations. Low-level processing operates on image arrays of raw data and thus adopts a bottom-up, or data-driven approach to image analysis. High-level processing is concerned with the manipulation of high level, abstract data representations and thus favours a top-down, hypothesis-driven approach. Segmentation can employ either or both of these approaches.

In the restricted domain of industrial machine vision systems the exploitation and imposition of scene constraints, together with the applicability of binary imaging, generally have the systemic benefit of reducing the segmentation task to a fairly trivial level. However this statement cannot be blindly applied to other computer vision applications. In fact, the task of autonomous segmentation is one of the most difficult tasks in computer vision and it continues to be the subject of much research effort, particularly in relation to multi-spectral and colour imagery.

There are two main approaches to segmentation:

(a) pixel-based or local methods;
(b) region-based or global approaches.

These approaches are complementary and should produce the same results, however in practice this is rarely the case. The pixel-based approach seeks to detect and enhance edges or edge elements within an image and then link them to create a boundary which encloses a region of uniformity. The region-based approach seeks to create regions directly by grouping together pixels which share common features into areas or regions of uniformity.

These two main approaches are sometimes identified as either discontinuity or similarity methods (edges are abrupt discontinuities in pixel grey-level values and image regions require some similarity criterion for creation).

### 5.1.1 Formal definition

The aim of segmentation is to divide the image into regions which have a certain uniformity. To assist in this division a *uniformity predicate* can be defined [1, 2].

Consider an image array of m columns by $n$ rows, figure 5.1a.

Let $R$ denote this complete array of image pixels, that is, the set of pairs $\{i,j\}$ where $i = 0,1,2, \ldots, (m-1)$ and $j = 0,1,2, \ldots, (n-1)$.

Let $R_a$ be a non-empty subset of $R$ consisting of contiguous image pixels.

A uniformity predicate, $P(R_a)$, is a logical statement which assigns the value True or False to $R_a$, depending only on the properties related to the

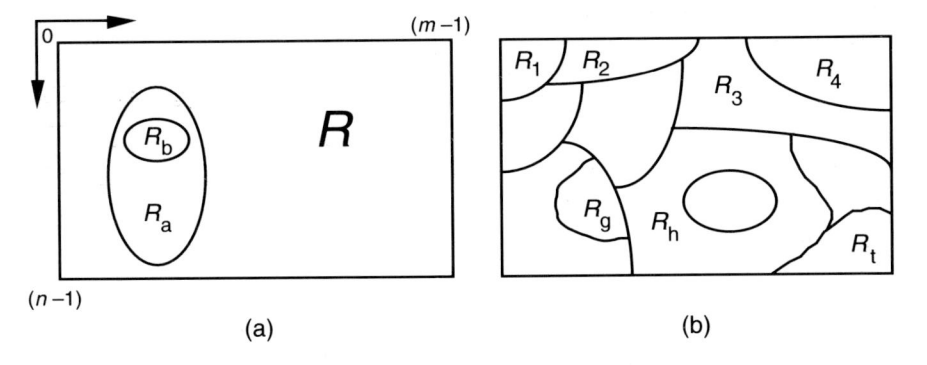

Figure 5.1 Uniformity predicate for segmentation

intensity matrix $f(i,j)$ for the points of $R_a$. Furthermore $P$ has the property that if $R_b$ is a non-empty subset of $R_a$, then

$$P(R_a) = \text{True implies that } P(R_b) = \text{True}$$

A segmentation of the array $R$, see figure 5.1b, for a uniformity predicate $P$ is a partition of $R$ into disjoint non-empty subsets $R_1, R_2, R_3, ..., R_t$ and can be defined mathematically as [3]

(a) $\cup R_g = R$ for $g = 1, 2, 3, \ldots, t$.
(b) $R_g$ is a connected region; $g = 1, 2, 3, \ldots, t$.
(c) $R_g \cap R_h = \phi$ for all $g$ and $h$; $g \neq h$.
(d) $P(R_g) = $ True for $g = 1, 2, 3, \ldots, t$.
(e) $P(R_g \cup R_h) = $ False for $g \neq h$.

where $P(R_g)$ is the logical predicate over the points in set $R$ and $\phi$ is the null set.

These formal conditions can be interpreted as follows:

(a) The union (or sum) of all regions equals the whole image. All pixels in the image must be assigned to a region.
(b) The region is contiguous and connected.
(c) The intersection of any pair of adjacent regions equals the empty set. Each pixel belongs to a single region only, there is no overlap between adjacent regions.
(d) For each region the uniformity predicate is true. Each region must satisfy some particular uniformity criteria.
(e) For any pair of adjacent regions the uniformity predicate is false. If any two adjacent regions are considered, then they cannot be unified.

There are a number of image features or attributes which can be used in the segmentation process, these include intensity (grey-level values), colour parameters (RGB, hue, saturation), boundary and range information, texture and motion. Uniformity predicates can include any of these attributes and are an essential mechanism in controlling the partitioning of an image. They also determine what attribute is to be used as a measure of uniformity and they specify how much variation in the parameters may be tolerated while still regarding the region as uniform.

### 5.1.2 Towards good segmentation

Segmentation is a critical component of a computer vision system because errors in this process will be propagated to the higher level analysis processes and increase the complexity of the subsequent tasks. Ideally the segmented regions within an image should have the following characteristics [3]:

(a) regions should be uniform and homogeneous with respect to some particular characteristic;
(b) region interiors should be simple and without many small holes;
(c) adjacent regions should have significantly different values with respect to the characteristic on which they are uniform;
(d) boundaries of each segment should be simple, not ragged, and must be spatially accurate.

Most image segmentation techniques are *ad hoc* and domain-dependent. It is difficult to obtain quantitative data on the quality of segmentation as the results are open to subjective interpretation, the only simple criterion available being a measure of the percentage of pixels mis-classified. Good segmentation is somewhat analogous to the definition of 'real-time', the segmentation is good if it provides appropriate output for the solution of the problem under investigation.

Achieving all the above desired properties in practice is extremely difficult. Insisting that adjacent regions have large differences in value can cause adjacent regions to merge and boundaries to be lost. Over-dividing or under-dividing regions may cause them to correspond to more than one surface, or conversely surface variability may split a single real surface into several regions.

The problem of segmentation of natural images is basically one of emulating psychological perception and therefore does not lend itself to a purely analytical solution. Any mathematical algorithms must be supplemented by heuristics, usually involving semantics or descriptions about the class of images under consideration.

Sometimes it is appropriate to go beyond simple heuristics and introduce *a priori* knowledge about the image. In such cases image segmentation proceeds simultaneously with image understanding. In segmentation, *a priori* knowledge refers to implicit or explicit constraints on the likelihood of a given pixel grouping. Such assumptions often arise from restrictions placed on the image as a consequence of domain-dependent considerations.

Even in a back-lit binary image, for example, where segmentation into regions of background and foreground is a trivial task, it is still necessary to label holes that occur within objects. They have the same pixel intensity values as the general background, but have a quite different significance. Holes can be uniquely identified by observing the hypothesis that they consist of regions of background intensity which is entirely enclosed by foreground intensity.

## 5.2 Pixel-based or local methods

The first clues about the physical properties of a scene are provided by the changes in intensity within the stored image. The geometric structure, sharpness and contrast of these intensity changes convey information about the physical edges in the scene. Pixel-based methods rely on the ability to detect the presence of such changes through either points (pixels), incremental edge elements ('edgels'), edge segments (lines), or connected line segments (boundaries) within an image array. Thus only point-wise or nearest-neighbour local information is used, and no account is taken of the general properties of the whole region.

Humans make great use of boundary information in image interpretation and can often recognise objects from only the crudest outline representation; consider the ease with which cartoon sketches can be recognised, for example. Finding edges that correspond to real physical changes within the scene can be quite complex because the presence of noise often produces disconnected pixel or edge elements as well as other apparent discontinuities due to texture variations. Thus many grey-level pixel changes have no real edge significance, and vice versa.

Edge-based segmentation can be summarised as a series of steps (see figure 5.2). First, the image pixel grey-level values are smoothed to overcome noise irregularities. Next, the candidate edge elements are enhanced using a local edge operator; the output image at this point contains unhelpful data (due to the noise) which can be removed by thresholding the image into a binary representation consisting of edgels and

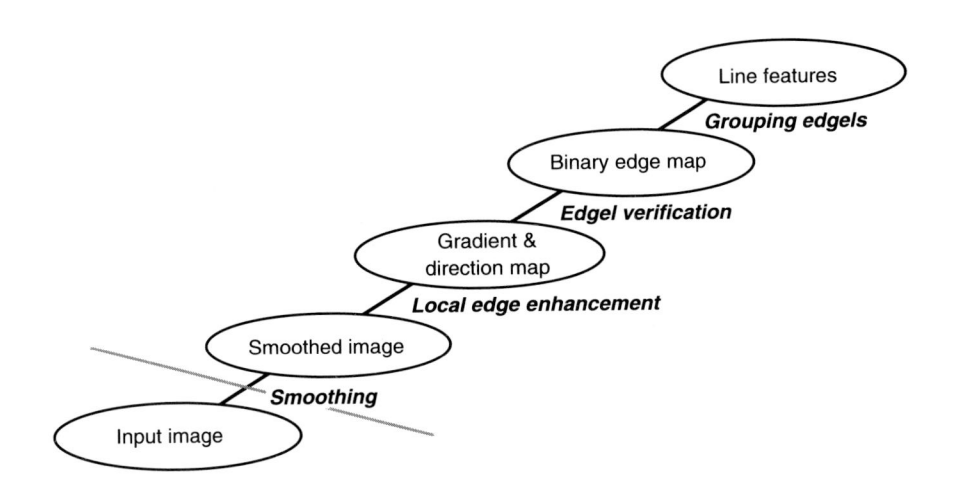

Figure 5.2 The edge-based segmentation process

non-edgels. Finally the individual edgels are grouped together in order to form a line feature.

The smoothing operation identified here is really part of the preprocessing operation discussed in Chapter 4; in addition, the discussion of preprocessing also encompassed edge enhancing filters which are the initial edge detecting operations from the viewpoint of segmentation.

### 5.2.1 Edge detection

*Gradient operators*

Every algorithm for edge detection is based on some model of the edge transitions occurring within an image. Commonly the transition in intensity is described by a step function and the image intensities on either side of the edge are assumed to be constants. In reality, the edge profile tends to be a slope because of noise and blur, and the performance of the edge detecting algorithms can be considerably improved by basing the algorithms on more realistic assumptions.

Various models for edge transitions have been illustrated previously in figure 4.13. An edge is a boundary between two regions which have different grey-level values. The basic approach is to calculate an intensity gradient in two orthogonal directions, $x$ and $y$. The partial finite differences in this case are given by

$$\Delta_x f(x,y) = f(x,y) - f(x-1,y)$$
$$\Delta_y f(x,y) = f(x,y) - f(x,y-1)$$

The digital approximation to the resultant gradient of $f(x,y)$ will be

$$\Delta f(x,y) = \sqrt{\{\Delta_x f(x,y)^2 + \Delta_y f(x,y)^2\}}$$

Since this expression may be computationally expensive, the digital gradient is generally considered to be either the sum of the absolute values of the two directional components or the maximum of these two components, i.e.

$$|\Delta f(x,y)| \approx |\Delta_x f(x,y)| + |\Delta_y f(x,y)|$$

or

$$|\Delta f(x,y)| \approx \max\{|\Delta_x f(x,y)|, |\Delta_y f(x,y)|\}$$

Figure 5.3 shows that the first derivative (gradient) of the grey-level distribution of pixels yields a positive peak at dark to light transitions and a

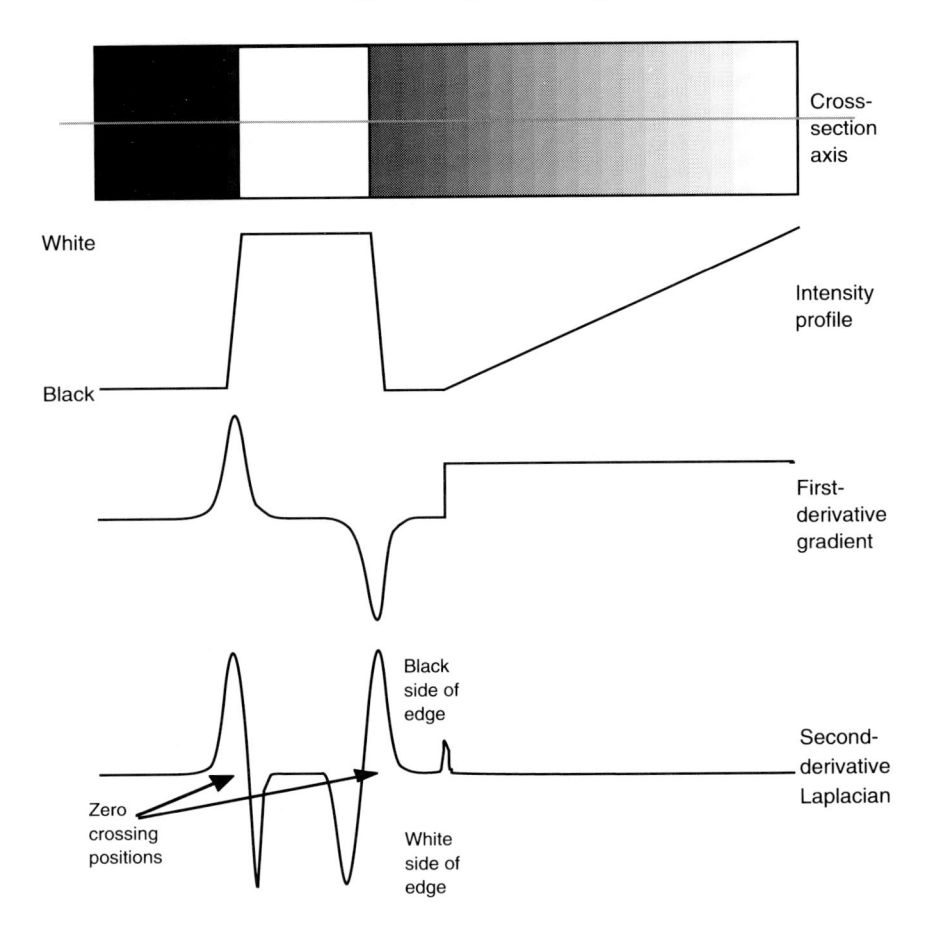

Figure 5.3 Edge derivatives

negative peak at light to dark transitions. The second derivative (Laplacian) yields a zero value in the middle of a double, bidirectional peak for either type of transition which can be used to identify the *position* of the edge. The second derivative also provides information about the intensity on either side of the edge, the positive value being associated with the darker side of the transition.

The usual way to look for intensity discontinuities is to run a mask or window over the image, as outlined in sections 4.1.1 and 4.4.2. The coefficients within the window will determine the type of discontinuity detected. Both single pixels and lines can be detected with the appropriate mask. However in real-world images, isolated points and thin lines are rarely encountered and the most common approach to pixel-based segmentation is edge detection, where edge-enhanced images are post-processed by thresholding and thinning to identify edgels.

The edge enhancing filters of Roberts and Sobel have already been introduced in Chapter 4. The Roberts operator uses two $2 \times 2$ masks to detect intensity changes across two orthogonal directions at 45° to the principal axes. The Sobel operator detects changes in the horizontal and vertical directions, at the same time using weighted mask values to effect some degree of smoothing.

The Laplacian operator is a second-order derivative that can also be used for detecting intensity changes. The general expression is

$$\nabla^2 f = \partial^2 f / \partial x^2 + \partial^2 f / \partial y^2$$

The standard digital version of the two-dimensional Laplacian operator is

$$\nabla^2 f(x,y) = 4f(x,y) - f(x-1,y) - f(x,y-1) - f(x,y+1) - f(x+1,y)$$

The Laplacian mask values have already been given in section 4.4.2. The Laplacian operator has zero response to gradual changes in intensity, as seen in figure 5.3, but does respond to changes immediately on either side of the edge with positive and negative values corresponding to the dark and light side of the edges respectively. Edge detection is therefore effected by isolating the points where the zero crossings occur. However, it should be noted that the Laplacian also responds strongly to noise in the image.

All methods which are gradient, or Laplacian, based are intrinsically noise sensitive, since the main cue for an edge is considered to be the grey-level difference that may be measured between adjacent pixels. Therefore some pixels having a difference in grey-level values from the background will be unavoidably extracted and identified as edge elements.

One approach, proposed by Marr and Hildreth [4], to extending the second-derivative edge detection process, while incorporating a smoothing function, is to convolve the image with a Gaussian smoothing function prior to the application of the Laplacian operator. Equation (4.2) can be re-written using polar coordinates, as

$$G(r) = \frac{1}{2\pi\sigma^2} \exp\left\{\frac{-r^2}{2\sigma^2}\right\} \quad \text{where } r^2 = x^2 + y^2$$

The second derivative (or Laplacian) of this function yields the 'Laplacian of Gaussian (LOG)' function:

$$\nabla^2 G(r) = \left\{\frac{1}{2\pi\sigma^4}\right\} \cdot \left\{\left(\frac{r^2}{\sigma^2}\right) - 1\right\} \exp\left\{\frac{-r^2}{2\sigma^2}\right\}$$

Convolving the Laplacian of Gaussian with an image enhances the edges within it and yields information on the direction of these edges through the orientation of the zero crossings. Remembering that the Gaussian response depends on the selected value of $\sigma$, it is possible to select the scale of features detected by the operator by adjusting $\sigma$. It is also clear that in the digital implementation the convolution mask size will vary accordingly. Thus, for example, when $\sigma$ is chosen as 3 a typical mask size is $23 \times 23$, necessitating a much longer processing time than for smaller $\sigma$ values.

'Directional operators', or 'compass-gradient operators', are similar to those discussed above, but here the image is 'convolved' (effectively correlated – see discussion in section 4.1) with a set of templates that represent ideal step edges in specified orientations. Therefore a change in intensity that is extended along one orientation in the image gives rise to a maximum response from the template whose orientation most nearly matches the direction perpendicular to the orientation of intensity change. The edge magnitude is taken to be the value of that maximal response and the edge orientation is quantised to that of the corresponding template. A variety of directional operators are available, for example Kirsch [5] and Prewitt [6], each of which utilises eight masks having discrete orientations of 0°, 45°, 90°, 135°, 180°, 225°, 270° and 315°; – figure 5.4 shows the set of templates due to Prewitt.

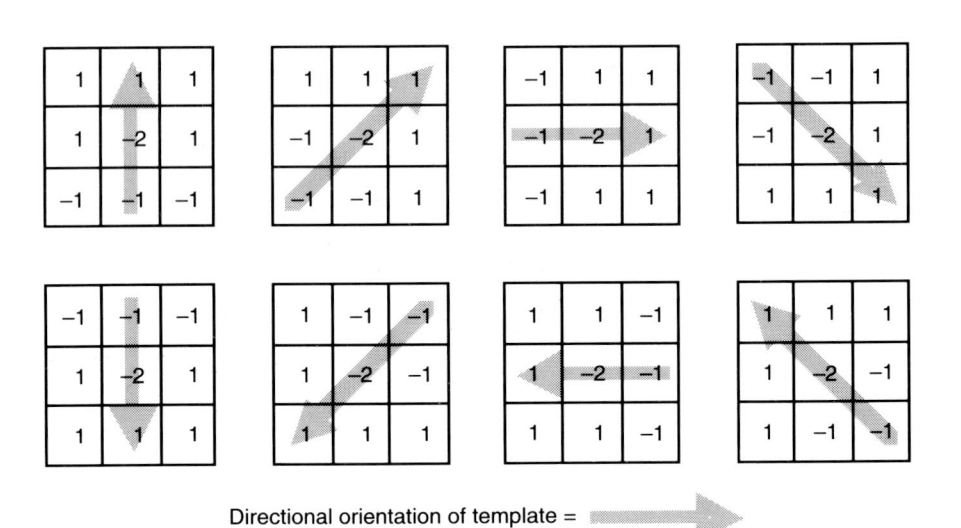

Figure 5.4  Templates for the Prewitt compass-gradient operator

*Multi-resolution operations*

Most of the early edge detectors used fixed size neighbourhoods. The window size in these operators therefore becomes an important factor in determining their success or failure to detect edges appropriately. In general the window size should be sufficiently large to cover the extent of the slope of the edges to be detected. However, there is a trade-off between large windows that are immune to noise and small windows that have higher resolving power. Therefore the idea of applying different size masks to the image has gained popularity.

The basic concept is that the smaller windows can detect finer changes in intensity profile, and are not affected by gradual changes in intensity, while the larger windows will ignore the fine detail and detect coarser changes in intensity profile. The larger windows are also sensitive to gradual changes in the light intensity. The analysis of intensity changes across such multiple resolution masks can however be very complex, and, at all mask resolutions, some of the detected 'features' may not correspond to significant physical discontinuities in the real-world scene, making the results even more difficult to interpret.

The potential application of different mask sizes naturally leads onto a discussion of image operations at different spatial resolutions. One data structure which assists these manipulations is the 'resolution pyramid' (see figure 5.5). It is a data structure in which certain operations may be

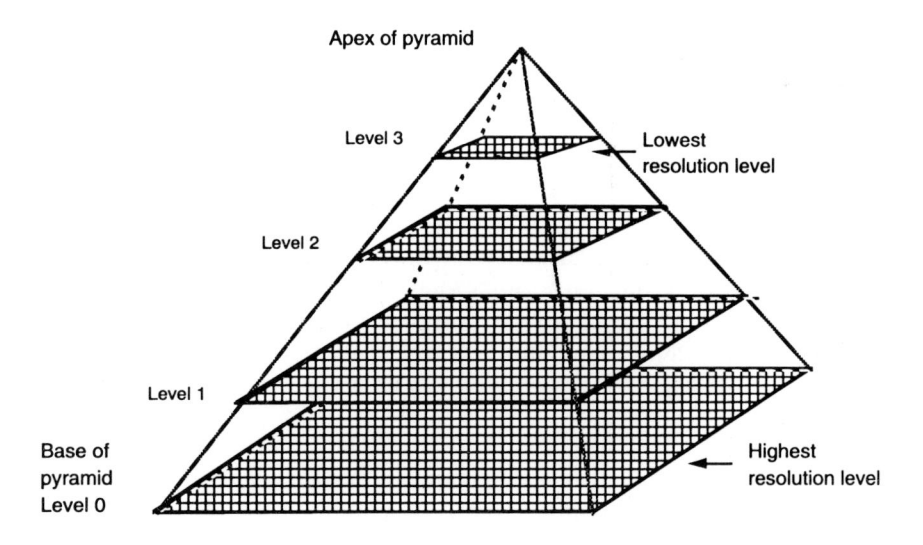

Figure 5.5 An image resolution pyramid

accomplished at a low image resolution, and refined at ever increasing resolutions, until the highest resolution of interest is reached.

Towards the top of the pyramid, computations are fewer and faster because of the reduction in dimensionality, and also unwanted detail (e.g. noise) may not be resolvable. One simple method of creating coarser resolution images is to divide the image into non-overlapping neighbourhoods of equal size and shape and replace each of them by their average pixel value.

In the Canny edge detector, edge features are first detected at a discrete set of image scales or resolutions [7]. The highest resolution image descriptions are then used to predict the results of the next level in the pyramid (next lower resolution). If there is a substantial difference between the actual description at the new level and that predicted from the lower level, it is assumed that there are important changes taking place at the higher level (lower resolution) that were not detected at the original highest resolution image. These features, detected at the lower resolution, are then added to the final output edge detected representation.

Clearly, this is a very sophisticated edge enhancement operator, but it produces very complete edge maps which only require minimal boundary detection. Although the Canny operator is complex, it relies upon locally derived data and therefore lends itself much more easily to hardware implementation than the conditional operations involved in boundary detection. In recent times this thinking has led to a reduction in interest in simple edge operators, like Roberts and Sobel, which require extensive subsequent boundary detection post-processing, in favour of complex edge finders such as those due to Canny and Marr and Hildreth.

### 5.2.2 Boundary detection

When the various edge operators described above are applied to an image they detect local edge elements, or edgels. However an image of disconnected edge elements is somewhat uninformative and additional processing is needed to group these elements into straight and curved boundaries. These closed boundaries provide patterns and structures which are more appropriate for the recognition and classification of any objects in the scene.

The goal of boundary formation, or edge linking, is to make a coherent one-dimensional edge feature from many individual local edge elements. Boundary formation techniques have to handle instances where edge elements do not correspond to any meaningful scene boundary, and conversely they have to cope when no edge elements are present even though a boundary should exist. The methods vary in the amount of *a priori*

## Boundary refining

One of the simplest techniques for linking edgels is to systematically analyse a small neighbourhood of pixels within an image that has already undergone edge detection. At each point within this edge map the gradient magnitudes and directions are checked and if these fall above a specified threshold the appropriate pixels can be linked.

*A priori* techniques utilise knowledge about the likely position of a boundary in order to guide the search for the real boundary, and hence refine the initial estimate of the most likely position. One such technique utilises a local neighbourhood search carried out at regular intervals along directions perpendicular to the *a priori* boundary [8]. An edge operator is applied at points along each of these perpendicular directions (see figure 5.6a), and for each such direction, the edge giving the maximum edge strength is selected as the one to link into a boundary. If sufficient edge elements are selected, then their positions within the image matrix are linked with some low-degree polynomial function and this curve is used to represent the actual boundary. This approach effectively samples the boundary and then reconstructs it from the samples. This enables it to cope with missing and spurious boundary points.

The '*divide and conquer*' technique is useful where a low curvature boundary is known to exist between two edgels [9]. Initially, a straight line is considered to link these two points (see figure 5.6b), and a search is carried out along lines normal to that which links the two points. If the search yields a point of maximum gradient and falling within some threshold distance away from the straight line then this point becomes a break point on the

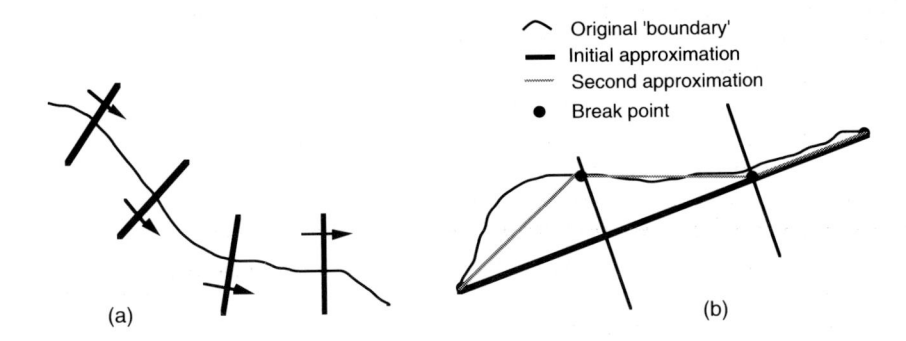

Figure 5.6   Techniques of boundary formation

boundary. Two new straight lines are thus formed and the process repeated on the two new lines. This technique iterates until no new break points are found and thus reconstructs the boundary from 'samples' using piecewise-linear approximations.

*Hough Transform*

The Hough Transform technique is used to find particular boundary shapes which are expected to occur within an image, provided these shapes can be described in some simple parametric form. It is most often used to detect straight lines, circles or ellipses but can be extended to look for any parameterised curve [10]. Its main advantages are that it is relatively unaffected by noise or missing portions of the boundaries of objects that it is trying to detect, but it is relatively expensive in terms of computational effort required.

In essence the Hough Transform produces a set of parameters which describe a boundary curve of the expected type that represents the best fit to the set of edge points in the given image. This is done by transforming every edgel position in the image 'space' (as defined by $x$ and $y$ axes) into a corresponding curve within a 'parameter space', or 'Hough space' (as defined by gradient, $m$, and intercept, $c$, axes in the case of the straight line Hough Transform, for example). A point in the Hough space where many curves intersect represents a simultaneous solution to the parametric equation for all of the edgels in the image space whose coordinates gave rise to those curves. This indicates a strong likelihood of a boundary shape of the expected type having been detected in the image space. The coordinates of the point of intersection in the Hough space correspond to the parameters of the curve detected in the image space; the greater the number of intersections, the greater the confidence that the detected boundary shape is genuine. It is this 'voting' effect that gives the Hough Transform its characteristic immunity to noise and discontinuous boundaries.

For example, consider the case where a straight line is to be detected. A point, $(x_i, y_i)$, in input space presumed to lie on this line (conventionally defined by the standard linear equation, $y = mx + c$) produces a locus of points in parameter space for all possible lines upon which it could lie (thus defined as $c = -mx + y$), as shown in figure 5.7. A second point in input space, $(x_j, y_j)$, similarly produces a locus of points in Hough space. Thus when all points of interest in input space (previously detected edgels) have been transformed to loci in Hough space, the intersections of these loci give a vote as to the best set of parameters for the unique line in the input space which will join all given points. The position of maximum intersection yields the parameters ($m_b$ and $c_b$) for the best fit solution.

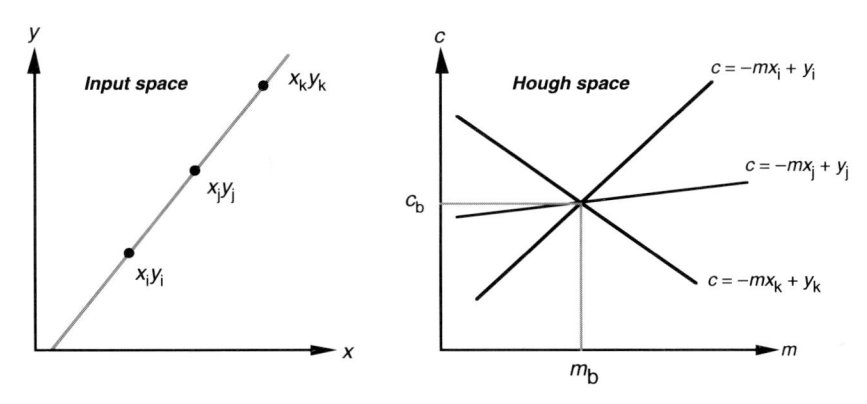

Figure 5.7 The Hough Transform used for straight line detection

One problem with this approach is that when implemented in discrete digital form a very large array of Hough space 'accumulators' is required to store all possible votes since the range of values should extend from minus infinity to infinity (e.g. a perfectly vertical line in input space has an infinite gradient). One way to overcome this limitation is to utilise polar coordinates $(r,\phi)$ instead of Cartesian form. A straight line can thus be defined by the length, $r$, and angle, $\phi$, of the normal vector connecting it to the origin (see figure 5.8). These parameters are related to the $x$ and $y$ coordinates by the expression

$$r = x\cos\phi + y\sin\phi$$

The three points shown in the input image are mapped into the Hough space (now in $(r,\phi)$ form). Each point in the input image, a, b or c,

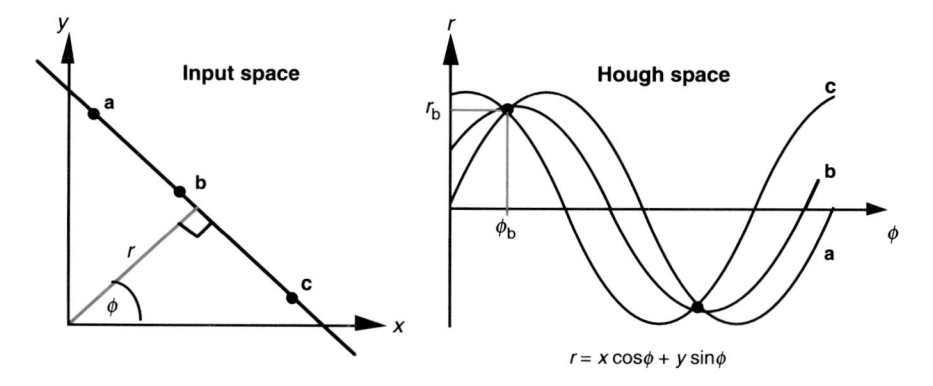

Figure 5.8 The Hough Transform – polar representation

transforms to a sinusoidal curve which is plotted over the range 0 to $2\pi$ radians. The position of maximum intersection can again be found and these unique values of $r_b$ and $\phi_b$ used to define the best straight line detected in the input image which joins the given pixel points, or edgels. As expected all the curves exhibit a symmetrical positive and negative response. Therefore the amount of computation can be halved by plotting curves in the Hough space only over the range 0 to $\pi$ radians.

If objects within a scene are known to contain circular shapes then the Hough transform can be used to detect circles, as defined by the expression

$$(x-a)^2 + (y-b)^2 = r^2$$

where $a$ and $b$ are the coordinates of the centre of the circle,
and    $r$ is its radius.

In this case there are three parameters present which leads to a three-dimensional Hough space. The principle is identical to that described above, but with an appropriate increase in the size of the accumulator array and therefore in the processing time.

Clearly, the complexity of the Hough Transform depends on the number of points in the input image and the number of coefficients required to define the functional representation of the curves to be detected. If a curve is given by the function

$$f(a_1, a_2, a_3, \ldots a_n, x, y) = 0$$

where $a_1, a_2, a_3, \cdots, a_n$ are the parameters which define the curve to be detected

then for each point in the input image, $(x_i, y_i)$, its generalised Hough Transform is a surface in $n$-dimensional space. The complexity of the Hough Transform increases very rapidly with the number of parameters needed to define the curve.

When applying the Hough Transform it must be remembered that the straight lines which have been detected are infinitely long and the circles are complete circles; the real image will contain finite lines and may only contain arcs. The problem of boundary formation requires the linking of appropriate edgels to form a complete real boundary. This can be accomplished by examining all points which fall on the best-fit curve and calculating the distance between disconnected pixels. A threshold can be set for this distance, and any points falling above the threshold are regarded as being disconnected, i.e. a gap at a point in a boundary is significant if the distance between that point and its closest neighbour exceeds the chosen threshold.

## 5.3 Region-based or global methods

As stated earlier, the region-based approach to segmentation seeks to create regions directly by grouping together pixels which share common features into areas, or regions of uniformity. This can be a relatively straightforward task for a high contrast image of an uncluttered scene. However, in the majority of practical cases it becomes a somewhat more complicated exercise. This means that they can be too computationally expensive for typical machine vision applications, being the last resort after other, simpler forms of segmentation have been found wanting. Despite this, region-based methods are of interest because they are generally less sensitive to noise than boundary-based approaches.

### 5.3.1 Region merging and splitting

'Region merging', also referred to as 'region growing', is a method whereby the image is divided into arbitrary elementary regions (often starting at the level of individual pixels), then merging these elementary regions according to some specified criteria until no more can be merged. The basic techniques are simple to describe and their implementation is eased by the use of good data structures (e.g. 'quad tree' – see figure 5.10) to handle the large number of initial regions and in the determination of adjacent regions.

The criteria for merging regions, i.e. the uniformity predicate identified above in section 5.1.1, is crucial in determining the final segmentation and it will depend on the type of data available and hence on the application under consideration. Similarly it can be measured over the entire region (using averaging techniques), or along the boundaries of two regions (using connectivity criteria), or any other combination. In addition it is important to decide when to stop the merging process, for example when a sufficient number of regions have been produced, and this should be contained within the uniformity predicate.

Figure 5.9 illustrates the region merging technique; in this simple example the grey-level values fall between 1 and 10 and the uniformity predicate used is for regions to merge when the difference in grey-level value intensity between adjacent regions is 1. Figure 5.9a shows the original distribution of grey-level values, then a first pass over the image identifies all regions with discrete grey-scale values and hence a primary region map is produced, as in figure 5.9b. It can be seen that regions labelled 9 and 10 are candidates for merging, as are regions labelled 5 and 6, and 6 and 7, etc.

All such possible mergers, where the grey-level difference of adjacent regions is 1, are identified in figure 5.9c. When all these mergers have taken place the final segmented image, figure 5.9d, displays one light and two dark

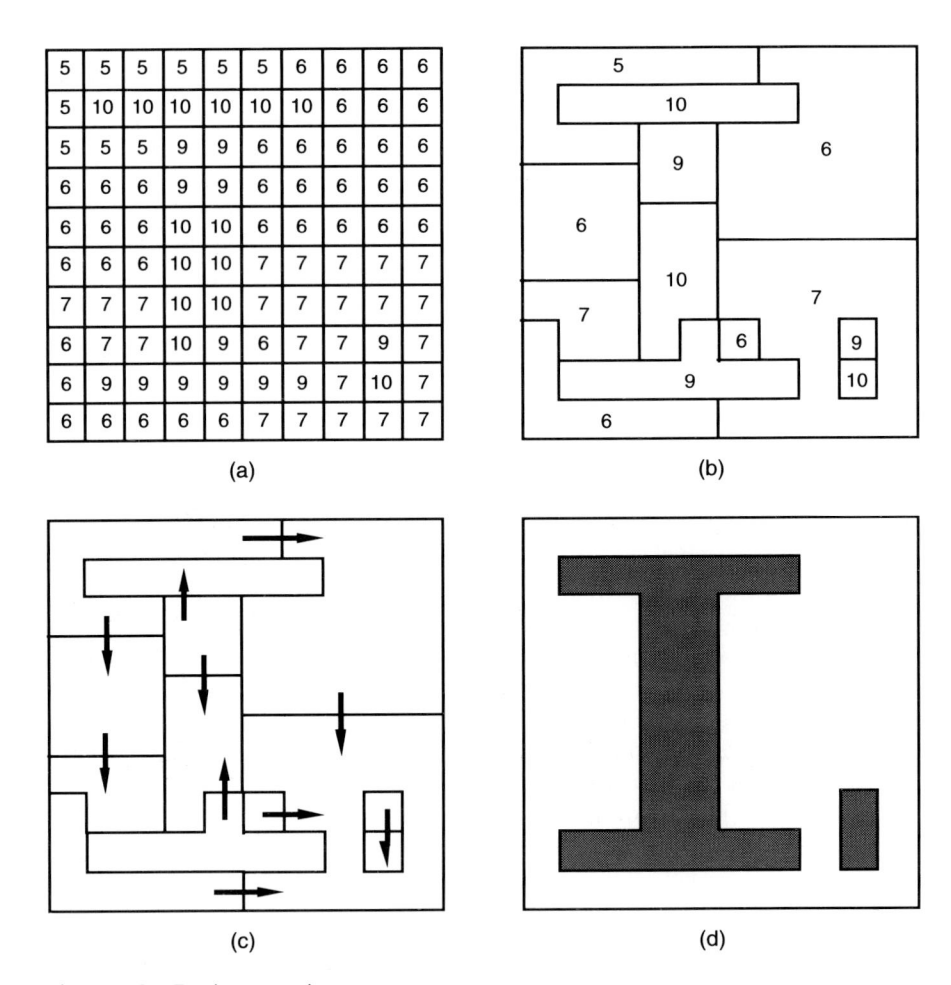

|   |   |   |   |   |   |   |   |   |   |
|---|---|---|---|---|---|---|---|---|---|
| 5 | 5 | 5 | 5 | 5 | 5 | 6 | 6 | 6 | 6 |
| 5 | 10| 10| 10| 10| 10| 10| 6 | 6 | 6 |
| 5 | 5 | 5 | 9 | 9 | 6 | 6 | 6 | 6 | 6 |
| 6 | 6 | 6 | 9 | 9 | 6 | 6 | 6 | 6 | 6 |
| 6 | 6 | 6 | 10| 10| 6 | 6 | 6 | 6 | 6 |
| 6 | 6 | 6 | 10| 10| 7 | 7 | 7 | 7 | 7 |
| 7 | 7 | 7 | 10| 10| 7 | 7 | 7 | 7 | 7 |
| 6 | 7 | 7 | 10| 9 | 6 | 7 | 7 | 9 | 7 |
| 6 | 9 | 9 | 9 | 9 | 9 | 9 | 7 | 10| 7 |
| 6 | 6 | 6 | 6 | 6 | 7 | 7 | 7 | 7 | 7 |

(a)　　　　　　　　　　　　　(b)

(c)　　　　　　　　　　　　　(d)

Figure 5.9　Region merging

grey level regions which may be interpreted as two objects in the scene, characters from a document reader, etc.

A similar result could have been achieved if the starting point had been to identify one pixel in the original image as a 'seed', for example one of the pixels labelled 10, and then apply the uniformity predicate to 'grow' from this single pixel region into the major dark area, as identified in the resultant segmented image.

The techniques classified as 'split and merge' first split the image in some uniform way into a set of arbitrary regions, then a uniformity predicate is applied either to subdivide the region further, or merge adjacent regions.

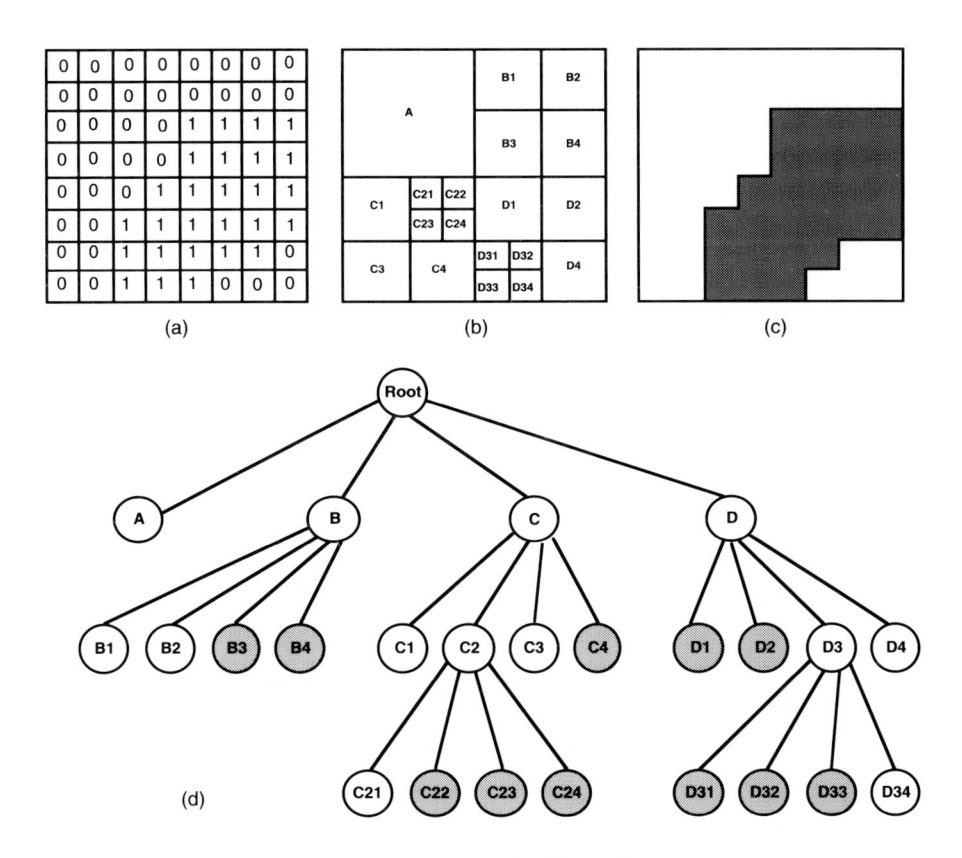

Figure 5.10 The quad-tree representation of a binary image

The quad-tree data structure is often used in conjunction with the split and merge segmentation techniques [11]. The subdivision of the image can be viewed graphically through a 'tree' structure in which the 'root' of the tree corresponds to the whole image and the 'leaves' of the tree correspond to individual pixels. The use of a quad tree as a form of image representation is illustrated in figure 5.10.

Tree notation adopts a set of curiously mixed metaphors, each 'node' (or 'parent') of the tree (represented by a circle in figure 5.10) having either four 'children', or none. Thus, descending one level in the hierarchy of the tree represents dividing the image into four sub-images, hence the term 'quad tree'. If all four children of a parent would have the same value, then their parent node can represent them perfectly well and the children do not exist in the finished tree (or the tree is 'pruned'). If the children have different values then they do exist and the non-terminal parent node adopts ('inherits') a data value equal to the average of its four children. This subdivision extends to the lowest level of the tree when the children represent leaves, or individual pixels. As such, the quad-tree representation of an image is a form of data compression, since it is configured as a hierarchical data structure with redundant data eliminated; – note that figure 5.10d only uses 25 nodes to represent 64 pixels.

A split and merge algorithm can be summarised as follows:

(a) split any region into four quadrants if the uniformity predicate is false;
(b) continue this subdivision for all new sub-images until the stopping criteria is reached (usually single pixels);
(c) merge any adjacent regions for which the uniformity criteria is true;
(d) stop when no further merging is possible.

The region merging process is similar to that described above except that the initial regions are derived in a different manner. The merging process begins at the leaves of the tree, if four leaves from a parent node satisfy the uniformity predicate then they are merged and deleted from the graph. The parent node then inherits the characteristics of the leaves, for example average grey-level value. Merging then continues at the next highest level in the tree structure until no more merging is possible.

In figure 5.10a the original image is represented as a binary array. Figure 5.10b illustrates the subdivision or splitting process into four regions which continues if the uniformity predicate is false. When the splitting is complete the merging process begins and produces figure 5.10c, which is the complete segmentation of the image into two regions. Finally, the quad-tree representation is given in figure 5.10d; shaded circles represent nodes that have been labelled as the dark region of the image, and vice-versa.

### 5.3.2 *Thresholding*

Thresholding is one of the most important techniques for segmentation and is a widely used tool for machine vision systems. It lends itself to simple implementation in hardware, which therefore results in high throughput rates, as generally required by industry.

Thresholding has already been discussed briefly in section 3.1.2 and at greater length in section 4.2.4 from the viewpoints of image representation and preprocessing respectively. From the segmentation viewpoint, thresholding is a method of producing regions of uniformity within an image based on some threshold criterion, T. The thresholding operation can thus be thought of as a test involving the function T and defined as

$$T = T\{x, y, A(x,y), f(x,y)\}$$

where $f(x,y)$ is the grey level of the pixel at $(x,y)$,

and $A(x,y)$ denotes some local property in the neighbourhood of this pixel.

A thresholded image $g(x,y)$ is defined as

$$g(x,y) = \left\{ \begin{array}{ll} 1 & \text{if } f(x,y) \geq T \\ 0 & \text{if } f(x,y) < T \end{array} \right\}$$

The value of the function T can be defined in one of three ways, identified above, and defined as:

(a) *Global threshold*: $\quad T = T\{f(x,y)\}$
where T is dependent only upon the grey-level value of the pixel at $x,y$.
(b) *Local threshold*: $\quad T = T\{A(x,y), f(x,y)\}$;
where T is dependent upon a neighbourhood property of the pixel as well as its grey-level value
(c) *Dynamic threshold*: $\quad T = T\{x,y, A(x,y), f(x,y)\}$;
where T is dependent on the pixel coordinates, in addition to the other two criteria

In practice the global thresholding technique is often used although the selection of the most appropriate threshold value, T, to use for a given application is the main dilemma for the system designer. This choice will be simplified if the image is of good contrast and the highest quality image has been captured (see section 4.2.4). This echoes earlier comments about the need to manipulate constructively the interaction between the sub-system modules as identified in the generic machine vision system model. Clearly, if correct illumination and image acquisition strategies are followed for a specific application, then the segmentation process can be trivialised to simple thresholding and the ideal global threshold value can be easily evaluated.

For a simple global threshold, where the image histogram is bimodal or has easily identifiable peaks and valleys, the selection of T is fairly straightforward (as in figure 4.8) and the selection process can be automated. However, if the image is noisy or there is a considerable spread in grey level values, then the selection of T is more problematic. The images shown in figure 5.11 illustrate the effect of differing threshold value applied to a typical grey-scale image; it is not easy to identify the correct, or optimum segmentation. In such cases optimal thresholding techniques rely on statistical analysis, e.g. probability density, histogram entropy, or minimum error analysis with the attendant processing overhead [12].

One approach to improving the segmentation process is to consider a histogram made up of only those pixels which lie at or near an edge or boundary of objects within the image. Such a histogram should contain sharper peaks and lower valleys. The main problem here is to decide which

Figure 5.11 The problem of segmentation through thresholding: (a) the raw 6-bit grey-scale image; (b) a single threshold set at 10; (c) a single threshold set at 20; (d) dual thresholds set at 20 and 45 (see section 4.2.4)

pixels lie on or near a boundary. To assist this decision, the Laplacian of Gaussian operator $\nabla^2 G$, as described in section 5.2.1, is first applied to the image so that the edge points will be clearly identified. Once identified the average grey-level value for these edge points is calculated and used as the global threshold value. Although this calculation may take some time, it is only carried out once during the system calibration and initialisation procedure and not before every threshold operation.

# References

1. T. Pavlidis, *Structural Pattern Recognition*, Springer-Verlag, Berlin, 1977.
2. R. Gonzalez and R. Woods, *Digital Image Processing*, 2nd edn, Addison-Wesley, London, 1992.
3. R. M. Haralick and L. G. Shapiro, 'Image segmentation techniques', *CVGIP 29*, pp. 100–132, Academic Press, New York, 1985.
4. D. Marr and E. Hildreth, 'Theory of edge detection', *Proceedings of the Royal Society of London*, **B207**, pp. 187–217, 1980.
5. R. Kirsch, 'Computer determination of the constituent structure of biological images', *Computers and Biomedical Research*, **4**(3), pp. 315–328, June 1971.
6. J. Prewitt, 'Object enhancement and extraction', in B. Lipkin and A. Rosenfeld (Eds), *Picture Processing and Psychopictorics*, Academic Press, New York, 1971.
7. J. Canny, 'A computational approach to edge detection', *IEEE Transactions on Pattern Analysis and Machine Intelligence*, **PAMI-8** (6) pp. 679–698, 1986.
8. R. Bolles, 'Verification vision for programmable assembly', *Proc. 5th IJCAI*, pp. 569–575, August, Morgan Kaufmann, Cambridge, Massachusetts, 1977.
9. P. Selfridge et al., 'Segmentation algorithms for abdominal computerized tomography scans', *Proc. 3rd COMPSAC*, pp. 571–577, Nov., IEEE Comp. Soc. Press, Los Alamitos, California, 1979.
10. D. H. Ballard, 'Generalising the Hough transform to detect arbitrary shapes', *Pattern Recognition*, **13**, pp. 111–122, 1981.
11. S. L. Horowitz and T. Pavlidis, 'Picture segmentation by a tree traversal algorithm', *Journal of ACM*, pp. 368–388, 1976.
12. F. Zahedi and R. Thomas, 'Hybrid image segmentation within a computer vision hierarchy', *IJEEE*, **30**, pp. 57–64, 1993.

# 6 Feature Extraction

The feature extraction aspect of image analysis seeks to identify inherent characteristics, or features, of objects found within an image. These characteristics are used to describe the object, or attributes of the object, prior to the subsequent task of classification. Feature extraction operates on two-dimensional image *arrays* but produces a *list* of descriptions, or a 'feature vector' (note the change in information format indicated on the generic model, section 1.4.1).

## 6.1 Image features

Many of the features of interest are concerned with the shape of the regions. Although *shape* is a familiar term to the layman, its meaning in terms of computer vision is by no means fully determined. However, following on from the previous process of segmentation, the general shape of an object (or region within an image) can be represented by features gleaned from either the external, or boundary, properties of the shape or from the internal, or regional, properties. For example, visual inspection of buttons could utilise features of perimeter, area and number of holes which could be obtained from a binarised back-lit image.

On the other hand, the surface properties, such as texture or brightness, of the objects under investigation might be the best way of distinguishing one from another, or good from bad. For example, when inspecting paper, features of interest are likely to be the degree of texture imposed by the mesh upon which the paper is dried, or the number of dark specks introduced perhaps by recycled raw material. Painted finishes are prone to the development of cracks, runs or an 'orange peel' texture, any of which is likely to result in unacceptable appearance and a hence rejection. This type of feature can only be extracted from a grey-scale image of a carefully lit scene.

Yet another class of features is concerned with the geometric *structure* of regions within the scene. Pioneering work on microscope images of rock samples at the Paris School of Mines led to the notion that geometric structure is not purely objective, but rather is the result of interaction between the phenomenon and the observer. This means that the idea of structure within images is highly dependent upon the purpose to which the interpreted image is to be put. 'Mathematical morphology' addresses this issue by introducing the concept of *structuring elements*. These are actively selected by the investigator to [1]

> '... *interact with the object under study, modifying its shape and reducing it to a sort of caricature which is more expressive than the actual initial phenomenon.*'

This has developed into a powerful and widely applied methodology which has a formal and rigorous mathematical basis in set theory and topology, and draws its terminology from there [2]. Morphology is not just limited to feature extraction – it provides a set of tools which may assist any of the image processing stages up to, and including, feature extraction.

## 6.1.1 Design of the feature extraction process

It is generally recognised that the choice of features to be derived from the image radically influences the effectiveness of the subsequent classification or interpretation stages of a machine vision system. Therefore, a set of features should be identified which can help to uniquely identify key differences between the (hopefully limited) range of objects that will be encountered within the constrained environment of the machine vision system. This strongly application-specific nature makes it difficult to lay down hard and fast rules for the selection of features from the wide range of methodologies available. However, it can be generally noted that features should ideally be completely independent of the translation (location), scale (size) or orientation (rotation) of the objects, so that variations in position and pose can be accommodated.

Although some applications will demand the use of features derived from grey-scale images, there are many useful feature descriptors that can be obtained more easily from the thresholded binary representation. The complexity of feature extraction solutions for machine vision problems can also be minimised by appropriate systemic design of the scene constraints and image acquisition sub-systems. For example, it may be possible to control the presentation of industrial work-pieces using devices such as conveyor belts, vibratory bowl-feeders and jigs so that the tolerance on orientation, translation and scale is small.

However, it has to be noted that one of the highly desired advantages of vision-equipped automation is the ability to be flexible by allowing manipulation of a variety of work-pieces, recovery from misfeeds, etc. Therefore, the vision system designer has to be careful when striking the balance between a minimal 'cost', but dedicated solution, and a more expensive, but equally more robust and flexible one.

## 6.2 Image codes

In Chapter 3 it was noted that a single 512 × 256× 8-bit resolution image requires 128 kbytes of storage. Until the feature extraction stage all of the data carries equal weight, giving rise to the high degree of dependence of low-level processes upon the two-dimensional topology of the image. However, it was noted above that feature extraction begins to change the data format from spatially correlated arrays to textual descriptions of structural and/or semantic knowledge. This corresponds to re-coding the image data in a significantly more compact form.

Image coding exploits segmentation and attempts to efficiently describe the properties of significant regions identified within the image in such a way that they can be reconstructed from the coded form with 'sufficient' accuracy. This implies that the coding process will facilitate the measurement or assessment of object features. It should come as no surprise, therefore, to learn that basic coding techniques are fundamental to many forms of feature extraction. While a detailed treatment of the subject of image coding [3–5] is beyond the scope of this text it is essential to consider some of the simplest methods of image compression, in order to understand how feature extraction proceeds (see also Chapter 9 for applications of more sophisticated image coding algorithms). For such an initial treatment it is most convenient to consider the coding of binary images and/or edge maps.

### 6.2.1 Run code

This code consists of a set of 'runs' whose values are the number of consecutive white and black pixels in the image. Referring to figure 6.1a and commencing at the top left-hand corner of the image, there are 3 white pixels followed by 2 black pixels, thus the first two numbers in the run code will be 3,2 (assuming that the first run is white, by convention). Continuing, the next run is 3 white pixels this time followed by 4 black pixels which run over onto the next row of the image. Thereafter there are 3 white followed by 2 black and finally 8 white. Thus the complete image run code is produced, as shown, and results in a considerably compressed form of the original binary representation. The information is now in a more accessible form. If there is only one object in the image then the area of the object can be found by summing every alternate number in the run code – the starting point depending upon whether the object of interest is black or white.

(a) Run code: 3234328

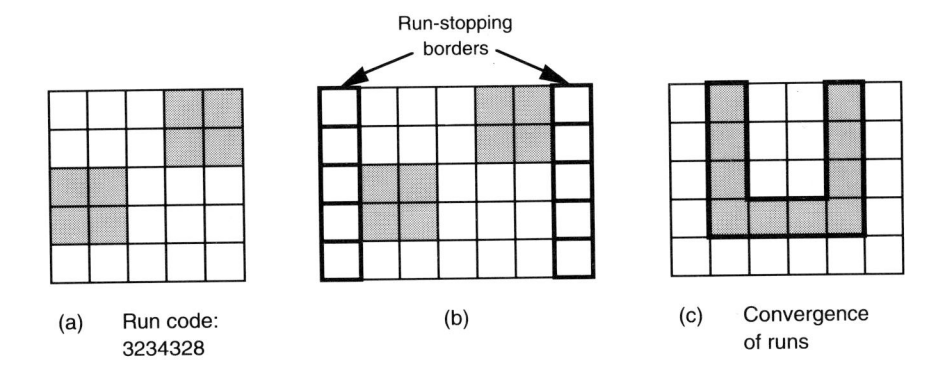

(b)

(c) Convergence of runs

Figure 6.1   Run code representation

Unfortunately, by converting the binary image into run code, the original row structure is lost so that runs overlap the image boundary. This causes problems when trying to distinguish between two or more objects. However, if a white border is imposed around the image, it has the effect of stopping black pixel runs from overlapping the image boundary and will assist subsequent analysis for feature extraction (see figure 6.1b). Production of the run code now involves totalling the number of bits until a change of state occurs, whereupon the total is stored as a run length and the total for the next run (row) starts at zero. The border can simply be added by inserting two white bits every eight pixels, but once again this depends upon *a priori* knowledge of which state represents background, and which foreground (object).

When there is more than one object within the image, tests are needed to check for object connectivity (overlapping runs in consecutive rows) and object convergence (one object which looks like two at the commencement of the run), see figures 6.1b and c.

Object features can be determined from the run code as follows: the area of each object can be found by summing the run lengths of appropriate state which are known to belong to that object. The maximum and minimum object widths can be found by identification of the longest and shortest runs (provided that the object has no holes). Perimeter can be estimated as the length of the top and bottom rows plus two units for the end point of every row in between – again provided that the object has no holes, unless the total (i.e. internal and external perimeter) is of interest.

Finally, it should be noted that the run code gives a non-isotropic, or orientation-dependent, image representation which may impose limits on its usefulness in many applications. Run-length connectivity analysis (effectively segmentation) is also not trivial, especially in the case of convergent objects. However, hardware implementation of run-length coding is very simple, making the technique always worth considering for real-time applications.

### 6.2.2 Chain code

A boundary may be represented by a 'chain' of connected steps of known direction and length, thus the 'chain code' is a concise way of recording a shape contour [6]. In a two-dimensional orthogonal image, array movement from one pixel to an adjoining pixel can only be undertaken in one of eight directions – thus the eight compass points can be used as direction vectors. By giving a number to each direction, as defined in figure 6.2, the outline of an object can be traced and coded as a sequence of numbers.

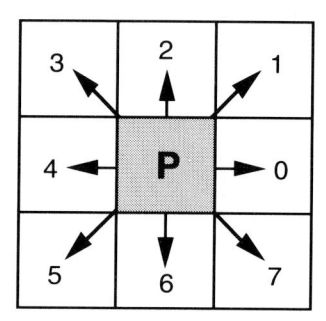

Figure 6.2 Chain code direction convention

In operation the chain code works best with binary edge maps. The sequence of illustrations given in figure 6.3a–i shows the chain code derived by raster scanning (i.e. scanning in strips from left to right and from top to bottom) an edge map until the first boundary pixel of the object is found.

The coordinates of this pixel are recorded as the start point of the contour, S – in this case (2,1). From this point on the cursor follows the contour, usually in a clockwise direction, the first step being in a south-east direction, coded as direction 7 according to the convention of figure 6.2. The second step is due east, coded direction 0; the third step is due south, coded as direction 6, etc. This procedure is followed until the final step, direction 1, returns the cursor to the start point, S. This yields the chain code 70654321, which has a chain length, or chain count, of 8 vectors.

If the starting point of the boundary cannot be regained then the contour is not complete (closed) and the code sequence is invalid and must be

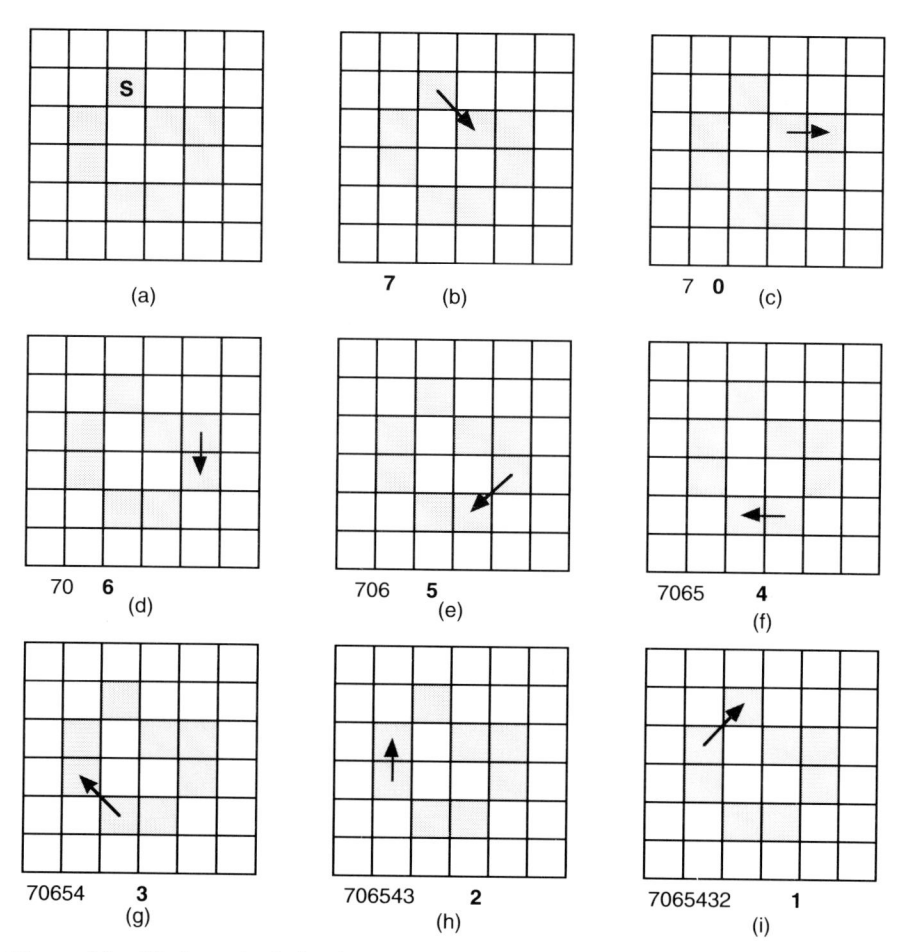

Figure 6.3 Chain code derivation

discounted. This could result from incomplete boundary refining, or objects which cross the edge of the field of view.

When a closed contour has been completely coded, raster scanning restarts and continues until another object is found or the end of the array is reached. It is therefore normal practice to delete boundary pixels from the edge-map array as they are coded so that each object is coded only once, and so that internal boundaries (due to holes) and subsequent objects might also be detected.

Finally, each closed contour must be checked to see if it represents another object (in which case it will be outside all other coded boundaries), or a hole in an object (whereupon it will be inside the boundary). This may be achieved by using Jordan's curve theorem which constructs lines from a

154                    *Applied Image Processing*

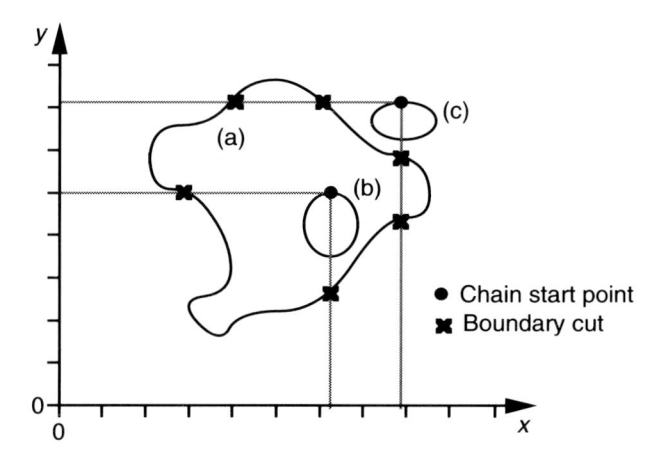

Figure 6.4   Illustration of Jordan's curve theorem: (a) the principal object boundary; (b) an internal boundary (hole); (c) an external boundary (hole)

point on the boundary to the edges of the image and checks if they cut any other boundaries [7]. If they cut an even number of times then the first boundary is outside the second, and if they cut an odd number of times then it is inside (see figure 6.4).

One of the problems with feature extraction by means of the chain code (see section 6.3.1) is concerned with inaccuracies introduced by sampling the original continuous image, since the chain vectors must be considered to link the centres of the pixels in the edge-map array. If vectors joining the centres of the *actual* boundary pixels are used, then features such as perimeter and area will be under-estimates of the real-world object dimensions. Conversely, if the centres of the pixels just outside the boundary are used then the derived features will be too large.

This problem can be significant when trying to recognise objects from their features because exact alignment of the sampling grid cannot be guaranteed each time an object is presented to the system. Therefore it is possible that significant errors could occur even when the very same object is viewed on subsequent occasions. Since this is actually a spatial resolution phenomenon, its effect (and others due to rotation and translation of objects) can be minimised by ensuring that the pixel size is always very small compared with the size objects of interest (thus making sure that the sampling frequency is high enough for meaningful results to be obtained). However, in cases where utmost accuracy is helpful, a variant of the chain code called 'crack coding' can produce better results.

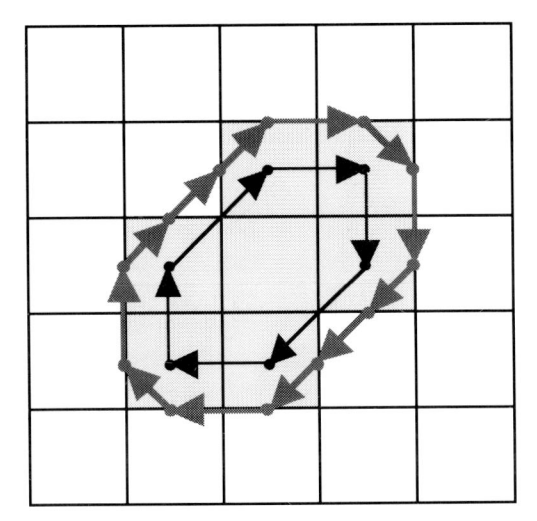

Figure 6.5  Benefits and penalties of crack coding. The crack code vectors (shaded arrows) clearly give a better representation of the object than the standard chain code vectors, but there are more vectors to process and more storage must be used

## 6.2.3 *Crack code*

The major difference between crack coding and chain coding is in the point at which the tracing of an object occurs, otherwise the principle is the same. Crack coding, as the name suggests, uses a point on the boundary *between* two pixels, rather than the pixel centres, for tracing vectors – hence the resulting code is a better representation of the true object contour. The major problem however is that the point between two pixels is not physically addressable in the original image array, and so additional post-processing is required on the edge map before crack coding can commence.

One approach is to double the resolution of the edge-map array and offset its origin so that each pixel in the original image is now surrounded by four additional pixels at the new, higher resolution (see figure 6.5). Now the 'cracks' between pixels can be addressed as the centres of these new 'pixels' which can be used for object contour tracing in the normal way. Clearly, this approach results in greater memory requirements and somewhat slower processing, although it has been shown to produce more accurate representations, especially of object vertices [8].

## 6.2.4 *Signatures and skeletons*

The signature of an object is a simple functional representation that can be used to describe and reconstruct it with appropriate accuracy. The 'polar

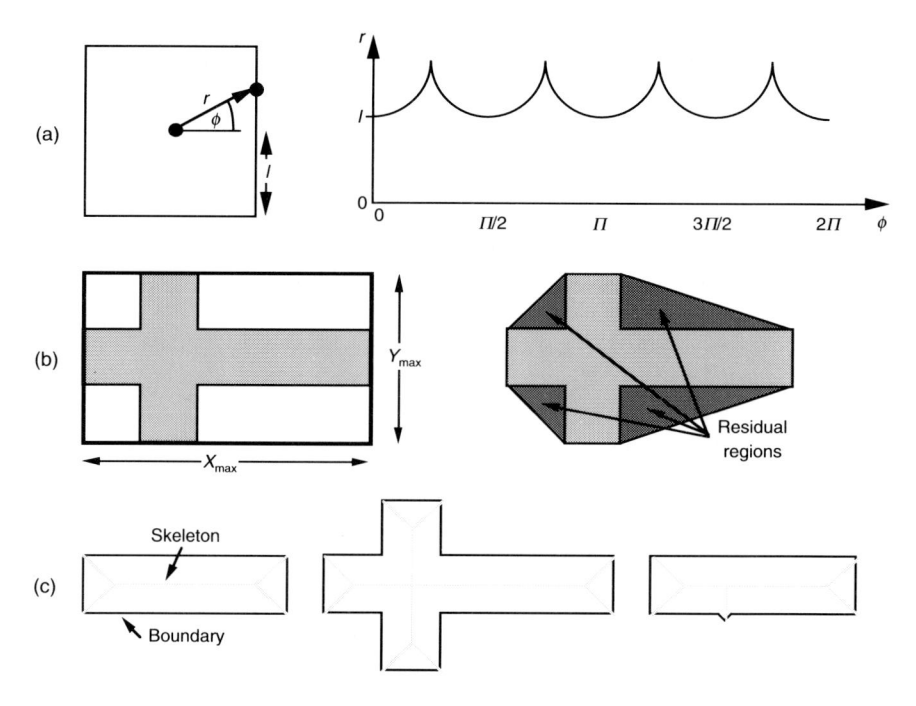

Figure 6.6  Various image coding strategies: (a) polar-radii signature for a rectangle; (b) minimum enclosing rectangle and convex hull; (c) illustrations of medial axis skeletons

radii signature' of an object shows the relationship between the distance from its centroid to points along its boundary as a function of angle – a rectangular object yields the simple two-dimensional graph shown in figure 6.6a. This is an orientation-invariant feature which allows an object to be compared with a standard prototype by cyclically shifting the signature of one with respect to the other in steps while checking for the best match [9]. For a circular object the graphical signature would simply be a horizontal line with the ordinate corresponding to the radius of the circle.

Another approach is to decompose the boundary into several segments; this is particularly useful when the boundary contains a number of significant concavities. The 'convex hull' approach, often referred to as rubber-banding, defines a line which touches the edge of the object at every point except that concavities are not followed. The number and shape of the regions left between the convex hull and the object itself are used to specify the shape.

A simpler, but less versatile technique involves the construction of a 'minimum enclosing rectangle' around the object whose sides are parallel to the horizontal and vertical axes of the image sampling system and just enclose the maximum extent of the object in each of those directions. It is not a rotationally invariant feature, but for very high aspect ratio objects, such as fibres, the diagonal of the rectangle ($\sqrt{X_{max}^2 + Y_{max}^2}$) approximates to the length of the object and is therefore invariant with rotation. Figure 6.6b illustrates firstly an object surrounded with its minimum enclosing rectangle and secondly the same object and its convex hull.

The 'medial axis transform (MAT)' generates a 'skeleton' of an object which can be used as a simple shape descriptor since it exhibits topological properties of connectedness, invariance to scaling and rotation, and preserves information from which to reconstruct the object [10]. The skeleton may be found by repeated application of a 'thinning' algorithm which removes layers of pixels from the object like peeling an onion, subject to the constraint that connectivity must be maintained (see section 6.5.3 for a morphological implementation). Figure 6.6c illustrates the skeleton for three simple shapes. While the skeleton of an object is intuitively easy to visualise, thinning algorithms tend to be computationally expensive and even minor disturbances on the object boundary can generate totally spurious 'bones' in the skeleton (compare the skeletons of the first and third shapes in figure 6.6c).

## 6.3 Boundary-based features

The techniques described in the previous section can be used to derive a number of simple, yet useful features from the boundary of an object, with which to assist object description. Some of these are listed here as

| | |
|---|---|
| Perimeter | The length of the traced outline |
| Area | The area enclosed |
| $X_{max}$, $X_{min}$, $Y_{max}$, $Y_{min}$ | The maximum and minimum co-ordinates reached by the outline |
| $X_{centroid}$, $Y_{centroid}$ | The $X$ and $Y$ coordinates of the centre of the area enclosed |
| Shape factor | Perimeter$^2$/Area |
| Chain count | The number of elements in the chain. |

### 6.3.1 Perimeter, area and shape factor

On studying the chain code convention definition illustrated in figure 6.2 it is clear that the *even* numbered vectors correspond to horizontal or vertical movements of unit pixel spacing, while *odd* numbered vectors, being at directions of 45° to the principal axes, correspond to movements of $\sqrt{2}$ times the unit spacing (assuming the pixels have a 1:1 aspect ratio). Therefore the object perimeter can be found by simply totalling the number of even components and adding $\sqrt{2}$ times the total number of odd components. Considering the binary image in figure 6.7 chain coded in the normal way, the derived chain code is

077076455453012334201

This chain contains ten even numbers and eleven odd numbers, thus the object perimeter, P, is given by

$$P = \text{Even count} + \sqrt{2}(\text{Odd count})$$
$$= 10 + 11\sqrt{2} = 25.56 \text{ units}$$

The calculation of area is based on finding the elemental contribution to area made by each vector in the chain coded boundary, as illustrated in figure 6.8. The value of the ordinate is required because the elemental area calculated is that between the vector and the $y$-axis. Since vectors 2 and 6 are vertical they make no contribution to the area value. By definition, the vectors 1, 0, 7 make a positive contribution to the area (because they point clockwise), whereas vectors 3, 4, 5 make a negative contribution to the area.

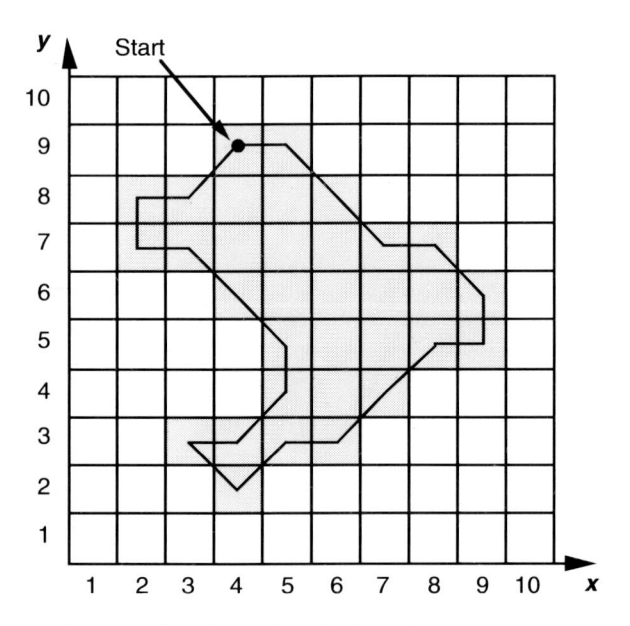

Figure 6.7  Perimeter estimation using chain code

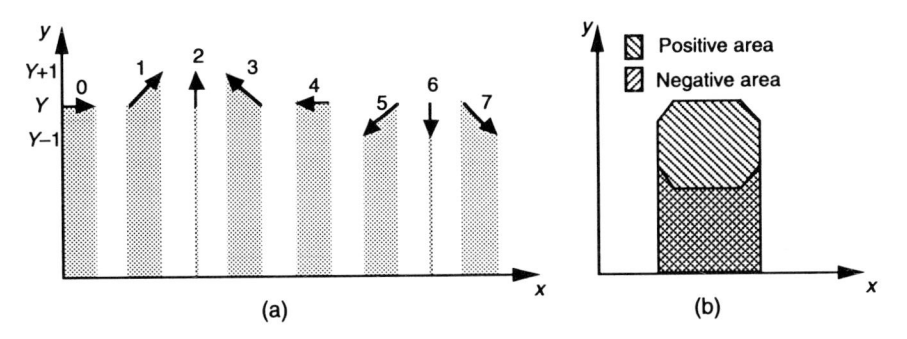

Figure 6.8 Area calculation from the chain code: (a) elemental contributions to area; (b) summary of area sign convention

The actual contribution to the area is determined by the area of the strip formed by the perpendiculars dropped from either end of the vector and the $y$-axis. Therefore vectors 0 and 4 contribute $Y$ square units while all vectors at 45° to the major axis (1,3,5,7) will contribute ($Y \pm 0.5$) square units, as appropriate.

The elemental contributions are summarised by the following equations:

| | |
|---|---|
| Chain code vector 0 | Area := Area + $Y$ |
| Chain code vector 1 | Area := Area + ($Y$ + 0.5) |
| Chain code vector 2 | Area := Area |
| Chain code vector 3 | Area := Area − ($Y$ + 0.5) |
| Chain code vector 4 | Area := Area − $Y$ |
| Chain code vector 5 | Area := Area − ($Y$ − 0.5) |
| Chain code vector 6 | Area := Area |
| Chain code vector 7 | Area := Area + ($Y$ − 0.5) |

The area of the object depicted in figure 6.7, whose chain code has already been determined, can now be estimated as follows:

| Vector code: | 0 | 7 | 7 | 0 | 7 | 6 | 4 | 5 | etc. |
|---|---|---|---|---|---|---|---|---|---|
| Ordinate: | 9 | 9 | 8 | 7 | 7 | 6 | 5 | 5 | etc. |
| $\Delta$Area: | +9 | +8.5 | +7.5 | +7 | +6.5 | +0 | −5 | −4.5 | etc. |
| Cumulative area: | 9 | 17.5 | 25 | 32 | 38.5 | 38.5 | 33.5 | 29 | etc. |

This is continued for the complete chain code to give the area as 21.5 square units.

Both the perimeter and area estimations described here yield results which are given in terms of pixel values (pixel count or square pixels respectively). For absolute measurements of size the magnification of pixel size should be calibrated in SI units, but for comparative purposes this is not necessary if the magnification remains fixed at all times.

The parameter termed 'shape factor', S, which can be expressed as the ratio (perimeter$^2$/ area) is therefore a dimensionless quantity. This renders it invariant of scale as well as rotation and translation making it a useful and effective feature that is simple to derive (see figure 6.9). Since it quantifies the relationship between a perimeter and the area it encloses, it is a useful measure of elongation of shapes.

The reciprocal of S is often referred to as the 'circularity' feature, since it rises to a maximum value (of $1/4\pi \approx 0.08$) for circular objects. For example, it is possible to distinguish between plain and shake-proof washers by setting a threshold on the circularity of the inner boundary, such that if it is substantially less than 0.08, the washer is plain.

The technique illustrated in figure 6.8 can easily be extended to calculate elemental contributions to 'first moments of area' about each of the Cartesian axes [11]. See section 6.4.2 for a discussion of how this can be used to locate the 'centroid' of an area. This is a key parameter in determining the position, orientation and identity of objects in silhouette analysis (also known as 'blob' analysis).

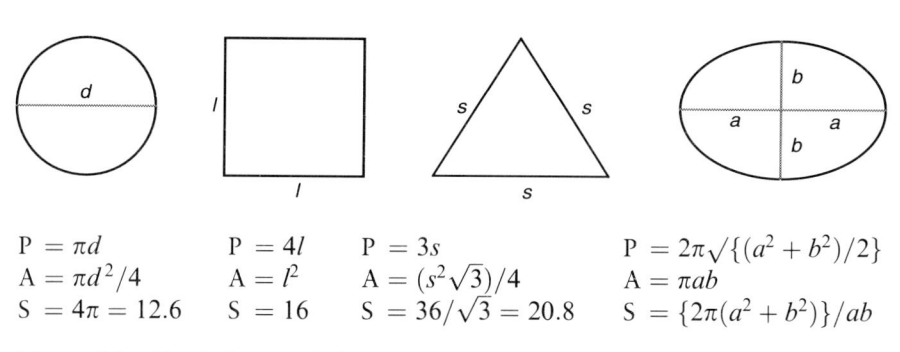

$P = \pi d$  
$A = \pi d^2/4$  
$S = 4\pi = 12.6$

$P = 4l$  
$A = l^2$  
$S = 16$

$P = 3s$  
$A = (s^2\sqrt{3})/4$  
$S = 36/\sqrt{3} = 20.8$

$P = 2\pi\sqrt{\{(a^2 + b^2)/2\}}$  
$A = \pi ab$  
$S = \{2\pi(a^2 + b^2)\}/ab$

Figure 6.9 Simple feature derivations for various shapes

## 6.4 Region-based features

For more complex objects the use of region-based features such as topology and texture may be considered. Although the *area* feature mentioned above is easily linked with boundary-based measurements, it is actually a region-based feature. In fact it is just the simplest of the family of region-based features which depend upon moments.

## 6.4.1 Topology and texture

Topological features give a global description of a region, or object, and are unaffected by scale changes or even distortion within the image (except tearing or folding). This is particularly useful since it allows, for example, an object to be described simply by the number of holes it contains regardless of its overall size. Consider a washer: it is generally, but not always, a circular object with one hole. The washer material itself represents one connected component and there is one hole. These topological properties can be combined to determine a feature called the Euler number, $E$:

$$E = C - H$$

where $C$ is the number of connected regions,
and $H$ is the number of holes.

Thus Euler number for the washer is 0, since there is 1 connected region containing 1 hole. Such topological features can be extended to include the number of object edges, faces, vertices, etc. [12].

In real-world images, surface properties such as smoothness, reflectivity, granularity, and regularity can be used to assist with object classification. Statistical analysis of the grey-scale histogram may yield feature measures such as distribution of surface intensities, relative smoothness and relative flatness. Simple frequency domain techniques such as searching for light–dark transitions along a particular row of the image will yield features of periodicity; this approach has been used to assess the tension in coil springs fitted on a car brake assembly. More generally, spectral analysis utilises the Fourier transform where the location of peaks in the frequency spectrum indicates the fundamental spatial frequency of the surface patterns.

Texture, as observed in wood grain, stone, cloth, grass, etc. is an important surface property which, although readily detected by the human visual system, is far from trivial for machines to identify; for example, regions having the same average grey level may have dramatically different textural appearance.

Analysis of texture begins by identifying basic texture elements ('texels') which repeat with some degree of predictability. Each texel will consist of a group of pixels which can have random, periodic or partially periodic distributions. On the whole, man-made materials feature regular, periodic textures whilst naturally occurring textures are random, so *a priori* knowledge of surface type can simplify solutions.

Texture analysis can be divided into two major groups of techniques, statistical and structural [13] – the former being used primarily for naturally occurring textures, having a random nature, while the latter is well suited to deterministic or periodic (usually man-made) textures. Statistical techniques include the 'autocorrelation function', where strips of the image are correlated with themselves after shifting by a defined 'lag'. Plotting this function against lag can identify the periodicity of the texture pattern. Image transforms such as Fourier can be useful for periodic and partially periodic patterns and have been successfully applied to terrain discrimination and detection of certain types of lung disease. Random texture can be assessed for 'coarseness' by determination of the average density of edge pixels per unit area in an edge map of the scene.

Structural techniques are generally best suited to analysis of 'strong' textures which have clearly defined deterministic placement rules for texels. The texels may be defined by shape, average grey level or homogeneity of local properties such as orientation or size. 'Weak' textures feature randomly placed texels and must be assessed by features such as edge density or 'relative extrema density', which represents the average number of pixels in a neighbourhood whose grey levels are local maxima or minima (defined by some suitable criteria).

Readers can discover more about the complex topic of texture analysis in Haralick [14] or Lipkin and Rosenfeld [15].

### 6.4.2 Moments

The application of moments provides a method of describing the properties of an object in terms of its area, position, orientation and other precisely defined parameters. The basic equation defining the moment of an object is given as

$$m_{ij} = \sum_x \sum_y x^i y^j a_{xy}$$

where the order of the moment is $i + j$,

$x$ and $y$ are the pixel coordinates relative to some arbitrary standard origin, and

$a_{xy}$ represents the pixel brightness.

Zero- and first-order moments can be defined as

$$m_{00} = \sum_x \sum_y a_{xy}$$

$$m_{10} = \sum_x \sum_y x.a_{xy}$$

$$m_{01} = \sum_x \sum_y y.a_{xy}$$

In a binary image the zero-order moment ($m_{00}$) is the same as the object area since $a_{xy}$ is either '0' (black) or '1' (white).

Any region-based feature will usually require a datum point from which further features may be derived. The 'centroid' (centre of area, or centre of mass) is a good parameter for specifying the location of an object; it is the point having coordinates $x'$, $y'$ such that the sum of the square of the distance from it to all other points within the object is a minimum. The centroid can also be expressed in terms of moments as

$$x' = \frac{m_{10}}{m_{00}} \quad \text{and} \quad y' = \frac{m_{01}}{m_{00}}$$

where $(x', y')$ are the coordinates of centroid with respect to the origin in use.

Referring back to section 6.2.1 the centroid can easily be calculated from the run code as illustrated in figure 6.10. Similarly, the chain code technique of section 6.2.2 can also be easily extended to obtain the centroid [11].

Moments $m_{02}$ and $m_{20}$ correspond to the moments of inertia of the object. It should be noted however that these basic moments are limited in their usefulness since they vary according to their position with respect to the origin and the scale and orientation of the object under investigation. A set of invariant moments would be of more use. These can be derived by first

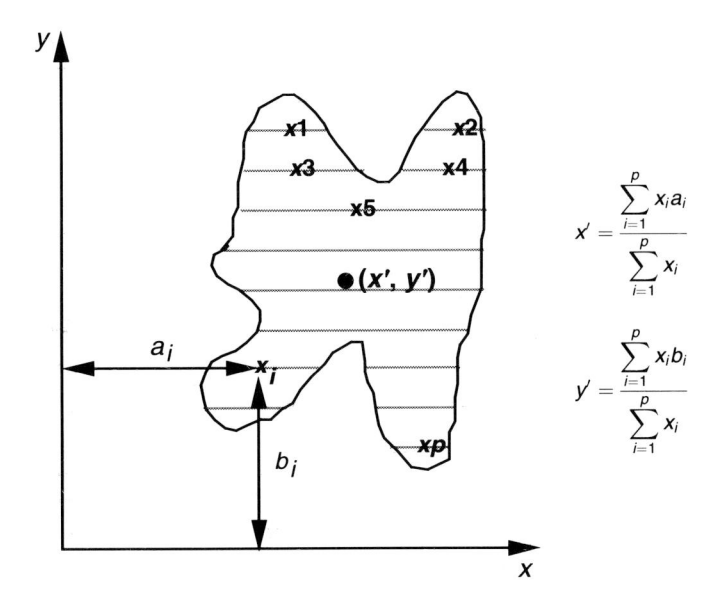

$$x' = \frac{\sum_{i=1}^{p} x_i a_i}{\sum_{i=1}^{p} x_i}$$

$$y' = \frac{\sum_{i=1}^{p} x_i b_i}{\sum_{i=1}^{p} x_i}$$

Figure 6.10  Centroid calculation using run code data. $x_i$ = length of the $i$th foreground run; $a_i$ = abscissa of centre of $i$th run; $b_i$ = ordinate of centre of $i$th run; $p$ = maximum number of runs

calculating the central moment, $\mu$, with respect to the centroid, as given by

$$\mu_{ij} = \sum_x \sum_y (x-x')^i (y-y')^j a_{xy}$$

then developing the normalised central moments, $\eta$, as

$$\eta_{ij} = \frac{\mu_{ij}}{(\mu_{00})^\lambda}$$

where $\lambda = \dfrac{(i+j)}{2} + 1$, and
$(i+j) \geq 2$ (first-order moments are always invariant).

From these normalised parameters a set of invariant moments, $\{\phi\}$, may then be defined. The equations governing all these operations up to order 3 are summarised in Table 6.1 [16]. The set of invariant moments makes a useful feature vector for the recognition of objects which must be detected regardless of position, size or orientation. They have been applied to character recognition, aircraft identification and scene-matching applications [17, 18].

Table 6.1  Central and invariant moments

---

*Central moments*

---

$\mu_{00} = m_{00}$

$\mu_{10} = 0 \qquad\qquad = \sum_x \sum_y (x-x')^1 (y-y')^0 a_{xy}$

$\mu_{01} = 0 \qquad\qquad = \sum_x \sum_y (x-x')^0 (y-y')^1 a_{xy}$

$\mu_{20} = m_{20} - x' m_{10} \qquad = \sum_x \sum_y (x-x')^2 (y-y')^0 a_{xy}$

$\mu_{02} = m_{02} - y' m_{01} \qquad = \sum_x \sum_y (x-x')^0 (y-y')^2 a_{xy}$

$\mu_{11} = m_{11} - y' m_{10} \qquad = \sum_x \sum_y (x-x')^1 (y-y')^1 a_{xy}$

$\mu_{30} = m_{30} - 3x' m_{20} + 2x'^2 m_{10} \qquad = \sum_x \sum_y (x-x')^3 (y-y')^0 a_{xy}$

$\mu_{03} = m_{03} - 3y' m_{02} + 2y'^2 m_{01} \qquad = \sum_x \sum_y (x-x')^0 (y-y')^2 a_{xy}$

$\mu_{12} = m_{12} - 2y' m_{11} - x' m_{02} + 2y'^2 m_{10} = \sum_x \sum_y (x-x')^1 (y-y')^3 a_{xy}$

$\mu_{21} = m_{21} - 2x' m_{11} - y' m_{20} + 2x'^2 m_{01} = \sum_x \sum_y (x-x')^2 (y-y')^1 a_{xy}$

*Derived invariant moments*
$$\phi_1 = \eta_{20} + \eta_{02}$$
$$\phi_2 = (\eta_{20} - \eta_{02})^2 + 4\eta_{11}^2$$
$$\phi_3 = (\eta_{30} - 3\eta_{12})^2 + (3\eta_{21} - \eta_{03})^2$$
$$\phi_4 = (\eta_{30} + \eta_{12})^2 + (\eta_{21} + \eta_{03})^2$$
$$\phi_5 = (\eta_{30} - 3\eta_{12})(\eta_{30} + \eta_{12})\{(\eta_{30} + \eta_{12})^2 - 3(\eta_{21} + \eta_{03})^2\}$$
$$+ (3\eta_{21} - \eta_{03})(\eta_{21} + \eta_{03})\{3(\eta_{30} + \eta_{12})^2 - (\eta_{21} + \eta_{03})^2\}$$
$$\phi_6 = (\eta_{20} - \eta_{02})\{(\eta_{30} + \eta_{12})^2 - (\eta_{21} + \eta_{03})^2\} + 4\eta_{11}(\eta_{30} + \eta_{12})(\eta_{21} + \eta_{03})$$
$$\phi_7 = (3\eta_{12} - \eta_{30})(\eta_{30} + \eta_{12})\{(\eta_{30} + \eta_{12})^2 - 3(\eta_{21} + \eta_{03})^2\}$$
$$+ (3\eta_{21} - \eta_{03})(\eta_{21} + \eta_{03})\{3(\eta_{30} + \eta_{12})^2 - (\eta_{21} + \eta_{03})^2\}$$

Orientation of an object may be defined as the angle of the axis of the minimised moment of inertia. This can be expressed in terms of the second-order central moments as

$$\theta = \frac{1}{2}\tan^{-1}\left[\frac{2\mu_{11}}{\mu_{20} - \mu_{02}}\right]$$

where $\theta$ is the orientation with respect to the $x$-axis.

The maximum and minimum distance vectors ($R_{max}$ and $R_{min}$ respectively) from the object centroid to its boundary are also used in description, as is the angle formed between them. The ratio of $R_{max} : R_{min}$ is another characteristic feature used as a measure of elongation or 'eccentricity'. It may also be expressed in terms of the second-order central moments as

$$\epsilon = \frac{(\mu_{20} - \mu_{02})^2 + 4\mu_{11}^2}{\mu_{00}}$$

where $\epsilon$ is the eccentricity of the object.

## 6.5 Mathematical morphology

Mathematical morphology has its origins in set theory and concerns the study of form and structure. Within image analysis it concerns the shape and properties of objects, or regions of an image, and how these may be changed, and useful features extracted.

While the formal mathematics of set theory utilises concepts of set inclusion, complement, union and intersection, a useful insight into the topic of morphology can be achieved using a more intuitive approach based upon plain English descriptions of the operations performed and the logical expressions OR, AND and NOT. In addition the familiar operation of moving a template, or mask, over an image and performing specific template–image comparisons at each template position will give similar results to the set theoretical approach.

### 6.5.1 Basic definitions

Despite the wish to make understanding of morphology more intuitive, it is true that most literature on the subject will use standard set notation. Therefore the following discussion will describe the operations and back this up with proper set notation. Note that points and vectors will be represented in lower-case italics, sets will be represented in upper-case italics. $\emptyset$ is the empty set, $\in$ means 'is a member, or element, of' and $\Rightarrow$ means 'implies that'. The expression $\{p \mid < condition(s) >\}$ means 'the set of points, $p$, such that the listed <condition(s)> is (are) true'. Figure 6.11 will serve to illustrate many of the following definitions more fully:

(a) The binary object can be identified as a subset of pixels $A$ contained or '*included*' within the total set of image pixels $N$. This is usually written as $A \subset N$. Therefore if $p$ is an element of $A$ then $p$ is also an element of $N$, or written mathematically $p \in A \Rightarrow p \in N$.

(b) The '*complement*' of the set, $A$, is the set which are not elements of $A$, that is, all the white background pixels in the figure; this may be denoted by $A^*$. Therefore if $p$ is an element of $A$ it cannot be an element of $A^*$, i.e. $p \in A \Rightarrow p \notin A^*$.

(c) The '*transposition*' of a set of pixels, $A$, is the set such that each pixel position is reflected about a defined origin. When the set, $A$, given in figure 6.11a is reflected about an origin taken to be the centre of the image, the set $A^t$ results (figure 6.11b, where $A^t$ equals $B$). Thus the set $A$ is the symmetrical set of $A^t$, with respect to its origin.

(d) The '*union*' of two sets $A$ and $B$ is denoted by $A \cup B$. Here the resulting set is all pixels such that each pixel is an element of set $A$ or an element of set $B$, i.e. $A \cup B = \{p \mid p \in A \text{ or } p \in B\}$. This is analogous to the logical OR operation and is illustrated in figure 6.11c, which shows the union of the images given in figures 6.11a and 6.11b.

(e) The '*intersection*' of two sets $A$ and $B$ is denoted by $A \cap B$. Here the resulting set is all pixels such that each pixel is a common element of both set $A$ and set $B$, i.e. $A \cap B = \{p \mid p \in A \text{ and } p \in B\}$. This is analogous to the logical AND operation and is illustrated in figure 6.11d, which shows the intersection of the two images given in figures 6.11a and 6.11b.

(f) The '*difference*' between two images is denoted as $A/B$, and in the language of set theory, is the resulting set such that each pixel is either an element of set $A$ or an element of set $B$, but not an element of both sets, i.e. $A/B = \{p \mid p \in A \text{ or } p \in B; p \notin A \cup B\}$. This is analogous to the logical XOR (exclusive OR) operation and is illustrated in figure 6.11e, which shows the difference of the two images given in figures 6.11a and 6.11b.

(g) '*Translation*' can be simply regarded as a vector addition. The translation of a set $A$ by $v$ is denoted by $A_v$. This is a set where each element is translated by a vector $v$. Figure 6.11f shows the result of the translation of a set $A$ (see figure 6.11a) by a vector (2, 2).

## 6.5.2 The hit–miss transform

Image morphology relies on the analysis of images using elementary patterns or 'structuring elements'. These structuring elements can be thought of as templates, but here the image–template calculations are evaluated through the algebra of sets, as defined above, instead of the multiplication operations used in the more familiar convolution approach of earlier chapters.

For example, figure 6.12b illustrates a $3 \times 3$ structuring element in which the reference pixel (origin) is defined to be the centre pixel, and which contains four foreground (black) pixels and five background pixels. This structuring element can be applied to locate upper right corners of objects, while ignoring the endpoints of lines (i.e. features of single pixel width). This involves moving the structuring element, or template, so that the reference pixel is stepped over the whole image. For each step a check is performed to see if ALL of the pixels of the template *match* the pixels of the underlying image (in a binary image, foreground template pixels must match with foreground image pixels AND background template pixels with background image pixels). At each position of the reference pixel where this is true, a foreground pixel is written to the output array at the corresponding position. Thus moving figure 6.12(b) over figure 6.12a as defined will identify corners in the image as required (see figure 6.12c).

In morphology this is known as the 'hit–miss' transformation of set $A$ with structuring element $B$, which is denoted $A \otimes B$ and is formally defined mathematically as

$$A \otimes B = \{p \mid B_{\text{fgd}} \subset A; B_{\text{bgd}} \subset A^*\}$$

where $A$ is the image set, $A^*$ is the complement of the image set,
$B$ is structuring element set,

168    *Applied Image Processing*

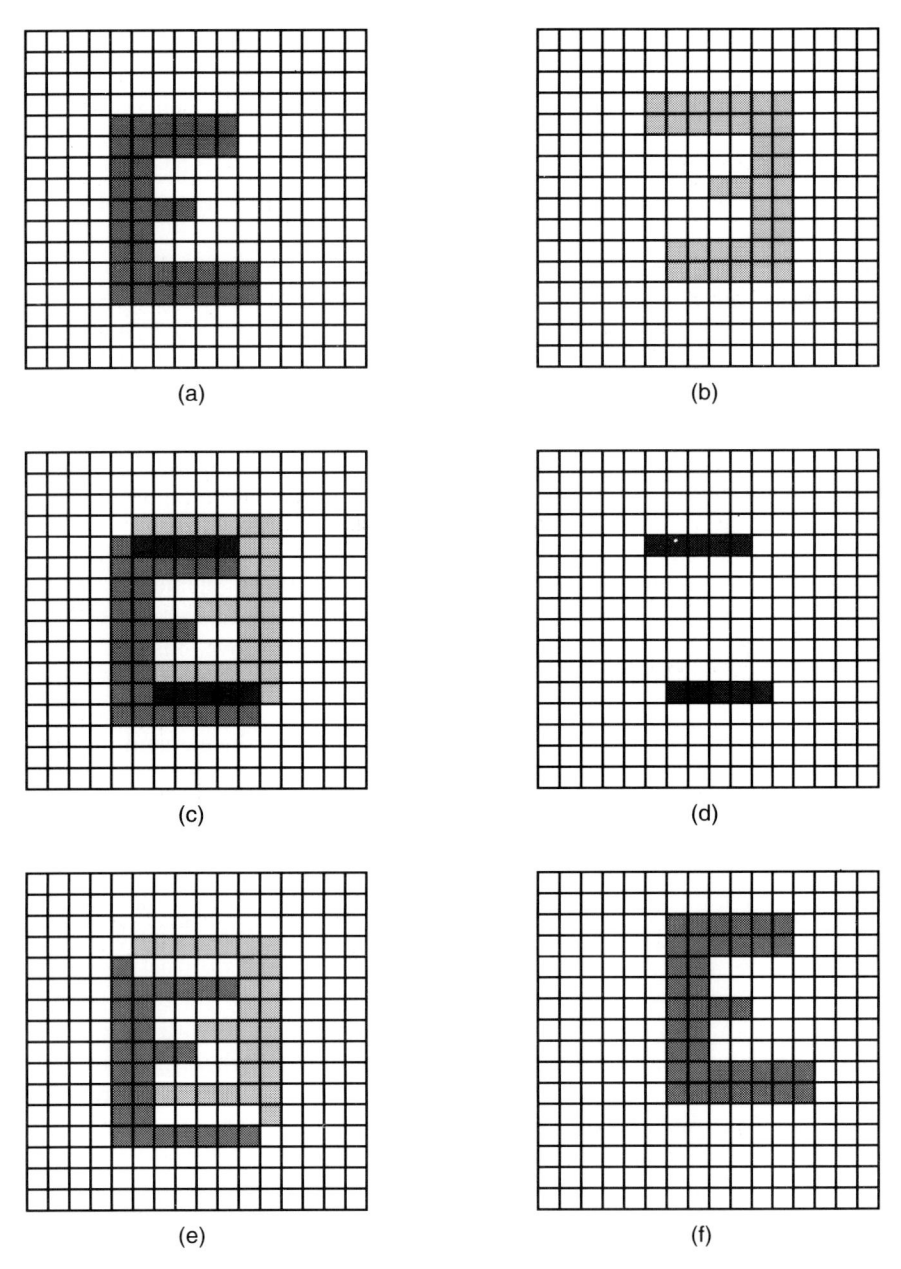

Figure 6.11   Illustration of some basic definitions in mathematical morphology: (a) original image, set $A$; (b) transposition of set $A$ to produce set $B$; (c) union, $A \cup B$ (equivalent to $A$ OR $B$); (d) intersection, $A \cap B$ (equivalent to $A$ AND $B$); (e) difference of images, $A/B$ (equivalent to $A$ XOR $B$); (f) translation, $A_v$, of set $A$ by vector $v$ (2, 2). Note: the foreground in (a) and (b) is shaded differently for emphasis.

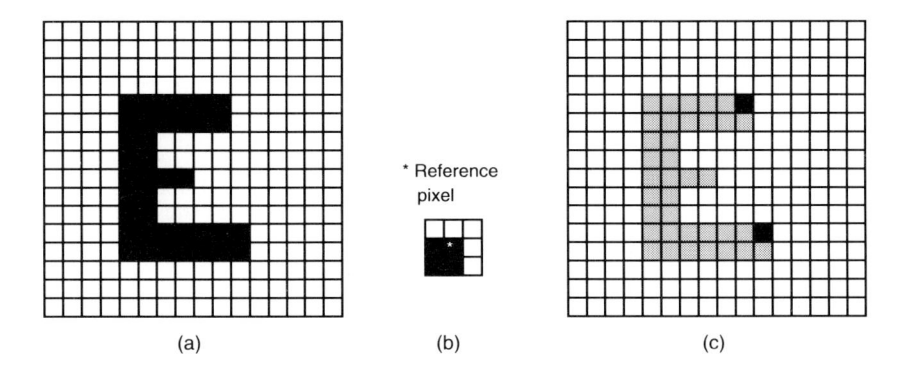

Figure 6.12 Corner finding using a structuring element: (a) the original image set; (b) the 'upper right' corner finding structuring element; (c) the output image set (with original object shown shaded for comparison)

$B_{\text{fgd}}$ represents the foreground points of the translated structuring element,

$B_{\text{bgd}}$ represents the background points of the translated structuring element, and

$p$ is the current position of the reference pixel of the structuring element.

So $A \otimes B$ defines the points where the structuring element $B$ 'hits' the set $A$ – i.e. its foreground AND background exactly matches the underlying image. Clearly, the hit–miss transform is logically analogous to correlation, and structuring elements are prototypes of features which need to be detected in the image. The reader might like to consider what structuring elements could be used to locate the other corners in the image, and also the endpoints of lines.

### 6.5.3 Erosion and dilation

Two operations which are fundamental to morphological analysis of images are '*erosion*' and '*dilation*'. Almost all morphological operations can be defined in terms of these two basic operations.

Erosion of set $A$ by structuring element $B$ is denoted $A \ominus B^t$ and is formally defined mathematically as

$$A \ominus B^t = \{p \mid B_p \subset A\}$$

where $B^t$ is the *transposed* form of the structuring element set, and

$B_p$ represents the structuring element centred at point $p$.

Note also that the more obvious $A \ominus B$ is NOT an erosion – it actually performs the 'Minkowski subtraction' of $B$ from $A$, but this will effect the same result if the structuring element has rotational symmetry. The relationship between erosion and Minkowski subtraction is therefore reminiscent of that between convolution and correlation, as discussed in section 4.1.1.

However, since the erosion operation is based on inclusion it is more simply conceived of as the output set obtained where the template is completely contained by the image set. Thus the eroded image of figure 6.13c is produced by stepping the structuring element, or template, over the input until ALL of the foreground pixels of the template fit over foreground pixels in the underlying image. At each position where this is true a pixel is written to the output array corresponding to the reference pixel position, $p$ (see figure 6.13c). This is equivalent to the hit–miss transformation of the image with the structuring element, where there are NO background pixels in the latter – i.e. $B_{\text{bgd}} = \emptyset$.

Notice how erosion enlarges holes in the object, shrinks its boundary, eliminates 'islands' and removes narrow 'peninsulas' that might exist on the boundary. Also consider the effect of eroding the object with a 3 × 3 grid of foreground pixels, $F$, and then finding the morphological difference between this and the original object. The result is a row of connected single pixels corresponding to the outermost extreme of the object foreground – a perfect boundary, known as the 'inner' boundary of the object. For the record this is expressed mathematically as

Inner boundary $= A/A \ominus F$

Dilation is the 'dual' of erosion, that is, the dilation of a set $A$ is equivalent of the erosion of the complement set $A^*$. (Note that this duality relationship is analogous to the duality of De Morgan's Laws in Boolean algebra). Therefore, dilation, denoted $A \oplus B^t$, may be defined as

$$A \oplus B^t = A^* \ominus B^t$$

This mathematical notion of duality means that dilation can be performed by eroding the complement set by the same structuring element. In practice, this means that a foreground object in a binary image can be dilated by eroding the background with the same structuring element. Alternatively dilation is formally defined mathematically as

$$A \oplus B^t = \{p \mid B_p \cap A \neq \emptyset\}$$

This means that an output pixel will be written at all points where the translated structuring element 'hits' the image set – i.e. they have a 'non-empty intersection'.

Stated more intuitively, the mathematical definition of dilation says that a foreground pixel will be written to the output set at all positions of the structuring element reference where ANY foreground pixel in the structuring element overlays a foreground pixel of the image set (i.e. part of an object). The result of this operation is illustrated in figure 6.13d.

Notice how dilation fills in holes in the object and expands its boundaries, filling in any narrow 'creeks' that might exist. The reader might like to consider how an 'outer' boundary of the object might be achieved using dilation.

In both figures 6.13c and 6.13d the shaded regions indicate the change that is produced by the morphological operator. In figure 6.13c the output image is indicated by the solid black area and in figure 6.13d it is indicated by the solid black area PLUS the shaded area. Notice that these operations significantly modify the *size* of the objects as well as their *shape*.

## 6.5.4 Opening and closing

Although dilation and erosion are dual operations it is not possible to reconstruct an image set by application of dilation after previously having eroded the image. The dilation operation will only be able to reconstitute the *essential* features of the structure of the object as modified by the structuring element. However, such a new set can be extremely useful in determining size and shape information. A pair (dual) of sequential operations may thus be formally defined as

'Opening', denoted $A_B$:    $(A \ominus B^t) \oplus B$;   i.e. erode then dilate

'Closing', denoted $A^B$:    $(A \oplus B^t) \ominus B$;   i.e. dilate then erode

The results of these operations are illustrated in figure 6.14b and c. In both cases the original input image is that given above in figure 6.14a and the structuring element is that given in figure 6.13b.

Opening and closing operations form the basis of boundary smoothing and noise elimination processes whether the noise is manifest as small holes within the object or as small protrusions external to it. Continuing with the geographical metaphor which seems so appropriate for the effect of morphological operations, opening smooths object 'coastlines', eliminates small 'islands' and cuts narrow 'isthmuses'. Thus it isolates objects which may be just touching one another, and is therefore a suitable precursor to studies of the distribution of particles sizes – for example, in analysis of wear particles in engine oil, ink particles in recycled paper, or cells in cytology. On

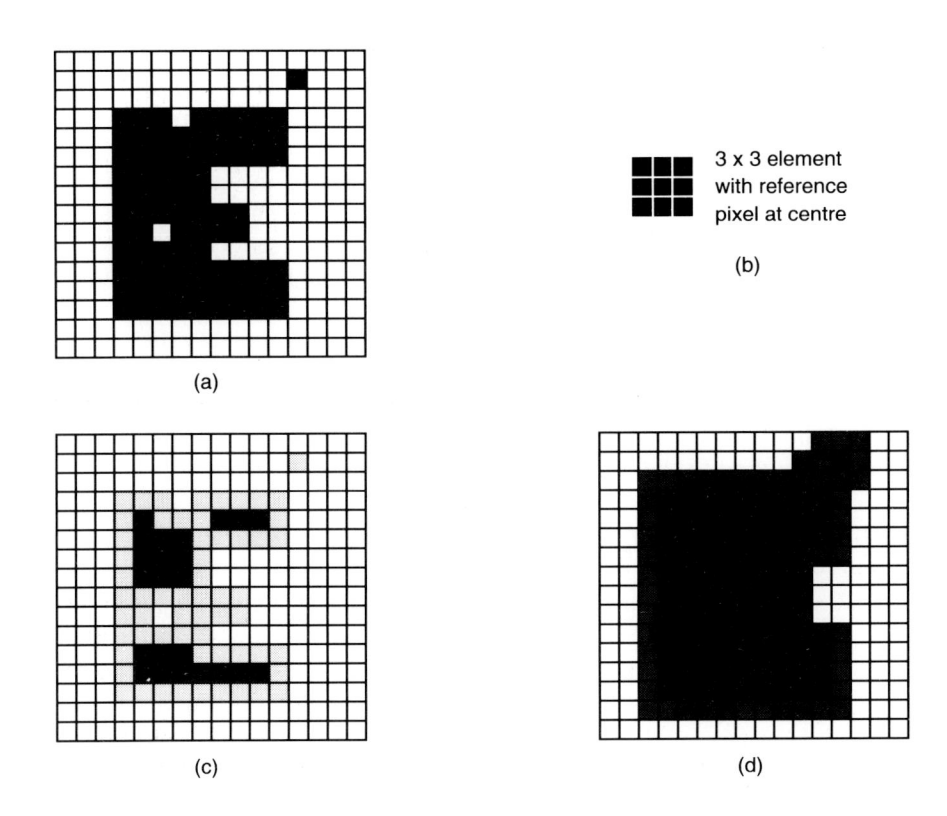

Figure 6.13  Erosion and dilation: (a) the original image (black = object); (b) the 3 × 3 structuring element; (c) the erosion of (a); (d) the dilation of (a)

the other hand, closing blocks up narrow 'creeks' and small or thin 'lakes' inside the object and links nearby objects. This simplifies the process of assessing the separation of particles.

Note that, while opening or closing significantly modifies the *shape* of the object(s), completing these operations restores the previously eroded or dilated object(s) to their original *size*. Application of the opening and closing techniques by the reader to suitable examples will clarify their use.

The focus of interest in this chapter is feature extraction and in section 6.2.4 reference was made to the skeleton of an object being one of the basic structural features worthy of identification. The skeleton of an object may be identified by the repeated application of some of the mathematical morphology techniques just described. Figure 6.15a–k illustrates the operations to produce the medial axis skeleton of a simple rectangular

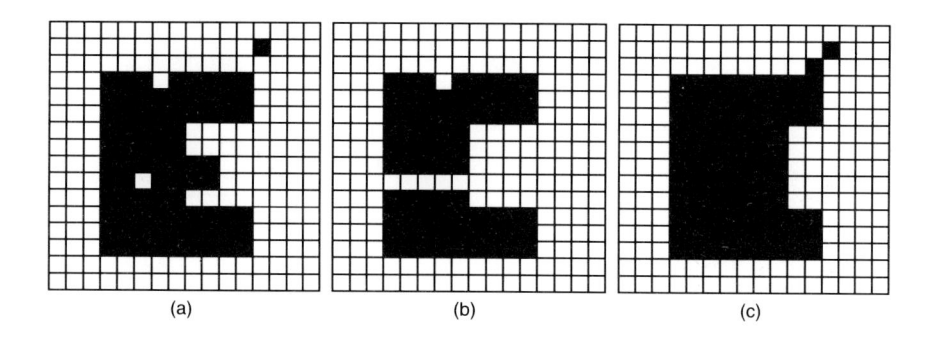

Figure 6.14 Sequential applications of erosion and dilation: (a) original image set; (b) opened version; (c) closed version

shape; the structuring element $S$ is also identified and its centre pixel is used as reference.

It can be seen that the production of the medial axis skeleton breaks down into an n-stage iterative process, where $n$ is the number of erosions performed on the original image, $A$, to produce the 'initial' image, $I(n)$, for that stage. Each stage consists of opening the initial image by $S$, then taking the morphological difference between the initial image and the opened version. This 'result' image, $R(n)$, is stored for collection at the end of the algorithm. Thus the processing of each stage can be mathematically summarised, using the notation previously defined, as

$$R(n) = (I(n)_S / I(n))$$

The algorithm will terminate when further erosion of the image set eliminates all foreground pixels – i.e. when $I(n) = \emptyset$. Then all the result images are collected up and combined together to form the final skeleton. Thus

$$\text{Medial axis skeleton} = \bigcup_{n=0}^{n_{max}} R(n)$$

where $n_{max} = 2$, in this case.

Mathematicians would summarise all of this much more succinctly as

$$\text{Medial axis skeleton of } A = \bigcup_{n=0}^{n_{max}} [(A \ominus nS)/(A \ominus nS)_S]$$

where $(A \ominus nS)$ indicates that $A$ has been eroded $n$ times by $S$, and the subscripted $S$ is the notation for opening by $S$.

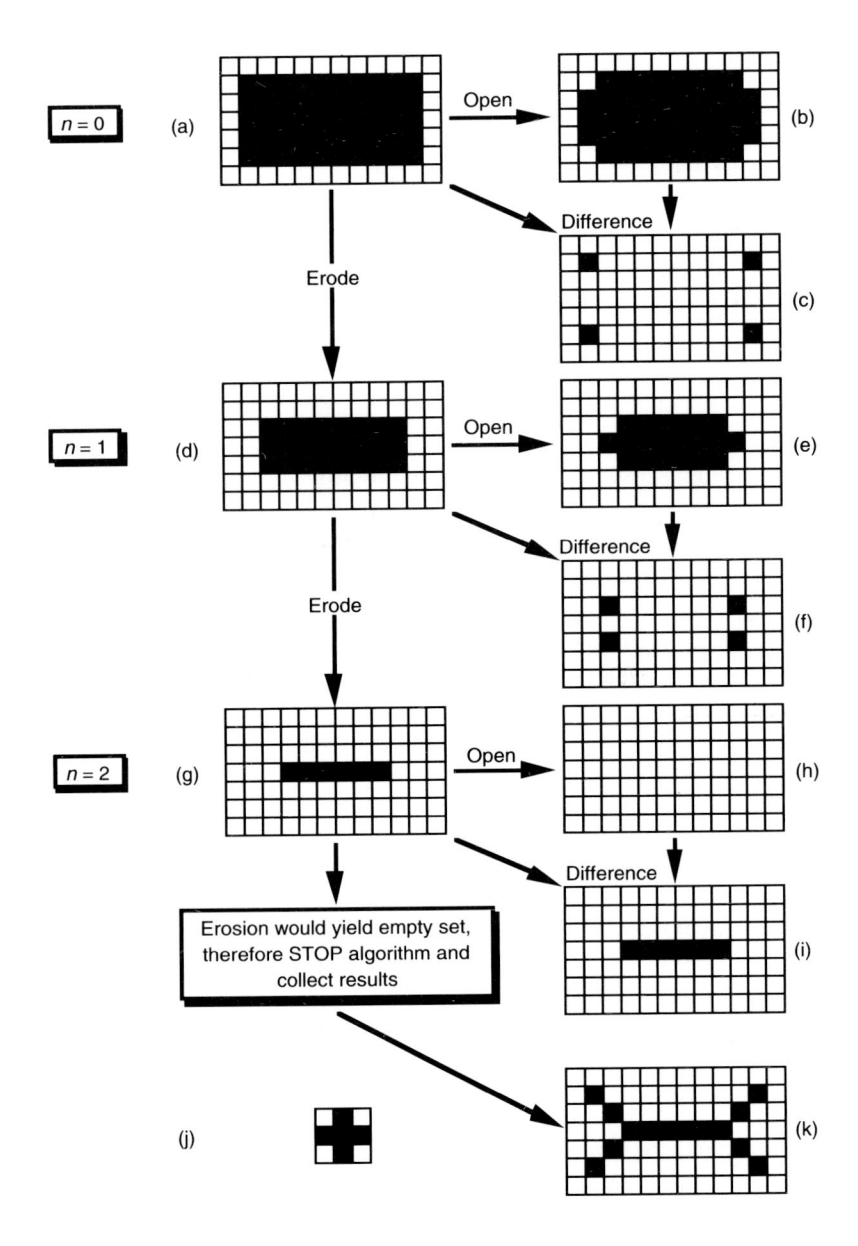

Figure 6.15  Obtaining the medial axis skeleton: (a) original image; (b) image (a) opened by structuring element (j); (c) difference of (a) and (b); (d) image (a) eroded by (j); (e) image (d) opened; (f) difference of (d) and (e); (g) image (d) eroded; (h) image (g) opened; (i) difference of (g) and (h); (j) structuring element; (k) result image – union of (c), (f) and (i)

## References

1. J. Serra, *Image Analysis and Mathematical Morphology*, Academic Press, New York, 1982.
2. G. Matheron, *Random Sets and Integral Geometry*, Wiley, New York, 1975.
3. A. Netravali and J. Limb, 'Picture coding: a review', *Proc. IEEE 68*, No. 3, pp. 366–406, 1980.
4. A. Jain, P. Farrelle and V. Algazi, 'Image data compression', in M. Ekstrom (Ed.), *Digital Image Processing Techniques*, Academic Press, New York, 1984.
5. Various authors, *Proc. IEEE 73*, No. 2, 1985.
6. H. Freeman, 'Computer processing of line-drawing images', *ACM Computing Surveys*, **6**(1), pp. 57–97, 1974.
7. R. Courant and H. Robbins, *What is Mathematics?*, pp. 267–269, Oxford University Press, London, 1941.
8. K. Dunkelburger and O. Mitchell, 'Contour tracing for precision measurement', *Proc. Int. Conf. Robotics & Automation*, pp. 22–27, IEEE, 1985.
9. J. Dessimoz and P. Kammenos, 'Software or hardware for robot vision', in I. Aleksander (Ed.), *Artificial Vision for Robots*, Chapter 2, Kogan Page, London, 1983.
10. H. Blum, 'A transformation for extracting new descriptors of shape', in W. Wathen-Dunn (Ed.), *Symp. Models for the Perception of Speech and the Visual Form*, MIT Press, Cambridge, Massachusetts, 1967.
11. P. Kitchin and A. Pugh, 'Processing of binary images', in A. Pugh (Ed.), *Robot Vision*, pp. 21–42, IFS (Publications), Bedford, 1983.
12. R. Gonzalez and R. Woods, *Digital Image Processing*, 2nd edn, Addison-Wesley, London, 1992.
13. R. Haralick, Statistical and structural approaches to texture', *Proc. IEEE 67*, pp. 786–809, 1979.
14. R. Haralick, *Image Texture Analysis*, Plenum Press, New York, 1981.
15. B. Lipkin and A. Rosenfeld (Eds), *Picture Processing and Psychopictorics*, Academic Press, New York, 1970.
16. M. Hu, 'Visual pattern recognition by moment invariants', *IRE Trans. Info. Theory*, **IT-8**, pp. 179–187, 1962.
17. S. Dudani, K. Breeding and R. McGhee, 'Aircraft identification by moment invariants', *IEEE Trans. Computers*, **26**(1) pp. 39–46, 1977.
18. R. Wong and E. Hall, 'Scene matching with moment invariants', *CGIP*, **8**, pp. 16–24, 1978.

# 7 Pattern Classification

## 7.1 Introduction

The term *pattern classification* refers simply to the process whereby an unknown object within an image is identified as belonging to one particular group from among a number of possible object groups. For example, in automatic sorting of integrated circuit amplifier packages there might be three possible types: metal-can, dual-in-line and flat-pack. The unknown object should be classified as being only one of these types.

The term *pattern* refers to a collection of descriptors or features derived from the unknown object (as described in Chapter 6). It is usual to refer to this derived group of quantitative feature measurements as a 'feature vector'. A *pattern class* is a group of patterns that share some common property. *Classification*, or recognition, is the process of correctly assigning unknown patterns to their respective pattern class. Ideally this classification process should be completely automatic and error free. However in practise it is not always a simple task to assign clear classification rules to ensure correct classification, particularly in a situation when the feature vectors themselves contain measurements with a degree of variability between objects in the same object class.

In general, one can identify three basic approaches to pattern classification. The first, statistical pattern classification, relies on defining a set of decision rules based on standard statistical theory. A set of characteristic feature measurements are extracted from the input image data which are then combined to define a feature vector identifying an object as belonging to one defined pattern class. The decision rules for this assignment are derived either from *a priori* knowledge of the expected distribution of the pattern classes or from knowledge acquired through a training process involving many initial measurements.

The second, syntactic pattern classification, utilises the underlying structure of the patterns themselves. In some cases the significant information lies not in the feature vectors themselves but in their interrelationships, thus it is the interconnection of features which assists the classification process. For classification it is necessary to quantify and extract structural information and to assess structural similarity of patterns; at the lowest level this structural information will be primitive elements or building blocks out of which more complex elements of the image are constructed. One approach to this structural pattern relationship is to adopt the syntax structure approach of a formally defined language.

A third approach uses an architecture which can be trained to correctly associate input patterns with desired responses. This alternative approach, using artificial neural networks, attempts to exploit knowledge of how biological neural systems store and manipulate information. Use of neural networks, or artificial neural systems, offers a new computing paradigm in which the network, through a process of learning from task examples, can store experiential knowledge and make it available for use at a later time.

Although more than one of these techniques may be used for a given classification problem it is the vision system designer who needs to select the most apposite method. The boundaries between the three approaches are not necessarily clear and obvious. All approaches share common features and objectives. For a given pattern recognition problem an analysis on the underlying 'statistical' or 'grammatical' structures may be essential to the vision system designer as well as a consideration of the suitability of a neural network implementation. Generally an increase in classification accuracy requires a corresponding increase in the computational effort and hence in the cost of the vision system solution.

## 7.2 Statistical methods

For the purposes of this section two general approaches to statistical pattern classification will be addressed. The decision theoretic approach seeks to identify a set of decision rules based on knowledge gained from an analysis of many previous measurements of known objects. Once the distribution of these feature vectors is known then a decision rule can be defined to separate the pattern classes [1].

The probabilistic approach utilises *a priori* knowledge of the expected distribution of these feature vectors and this knowledge is incorporated into the decision rules. Such an approach should yield a greater degree of classification accuracy since all feature measurements themselves are subject to variation, e.g. noise, and hence any measurement will itself be subject to a statistical variation [2].

This section begins with a discussion on template matching and feature analysis techniques within the context of optical character recognition. Although these techniques are examples of simple heuristic approaches to pattern recognition, they can nevertheless be viewed as belonging to the decision theoretic category.

## 7.2.1 Template matching

Template matching is the simplest pattern classification approach. Here a template is defined as an ideal representation of the object (or pattern) to be identified or classified within the image. The template matching process involves moving the template to every position in the image and evaluating the degree of similarity at each position; as such it is a 'correlation' exercise. Since correlation and convolution are mathematically identical for symmetrical templates (or windows), matching can be achieved using the convolution operation discussed in Chapter 4. This explains why the process of template matching is sometimes described as 'matched filtering'.

A 'global' template represents the complete object under investigation whereas a 'local' template utilises several local features of the object. One of the limitations of this approach is that no variation in scale or orientation is

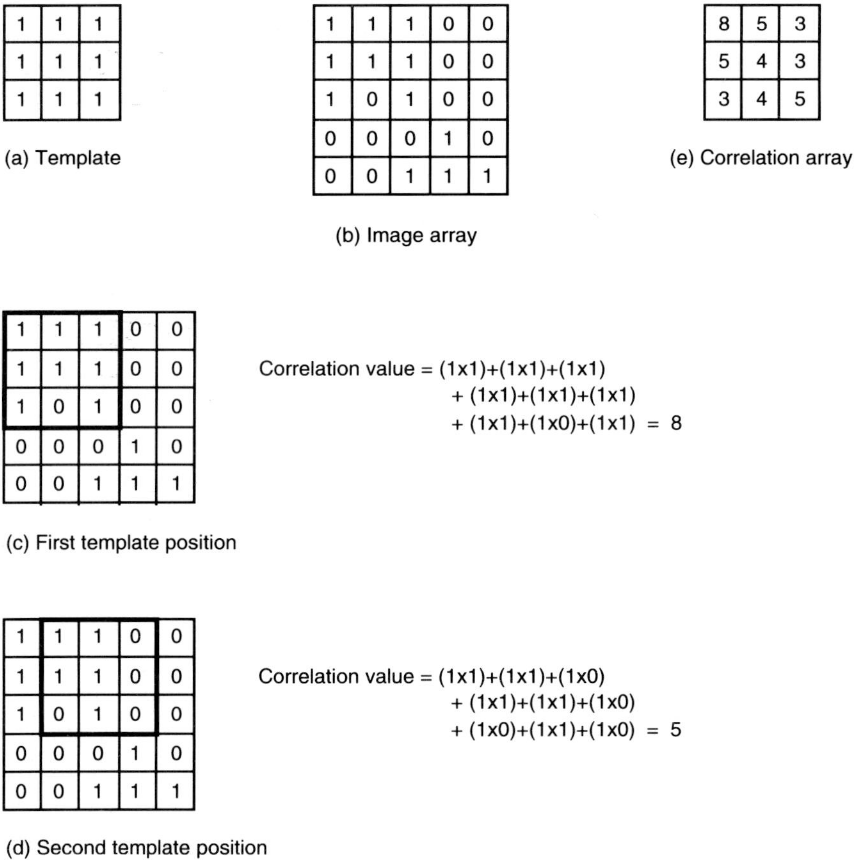

| 1 | 1 | 1 |
|---|---|---|
| 1 | 1 | 1 |
| 1 | 1 | 1 |

(a) Template

| 1 | 1 | 1 | 0 | 0 |
|---|---|---|---|---|
| 1 | 1 | 1 | 0 | 0 |
| 1 | 0 | 1 | 0 | 0 |
| 0 | 0 | 0 | 1 | 0 |
| 0 | 0 | 1 | 1 | 1 |

(b) Image array

| 8 | 5 | 3 |
|---|---|---|
| 5 | 4 | 3 |
| 3 | 4 | 5 |

(e) Correlation array

| 1 | 1 | 1 | 0 | 0 |
|---|---|---|---|---|
| 1 | 1 | 1 | 0 | 0 |
| 1 | 0 | 1 | 0 | 0 |
| 0 | 0 | 0 | 1 | 0 |
| 0 | 0 | 1 | 1 | 1 |

Correlation value = (1x1)+(1x1)+(1x1)
              + (1x1)+(1x1)+(1x1)
              + (1x1)+(1x0)+(1x1) = 8

(c) First template position

| 1 | 1 | 1 | 0 | 0 |
|---|---|---|---|---|
| 1 | 1 | 1 | 0 | 0 |
| 1 | 0 | 1 | 0 | 0 |
| 0 | 0 | 0 | 1 | 0 |
| 0 | 0 | 1 | 1 | 1 |

Correlation value = (1x1)+(1x1)+(1x0)
              + (1x1)+(1x1)+(1x0)
              + (1x0)+(1x1)+(1x0) = 5

(d) Second template position

Figure 7.1   Template matching

allowable between template and image, hence some preprocessing may be necessary to ensure these conditions exist prior to the classification process.

Consider the 3 × 3 template illustrated in figure 7.1a. This represents the pattern which is to be detected within the total image array shown in figure 7.1b. The template matching process commences with the template in the top left position, figure 7.1c, when the correlation between the template and array can be quantified by summing the products of corresponding pixel values within the template and image array respectively. This value 8 is then stored in a correlation array, figure 7.1e. The process is repeated as the template is scanned across the whole image array.

From the resulting correlation array it can be seen that the highest correlation value is 8 and hence the most likely position of occurrence of the object, as defined by the template, is at the first template position. Note also that there is not a perfect match, which would be signified by a correlation value of 9, but there is a high degree of probability that the object has been identified or classified at the origin (position 0, 0) of the correlation array.

If the orientation of the image is expected to vary, assuming all other parameters remain constant, then a range of separate image templates can be used in which each one corresponds to one of a number of discrete image orientations. Again the match occurs at the position corresponding to the highest value in the correlation array. This approach, however, can become expensive for large templates owing to the increase in the volume of computation. The use of small local templates, which focus on key salient features, can reduce the overhead. The template matching approach is most useful within well controlled machine vision environments and also in applications where the number of objects to be classified is not excessive.

### 7.2.2 *Template matching and optical character recognition*

One of the earliest applications of automatic pattern classification was the recognition of printed characters as found on bank cheques and letter postal codes. This is an example of an application in which the system designer has exerted maximum control over the scene constraints by allowing only a small set of stylised characters, and thus constraining the variability of the input image data.

Within the sphere of optical character recognition (OCR) one can identify three types of problem, dependent upon their degree of complexity. The simplest one is where a manufacturer produces an acquisition system which is designed to suit a particular character font; this is the case identified above. The two fonts in general use are known as OCR-A and OCR-B, illustrated in figure 7.2. OCR-A was an attempt at creating a stylised set of characters which brought together a number of differing fonts into a unified standard [3]. OCR-B was designed to produce a set of characters with visually acceptable shapes that were as near as possible to conventional characters. The second type of problem is where the OCR system has to recognise printed characters of any number of different fonts, and the third problem is that of recognising unconstrained characters, i.e. hand-written characters [4].

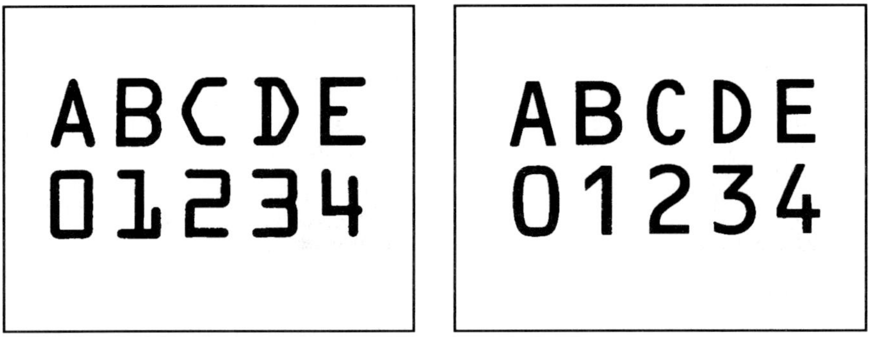

Examples of the OCR-A font

Examples of the OCR-B font

Figure 7.2  Standard optical character recognition fonts

Within simple OCR systems the data acquisition is performed by a scanning mechanism employing a vertical column of light sensors. This allows for variable vertical resolution, the horizontal resolution being a function of the scanning velocity. The template matching approach can be applied to such an OCR problem.

Figure 7.3a illustrates a typical standard font character after its acquisition. Figure 7.3b illustrates the use of a global template in which the unknown character is matched in turn against a complete set of replicas. In some marginal applications, global masks may fail because a significant portion of the character area is common to many characters. Thus figure 7.3c illustrates a variation of this approach which is applicable in such cases whereby a small number of pixels in the character area are selected and used to match with a stored set of data. This is arguably the matching of a small set of character features where the number and position of the pixels have been carefully chosen so that they can classify all the prototype characters uniquely. This type of 'peep hole' mask is limited to fairly simple classification problems and can be sensitive to noise, giving rise to changes in expected pixel values. An extension of this technique is the weighted area mask in which all the selected points in the character area are in some way summed together in differing proportions (weighted) in accordance with some predetermined value and then used in the correlation with the standard character set.

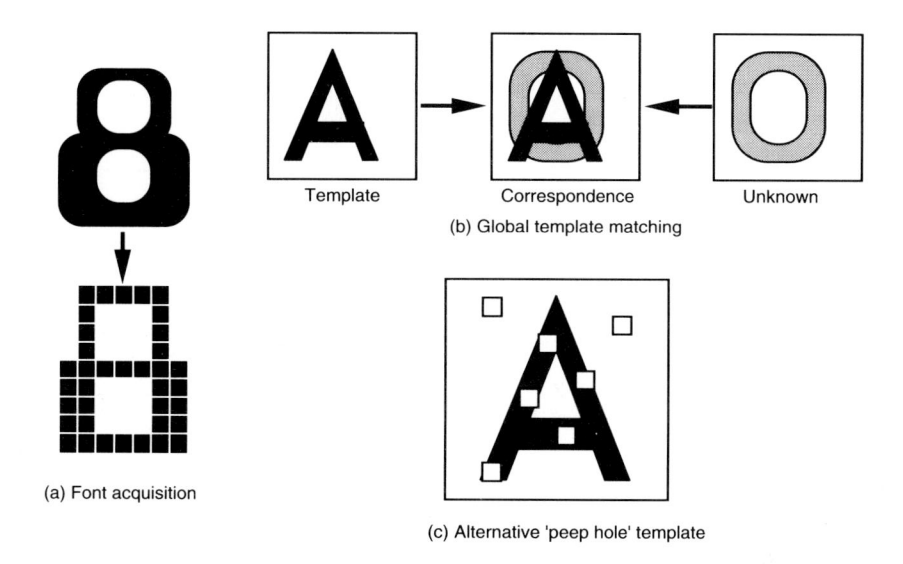

Figure 7.3  Template matching in optical character recognition

### 7.2.3 *Feature analysis*

A general technique based on the correlation of a group of individual properties or features which describe each different character class is known as feature analysis. To illustrate this technique consider the two examples given in figure 7.4. The first uses a stroke analysis approach in which the characters are classified from an analysis of their horizontal and vertical line structures, figure 7.4a. The figure illustrates that the numeral 6 is constructed from a combination of two vertical strokes, one short and one long, together with two short horizontal strokes. Such an approach is rather primitive but can be successful with standard character font numerals.

Figure 7.4b focuses on the geometric features of the characters. Here it is the general form of the character that is used and hence it does not depend on their exact size. Features such as line endings, corners, junctions, crossovers, etc. are utilised and the character is classified from the features it contains and their position relative to each other. The figure shows that numeral 4 contains four line ends, one crossover and one corner, while numeral 6 contains one line end, one junction and one circular curve.

(a) Stroke analysis

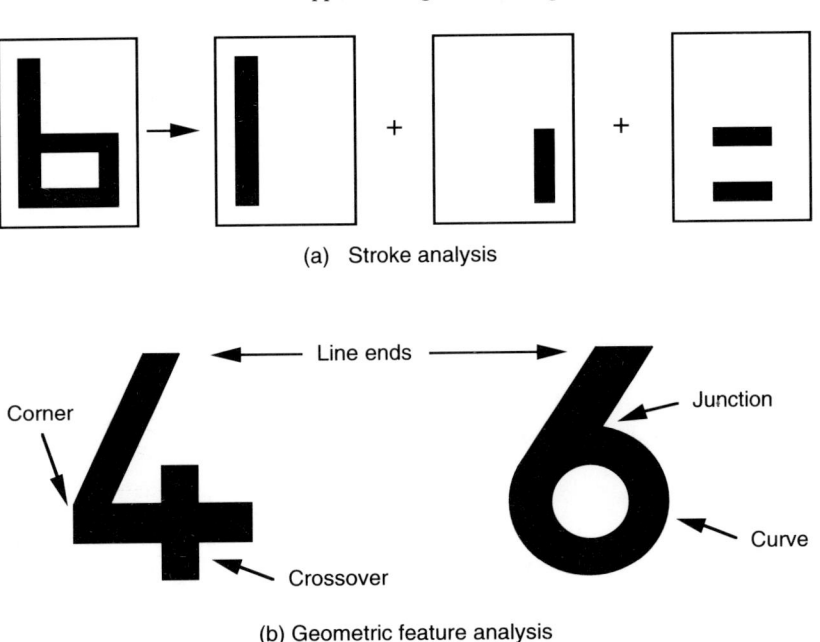

(b) Geometric feature analysis

Figure 7.4 Feature analysis

### 7.2.4 Decision theoretic approaches

In the decision theoretic approach, sometimes identified as discriminant function analysis, a 'weighting' vector is defined which will give an identifiable response with one class of patterns (objects) and a different response when operating on another class of patterns (objects)[5]. The principle is illustrated in figure 7.5a in which two appropriate object features have been extracted and are represented as points (feature vectors) in a two-dimensional graph (feature space). A decision surface can now be identified and defined which clearly separates these two feature vector clusters and hence can be used to classify the two objects. This approach is ideal for a simple problem where only two distinct types of objects are present and an accept or reject strategy is required. The diagram shows the simplest approach where a linear decision function of the form $(y = mx + c)$ is sufficient.

A linear decision function $d(X)$ can be defined such that

$$d(X) = X_1 - mX_2 - c$$

where $X_1$ and $X_2$ are two measured features,
    $m$ is the slope of the line and $c$ is its intercept of the $X_1$ axis.

Pattern Classification 183

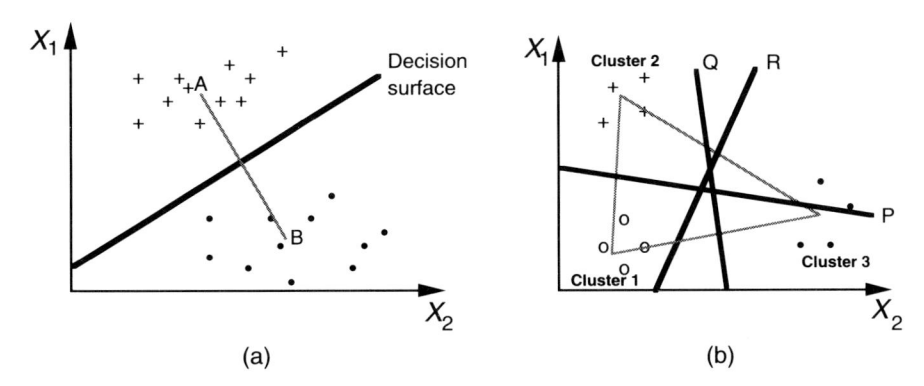

(a)

(b)

Figure 7.5  A linear decision surface

All points that lie on this line will satisfy the condition $d(X) = 0$. However any point that lies above this decision surface will satisfy the condition $d(X) > 0$ (i.e. any feature vector identified by a + symbol), while any point lying below this surface will satisfy the condition $d(X) < 0$ (identified by a • symbol). The key step in this approach is to identify and define the decision surface and it is often found by drawing the perpendicular bisector of the line joining the mean values of the two feature vector clusters, as illustrated by A and B in figure 7.5a. This is termed the minimum distance classifier.

The linear decision surface approach can be extended to a three-class problem by utilising three linear decision surfaces, as shown in figure 7.5b. Here the three classes (1, 2 and 3) are each represented by a cluster of measurements. Each decision surface then provides a distinction between any pair of clusters and a simple algorithm can be developed to deduce the correct classification of each unknown feature vector.

The real difficulty in using decision surfaces comes in obtaining an appropriate function for problems where the distribution of feature vectors necessitates a non-linear decision surface to be defined, such as figure 7.6a. Furthermore if three feature vectors are to be used then the decision surface becomes a decision plane in a three-dimensional graph. If the number of features increases then the problem becomes one of defining an equation for a 'hyperplane' in $n$-dimensional feature space.

Another variation is to identify an object by checking the degree of similarity which it has to other objects under investigation. Here classification can be achieved by first calculating the distance between the unknown object and all neighbouring objects within each of the clusters as in figure 7.6b. Secondly, a selected number, say three, of the smallest distances are then used to produce sample classes for the unknown (say these three minima indicate classes 2, 2, 3 respectively). Final classification is performed by tabulating the sample decisions thus produced (i.e. two votes

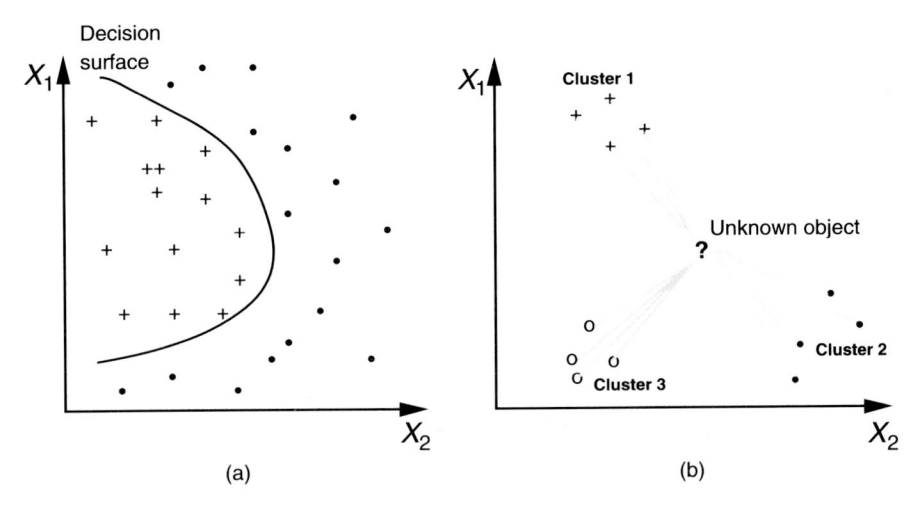

Figure 7.6 A non-linear decision surface

for class 2, one vote for class 3) and selecting the class with the highest number of contributions (class 2). This is referred to as the nearest-neighbour classification technique.

### 7.2.5 Probabilistic approaches – Bayes classifier

In the above decision theoretic approach no consideration is made about the probability of occurrence of a specified feature vector. For instance, consider the simple case incorporating feature vectors for a two-class problem, the designer has *a priori* knowledge that there is a 99% chance that feature vector A (object A) is present but only 1% chance that feature vector B (object B) is present. In such a case if every unknown object is assumed to be object A there will be a 99% success rate. Clearly the probability of occurrence of an object is important.

An extension of this approach would be to assign the classification using the relative probability of occurrence of the two pattern classes. Figure 7.7a illustrates two such probability distributions with respect to one particular feature value. These distributions can be designated class conditional distributions and they contain inherent information about the pattern classes under consideration.

In general:

$p(C_A)$ is defined as the (*a priori*) probability of occurrence of class A, and

$p(C_B)$ is defined as the (*a priori*) probability of occurrence of class **B**.

Then

$p(x/C_A)$ is the (class conditional) probability of feature value $x$ belonging to class **A**,

and

$p(x/C_B)$ is the (class conditional) probability of feature value $x$ belonging to class **B**.

Even with these distributions to hand it is still not an easy task to achieve a high degree of classification accuracy since any overlap between the two-

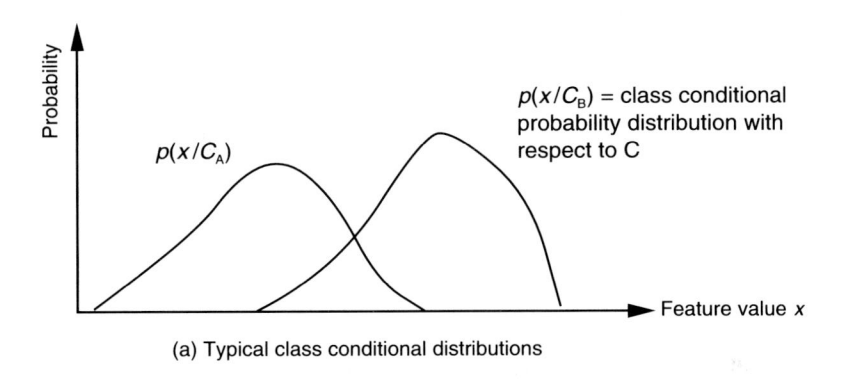

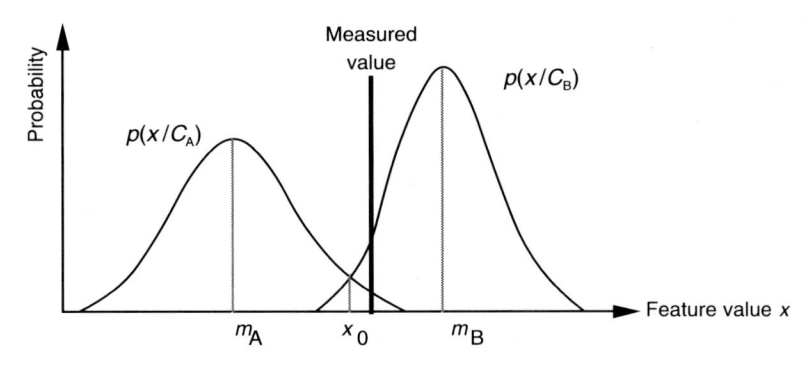

(a) Typical class conditional distributions

$p(x/C_B)$ = class conditional probability distribution with respect to C

(b) Example of two Gaussian distributions

Figure 7.7  Probability distributions

class conditional distributions will introduce some ambiguity in the classification when using only the single chosen feature value. However if these distributions are a good representation of the relationship between features and classes then this information can be exploited to maximum effect.

The probabilistic pattern classification approach relies upon an initial evaluation of the class conditional probabilities [6]. A given feature value is then obtained from the unknown sample, and this is used to determine the probability that the unknown belongs to a particular pattern class. This determination process is repeated for all pattern classes. The classification is made by selecting the class for which the unknown has the highest probability of membership. In practice it is not easy to obtain these class conditional probability distributions since it requires a very large number of pre-identified samples for each feature that can occur.

Thus the decision rule can be stated mathematically as

$$p(C_A/x) > p(C_B/x) \quad \text{for all } A \neq B$$

where $p(C_A/x)$ is the (*a posteriori*) probability that feature $x$ comes from class A, and

$p(C_B/x)$ is the (*a posteriori*) probability that feature $x$ comes from class B.

To evaluate the decision rule probabilities, use can be made of Bayes theorem which enables *a priori* probabilities to be converted into *a posteriori* ones. Bayes rule states that

$$p(P/Q)p(Q) = p(Q/P)p(P)$$

Hence the above decision rule can be rewritten as

$$p(x/C_A)p(C_A) > p(x/C_B)p(C_B) \quad \text{for all } A \neq B$$

The practical difficulty in obtaining reliable class conditional probability distributions means that it is often appropriate to assume that a Gaussian distribution exists since it approximates to many real-world situations. Figure 7.7b illustrates such a distribution for a two-class problem. Classes A and B have mean values of $m_A$ and $m_B$, and standard deviations $\sigma_A$ and $\sigma_B$ respectively. Applying the Bayes decision rule to this problem produces the following function for class A:

$$\text{Decision function, } d(x) = p(x/C_A)p(C_A)$$

$$= \frac{1}{\sqrt{2\pi\sigma_A^2}} \exp\left(-\frac{(x-m_A)^2}{2\sigma_A^2}\right) p(C_A)$$

and for class B:

$$d(x) = \frac{1}{\sqrt{2\pi\sigma_B^2}} \exp\left(-\frac{(x-m_B)^2}{2\sigma_B^2}\right) p(C_B)$$

The boundary between the two classes is a single point. If the two classes are equally likely to occur then $p(C_A) = p(C_B) = 0.5$ and the decision boundary is the value of $x_0$ for which $p(x_0/C_A) = p(x_0/C_B)$. Any feature value to the right of $x_0$ is classified as belonging to class B while any point to the left of $x_0$ is classified as A. Considering the measured value shown in the figure it is clear that $p(x/C_B)$ is greater than $p(x/C_A)$ and hence the unknown is assigned to class B. When the classes are not equally likely to occur, $x_0$ moves to the left if class B is more likely to occur and to the right if class A is more likely.

It is worth noting that for the two-class problem described above there are three possible outcomes to the classification process. The first two are the obvious ones of correct classification (e.g. $C_A \rightarrow C_A$) and incorrect classification (e.g. $C_A \rightarrow C_B$). However it is possible to design a system which includes a reject decision when the classifier has insufficient information on which to make a decision (e.g. $C_A \rightarrow$ Reject).

Any incorrect decision will incur a cost, or loss, to the user. This is another parameter that the vision system designer has to account for since it may be appropriate under certain conditions to accept a degree of incorrect classification. Such a cost can be defined as

$$\lambda(C_A/C_B)$$

that is, the cost incurred in assigning an unknown to class A when it belongs to class B.

The simplest way to use such a cost function is to incorporate the statistical knowledge about class membership such that a decision will incur a cost which depends on the conditional probability $p(C_A/x)$. Thus a 'conditional risk' or 'expected cost' function can be defined as

$$R(C_A/x) = \lambda(C_A/C_B) p(C_B/x)$$

A new decision rule can now be defined as one that classifies an unknown pattern so as to minimise the associated risk. This can be formally stated as

Assign $x$ to $C_A$ such that
$R(C_A/x) < R(C_B/x)$ for all B $\neq$ A

Returning to the earlier discussion on decision functions it should be remembered that in general terms there will be a defined decision function for each class of pattern to be classified. Thus for a ten-class problem, ten decision functions will be defined with each one giving a quantifiable result. In general the classification will be made by selecting the pattern class associated with the maximum resultant value. Hopefully one decision function produces a very much higher value than the others and hence classification is made with a high degree of confidence. However if this is not the case it is sometimes useful to introduce a 'confidence level' or 'rejection margin' directly into the decision rule. Such a rule would become:

Assign to $C_A$ such that
$$d_A(x) > d_A(x) + \gamma$$

where $\gamma$ is the confidence level.

The effect of increasing $\gamma$ is largely to reduce the number of classification errors. However the number of rejected samples increases dramatically for larger values of $\gamma$ and ultimately the system designer has to decide on the appropriate trade-off between error rate and rejection rate.

## 7.3 Syntactic methods

Syntactic pattern recognition relies on the concept that, within every pattern, there is an inherent structural relationship between its basic elements. The principle of the syntactic pattern recognition approach is to decompose a complex image pattern into a hierarchy of sub-patterns and to develop rules by which these sub-patterns are interrelated and hence can be recombined to form the high level pattern. Decomposition of the pattern is known as 'parsing' and the rules of combination are based on the syntax of formal language [7].

The lowest level pattern features are called pattern 'primitives'. These are the components which make up the high level pattern description. The methods by which these primitives are combined are known as the pattern 'grammar'. Pattern classification is performed in the following way; firstly a sample set of patterns is analysed to find the pattern primitives and the associated pattern grammar (the training phase). Secondly, an unknown pattern is presented to the vision system where it is decomposed into its primitives and then classified by parsing these primitives to see which known pattern it reproduces.

Selection of appropriate pattern primitives for a given vision problem is a difficult task. Considering the language analogy it is clear that any English sentence can be parsed into a number of phrases which in turn are made up from a number of speech elements, which in turn are made up of words composed from the 26 primitive alphabetic characters. Figure 7.8 illustrates

Sentence = Brighton is a seaside town

(Sentence)
(Noun)(Verb phrase)
Brighton(Verb)(Noun phrase)
Brighton is (Adjectival phrase)(Noun)
Brighton is (Article)(Adjective) town
Brighton is a (Adjective) town
Brighton is a seaside town

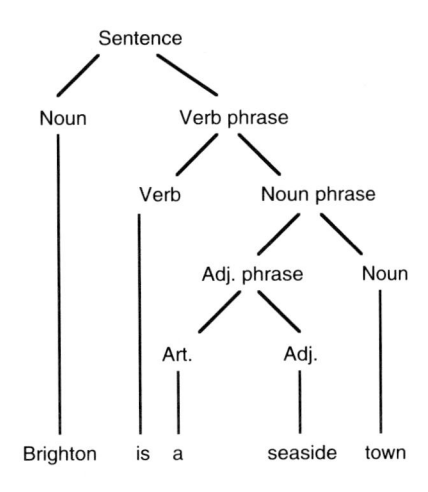

(a) Grammatical analysis

(b) Graphical representation

Figure 7.8  Parsing in formal language grammar

this parsing process where the results are displayed in both tabular and graphical form.

The formal grammar $G$ for such an operation can be defined as;

$$G = (\Sigma, N, P, S)$$

where $\Sigma$ is the set of primitives (e.g. words),
$N$ is the set of intermediate quantities (e.g. verb, adjective),
$P$ is the set of production rules (e.g. phrase, clause),
$S$ is the starting symbol or root (e.g. sentence).

This grammar may be used either to create a string of terminal symbols $S$ using $P$, or to analyse the symbol $S$ (given $\Sigma$ and $P$) to see if it was generated by this grammar and, if so, what structure was used. The method of combining $\Sigma$ and $P$ gives the grammar its structure.

Some examples using a picture description language are shown in figure 7.9 [8]. Any basic picture element is defined as having a tail and a head, and this picture element is defined as a picture primitive as in figure 7.9a. A number of pattern primitives are defined in figure 7.9b and a selection of derived operators which incorporate the primitives are shown in figure 7.9c. Figure 7.9d illustrates the application of the given picture description language to some alphanumeric characters in which the structure of each character is identified as a series of connected primitives and the formal

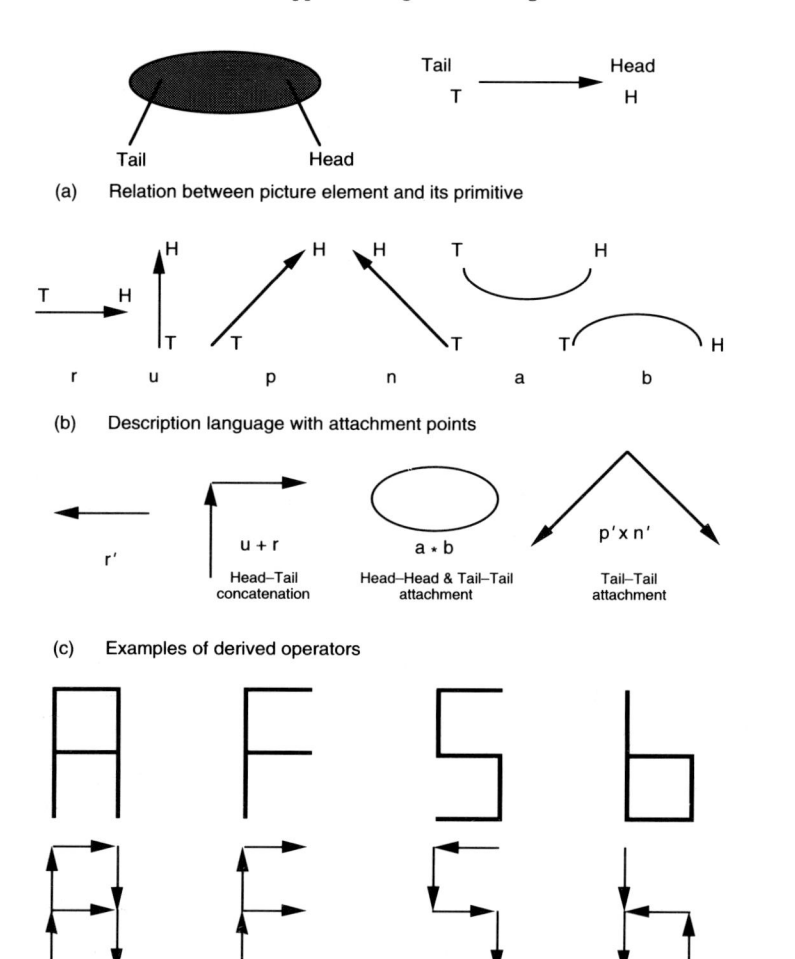

(a) Relation between picture element and its primitive

(b) Description language with attachment points

(c) Examples of derived operators

$A = u + \{(u + r + u') * r\} + u'$

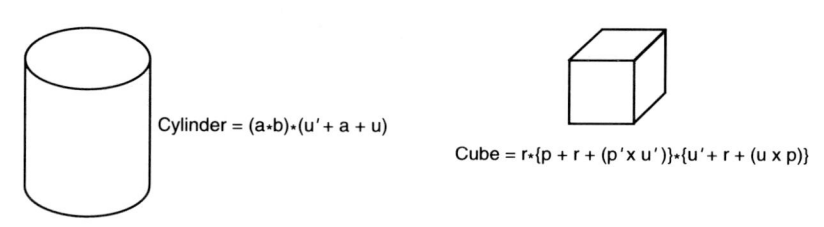

$F = u + (r \times u) + r$

$5 = r' + u' + r + u' + r'$

$6 = u' + \{r' * (u' + r + u)\}$

(d) Alphanumeric examples

Cylinder $= (a * b) * (u' + a + u)$

Cube $= r * \{p + r + (p' \times u')\} * \{u' + r + (u \times p)\}$

(e) Examples of cylinder and cube

Figure 7.9 Picture description grammar primitives and examples

grammar definition incorporates the parsing information. Finally other examples of a cylinder and a cube are given in figure 7.9e.

From these examples it can be noted that the pattern representation capability of formal grammars is somewhat limited. However syntactic pattern recognition encompasses a broader spectrum of approaches including representations as trees and attributed graphs. Many of these approaches are more popular in the areas of model-based vision and artificial intelligence [9, 10].

## 7.4 Neural network approaches

The human brain is composed of approximately $10^{10}$ nerve cells called *'neurons'*. The computational power of the brain lies in the interconnection, hierarchical organisation and firing characteristics of this multiplicity of neurons. Thus the brain may be described as a 'neural network'.

A neuron can be imagined as a cell body which receives electrical signals via many input nerve endings (*'dendrites'*) and which can supply electrical signals via a number of output nerve endings (*'axons'*). The input to the network is provided by sensory receptors, e.g. via the optic nerve from the retina in the eye, the input stimuli being in the form of electrical pulses. When electrical pulses appear at the output of a neuron it is said to have 'fired'. The dendrite–axon contact between one neuron and the next is called a *'synapse'*. The synaptic inputs can be either inhibitory, in which case the neuron does not fire, or excitatory in which case the neuron fires and the electrical pulse is transmitted onward.

Neural network approaches seek to use artificial neurons, constructed from electronic devices, to form large interconnected networks. Since the brain can be regarded as one enormous parallel processor, a variety of terms appear in the literature to identify this approach, namely 'neural network', 'neuro-computer', 'connectionist machine', 'parallel distributed processor', 'associative network', etc. Further, it is known that the human brain learns by experience and so the neural network approach uses a non-algorithmic computing paradigm which can be defined as [11]:

*'the study of a network of adaptable nodes (neurons) which, through a process of learning from task examples, store experiential knowledge and make it available for use'*

Pattern recognition utilising neural networks is currently the subject of much research and while the hardware implementation of artificial neurons is at an early stage of development, three basic attributes characterise all neural networks. These are:

(a) the characteristics of the individual neurons or nodes;
(b) the network topology, or interconnection of these nodes;
(c) the strategy for training the network.

### 7.4.1 Neural network nodes

The earliest neuron model was suggested by McCulloch and Pitts [12] and sought to utilise a simple summing device to add together the weighted sum of its inputs, the inputs being limited to binary values. An output signal would only be generated if the weighted input sum was greater than some specified threshold. The weighting and threshold values were fixed by the designer.

Although this early model served to illustrate the potential for implementation in combinational logic circuitry, it did not incorporate the idea of learning by association or experience which is at the heart of a biological neural system. The first device to incorporate such possibilities was the 'perceptron' [13] although it subsequently proved inadequate for practical pattern recognition problems. Extensions incorporating external inputs, inhibitory inputs, and weighted and non-linear combinations of

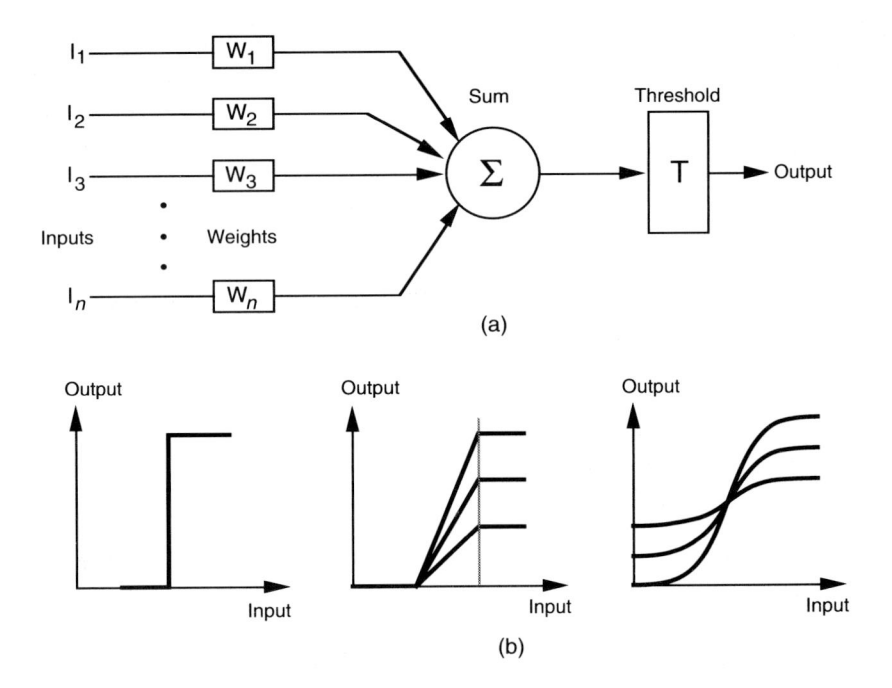

Figure 7.10 Neural network model

inputs were all developed [14, 15]. Even now, hardware models of neuron nodes are relatively simplistic and it seems likely that they must yet undergo considerable development, with consequent increase in complexity, if practically useful neural pattern recognition systems are to be fully realised.

The model shown in figure 7.10a is based on the perceptron. At the centre of the model is the summation unit which adds together the weighted sum of the input signals. The output of the summation unit goes to a threshold device which ensures the firing of the node is in accordance with some appropriate threshold activation criterion. This node activation criterion may differ for different classes of neural network. In its basic form it may be a discrete binary threshold where the output signal $O$ becomes either a logic '1' or '0', and the operation of the node is similar to that within a standard combinational logic circuit. For such a simple threshold activation function the output can be written as

$$O = \text{'1'} \quad \{\text{if } I_1W_1 + I_2W_2 + I_3W_3 + \ldots I_nW_n > T$$
$$\phantom{O =\ } \text{'0'} \quad \{\text{otherwise}$$

where $I_1, I_2, I_3 \ldots I_n$ are the node input signal values,
$W_1, W_2, W_3 \ldots W_n$ are the weights applied to each input, and
T is the threshold value for the node.

In other cases the node activation criteria may feature analogue, rather than digital, characteristics, as illustrated in figure 7.10b. The figure illustrates a simple threshold, a linear threshold characteristic and a sigmoid function, with the latter two showing various values of gain. These illustrations introduce the concept of time delay into the operation of the network as well as the concept of analogue node firing characteristics. Differences in the definition of the node activation criterion lead to different classes of neural networks.

When applied to a pattern recognition problem the node essentially acts as an 'associative network' with desired output responses being associated with a given set of input signals. This mode of operation can be simplistically summarised as follows:

(a) Apply a known set of input signals to the node (these can be features or feature vectors from a *known* image). The required output response from the node is now defined.
(b) Adjust the individual internal node weights in order to produce the desired output response given the applied input signals. This process is known as *'training'*. It is critical to the success of the system and is often difficult and time-consuming.
(c) Once trained any unknown object will be appropriately recognised and classified by its output response.

While the model depicted in figure 7.10a may, at first sight, lend itself to implementation using basic operational amplifier circuitry, a difficulty arises in defining how the weights themselves are adjusted during training. The network must be capable of self learning – i.e. the initial signal produced when known inputs are applied must be used as a correction factor to adjust the weights and hence produce a more desirable output (similar to an error signal in a control system). This is far from trivial. One technique has utilised the transfer of charge in MOSFET-based amplifier circuits [16], although with limited success. Most researchers currently use software simulation packages.

Hardware implementations of neural nodes is still the subject of much research with digital and analogue techniques for weight design both under current investigation [17, 18].

### 7.4.2 Network topologies

Using the basic neuron node model described above, a variety of node interconnection topologies can be devised. Figure 7.11 illustrates a selection of arrangements.

The neural node model given above in figure 7.10a is depicted by the symbol shown in figure 7.11a, the number of inputs being dependent upon the complexity of the network. For more sophisticated designs, two basic topologies exist:

(a) feed-forward networks;
(b) feedback networks.

Figure 7.11b illustrates a feed-forward network utilising three nodes, in which the three input signals $I_1$, $I_2$ and $I_3$ are connected to all nodes. Here only one 'layer' (column) of nodes is used so that the node outputs are also the overall network outputs.

Figure 7.11c illustrates an extension to this, whereby there are three layers of nodes which are used to feed-forward the three primary input signals to two primary outputs. In this network there is now a 'hidden layer' of nodes which are not accessible directly for signal measurement. In such a network, only the error at the primary outputs is available during the training phase when all weights, including those within the hidden layer, must be automatically adjusted. Feed-forward networks are also termed 'associative networks'.

The network shown in figure 7.11d is a single layer feedback network example in which the node outputs are connected back to their inputs via a delay device ($\delta$) to overcome timing problems during operation. Such a feedback network is termed 'auto-associative', since it associates the training pattern with itself.

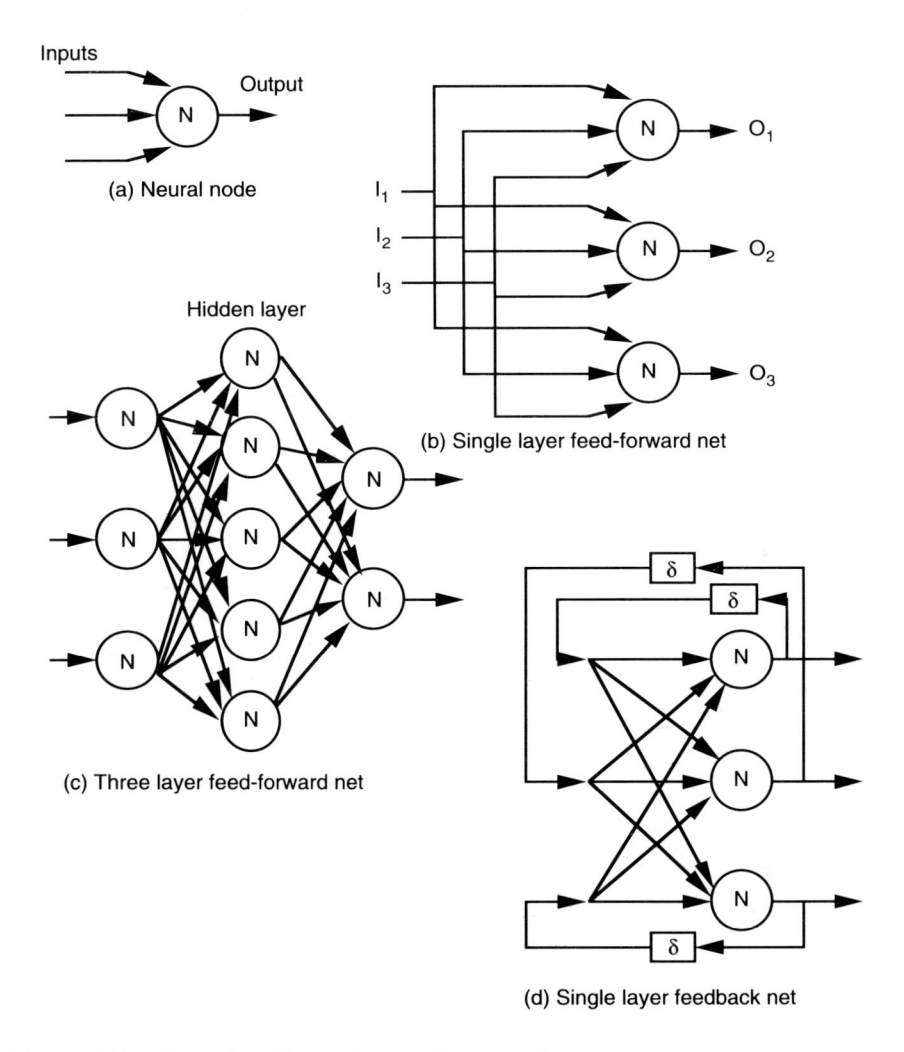

Figure 7.11  Examples of neural network topologies

One auto-associative network of particular interest is the Hopfield network [19–21] in which every node is connected to every other node in the network. In operation the network changes in a similar way to a clocked sequential logic network and can reach a 'stable' state which identifies the end of the training process. However, because of the 'memory' property of such networks, there may be more than one stable state and thus spurious stable states may be entered. Various training strategies have been developed to overcome this hazard [22, 23].

In a feed-forward network the output appears immediately after the input has been applied (analogous to the operation of a combinational logic circuit) whereas a feedback network is a dynamic circuit incorporating memory (analogous to the operation of a sequential logic circuit). Moreover, the introduction of the delay element permits the use of continuous time node activation characteristics, as in figure 7.10b.

### 7.4.3 Training strategies

As stated in section 7.4.1, when considering the basic operation of a node, a suitable training strategy needs to be adopted in order to produce the desired output response from the neural network. For a simple associative network the term 'training' is appropriate but in other more complex feedback networks the behaviour desired from the network is one of independent 'learning'. This is particularly relevant when a network is applied to a problem where the *a priori* knowledge of the required inputs and output is limited or non-existent. Although not all networks are actually required to exhibit this learning ability, the two terms 'learning' and 'training' are often used interchangeably.

Another distinctive and highly prized feature of neural networks is their ability to *generalise* – i.e. their ability to interpolate input patterns that have not previously been seen by the network. Many networks provide input to output mappings with very good generalisation capability, and this characteristic can ensure good performance in the presence of noise. This capability is fully exploited in the recognition of hand-printed characters, where no two samples are ever the same, but all instances are roughly similar [24]. Other text recognition problems which are best tackled by neural approaches relate to Chinese characters and cursive English script. In the research arena, neural networks have been successfully applied to the localisation of eyes and other facial features within photographs, irrespective of gender, skin colour or the wearing of spectacles [25]. There is a strong motivation for doing this type of investigation in the visual telecommunications arena, for awareness of the movements of key facial features would ultimately allow 'talking head' images to be reconstructed synthetically at the remote end. Head and mouth movements would be controlled by simple instruction strings from the near end over a very low bit-rate video-phone link (cf. section 9.9).

For pattern classification problems it is often appropriate to utilise an existing neural net structure, rather than design a new application-dependent one, and apply appropriate known training algorithms. In general all training or learning can be classified as either 'supervised' or 'unsupervised'.

In supervised learning the desired network output is known and so the output error signal (difference between desired output and measured output) is obtained by the system designer and used to adjust the weights within the network such that the error is minimised. A set of input and output training patterns is required for this learning mode. The system designer estimates the degree and direction of weight adjustment and reduces the error accordingly. This is not necessarily a straightforward process and most supervised learning algorithms perform some form of 'stochastic minimisation process'.

In unsupervised learning the desired output is not known and so there is no explicit error signal available with which to adjust the network weights. Hence learning is accomplished by observing responses to inputs which are effectively unknown. Unsupervised learning algorithms generally use input signals which are typical of the pattern classes under investigation, but are not in any way identified or classified as belonging to one group. Thus the network must discover for itself the possible existence of underlying patterns or input to output mappings – it must have its own way of judging successful outcomes.

Under these two broad learning categories one may identify appropriate learning algorithms for both main types of network topology. Typically, supervised learning in feed-forward networks use 'error back-propagation' techniques [26] while for feedback networks 'optimisation' techniques are appropriate [27]. Unsupervised learning in feed-forward networks requires 'self organising' techniques [28], while feedback networks may use 'adaptive resonance' approaches [29].

### 7.4.4 A memory-based classifier

A successful proprietary pattern classifier has been designed which utilises a weightless associative network structure [12, 30, 31]. At its heart are Random Access Memory elements which are trained during one memory write operation while the data (true or false) are supplied by the system designer and related to the desired response of the network. Although it is different in principle from the conventional neural network described above, it definitely exhibits properties such as generalisation; for example, it can distinguish a face which is smiling from one which is serious, despite there being other major differences between the faces being compared [32]. Another application area where such a network performs well is security, particularly the detection of intruders in images which contain a lot of noise. This is a valuable attribute since security systems must perform well under the type of poor lighting conditions which are often preferred by intruders!

An outline of the mode of operation and some of the properties of the memory-based classifier will serve to highlight and clarify its 'neural'

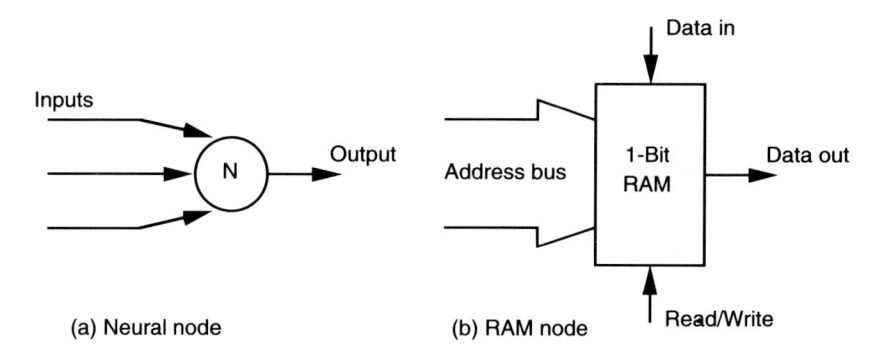

Figure 7.12  RAM-based node

(a) Neural node

(b) RAM node

behaviour. Figure 7.12 compares the essential features of the neural network node with a RAM-based node. The usual Write operation of the RAM is to store data in locations defined by the address bus pattern; in this example the data are a single bit which is connected to the data-in line (i.e. a 1-bit wide RAM is used). During a Read operation the stored data appear on the data out line when the appropriate address is applied.

Another view of this process is as follows:

(a) during the Write process, regard the address bus input as a known pattern and store a logic 1 data-in signal in this location to record the fact that a specific pattern has been applied;
(b) during the Read operation the data-out line is monitored when the same logic 1 will only appear if the same address has been applied to its address bus.

In this way the RAM can be thought of as a node which recognises an input pattern on which it has been trained. There are no weights associated with this node, although the data-in signal satisfies a similar role. In this simple node there is no generalisation property and in this respect its performance is unlike the neural node discussed above.

However a device with perceptron-like characteristics can be produced by interconnecting a number of RAM elements as shown in figure 7.13 – such a network being termed a 'discriminator'. Three RAM elements are shown in the figure with each one receiving three input address lines from the binary image grid representation (note that the data-in and Read/Write lines have not been drawn, for clarity). During the training (Write) phase, and assuming the input pattern shown, a logic 1 is placed on the data-in lines and stored in each RAM at the addressed memory location. During the recognition (Read) phase an unknown image will produce a pattern on the

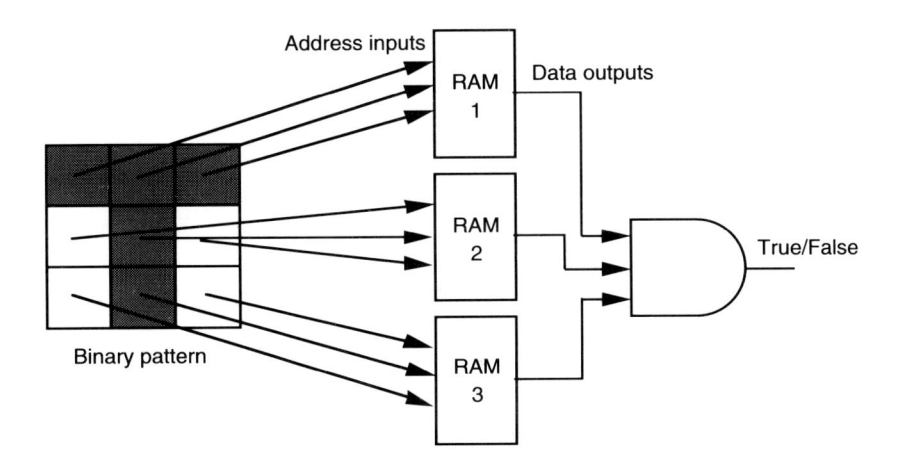

Figure 7.13  RAM discriminator

address bus; if this address pattern corresponds exactly to a previously trained address pattern then a logic 1 will be produced at each RAM output. These three outputs can be suitably processed through the three input AND gate shown to produce a True or False classification result.

It should be noted that the image features selected need to be in binary form and a direct connection to pixels of a binary image is appropriate. Referring to figure 7.14 a group or subset of pixels is termed an '$n$-tuple', and in this illustration 4-tuples are shown. The state of the $n$-tuple is defined by the assigned binary values, for example, in figure 7.14a the 4-tuple is 1001 while in figure 7.14b the 4-tuple is 1110. The figure also illustrates another

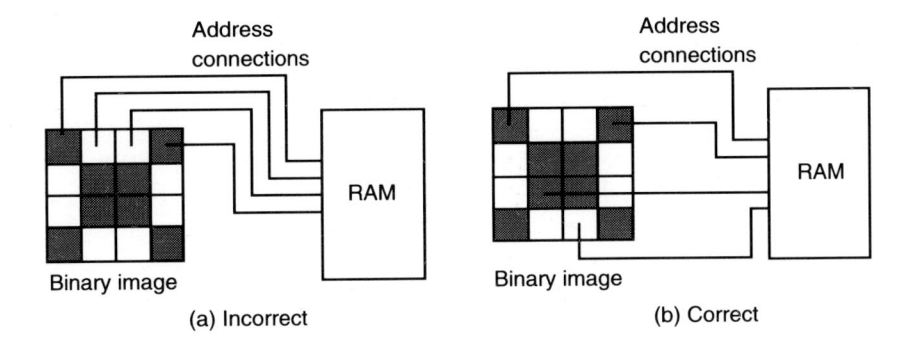

Figure 7.14  $n$-tuple connections

important design consideration, i.e. that it is statistically better to randomly select the n-tuple connections from the image. This makes the topology of figure 7.14b a better performer than that of figure 7.14a.

In the discriminator design, an optimal selection of $n$-tuple size and number of training set patterns are required. Too few training patterns results in poor classification performance, while too many give rise to saturation, where the discriminator will respond positively for almost any unknown input pattern. In addition, increasing the $n$-tuple connection above 4 does not significantly improve the quality of performance.

For the discriminator in figure 7.13 the 3-tuple state connected to $RAM_1$ is 111 while the 3-tuple states for $RAM_2$ and $RAM_3$ are both 010. Consider the three training patterns $T_1$, $T_2$ and $T_3$ shown in figure 7.15a to be applied to its inputs. As each training pattern is applied in turn a Write pulse is applied to each RAM and a logic 1 stored in the appropriate location. Each RAM has two unique address patterns applied to it, as shown in figure 7.15b, although their values are of no particular importance. However, figure 7.15c shows that since each RAM will respond positively to two training patterns there will be five other input patterns that will cause a True output.

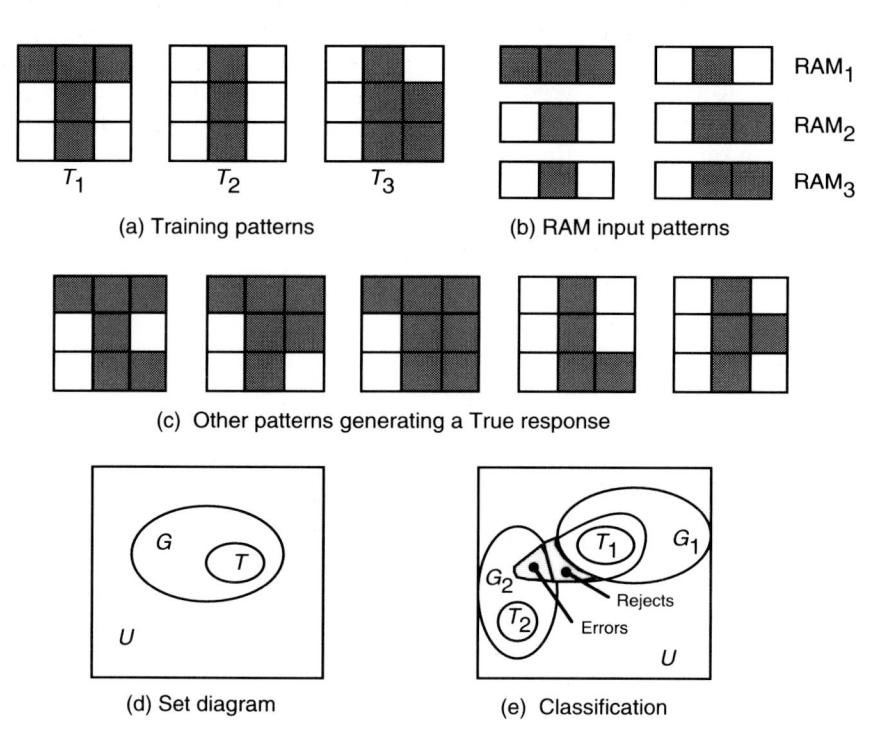

Figure 7.15 Training and generalisation within the discriminator

This property of being able to respond to patterns other than the training set is the generalisation property identified in the perceptron model. For this simple $3 \times 3$ grid there are $2^{(3\times3)}$ or 512 possible binary patterns ($U$). For training, 3 input patterns ($T$) have been identified, and as each RAM sees two sub-patterns there are a total of $2 \times 2 \times 2$ or 8 patterns ($G$) which will produce a True output signal. This is summarised in the set diagram of figure 7.15d.

For a two class pattern recognition problem two discriminators are needed, each one being trained on a representative sample of training patterns for one of the object classes, see $T_1$ and $T_2$ in figure 7.15e. When unknown patterns (assumed to be class 1) are subsequently applied a True classification takes place if these patterns fall within the $G_1$ area in the set diagram. If the unknown patterns fall outside this area they are False and rejected since there is no recognition by either discriminator. However if they fall outside $G_1$ but come within $G_2$ they will produce a True output for the other class of objects and hence result in classification errors.

In a practical system the number of discriminators required will be equal to the number of pattern classes to be recognised. A multi-category classifier can be made by replicating the system of figure 7.13 and *summing* the outputs obtained to make majority classification decisions. Such a multi-category classifier is depicted in figure 7.16 which indicates the input mapping from a $512 \times 512$ image array into each discriminator. The outputs are then connected to a logic unit designed to output a classification response based on the maximum values obtained from each discriminator.

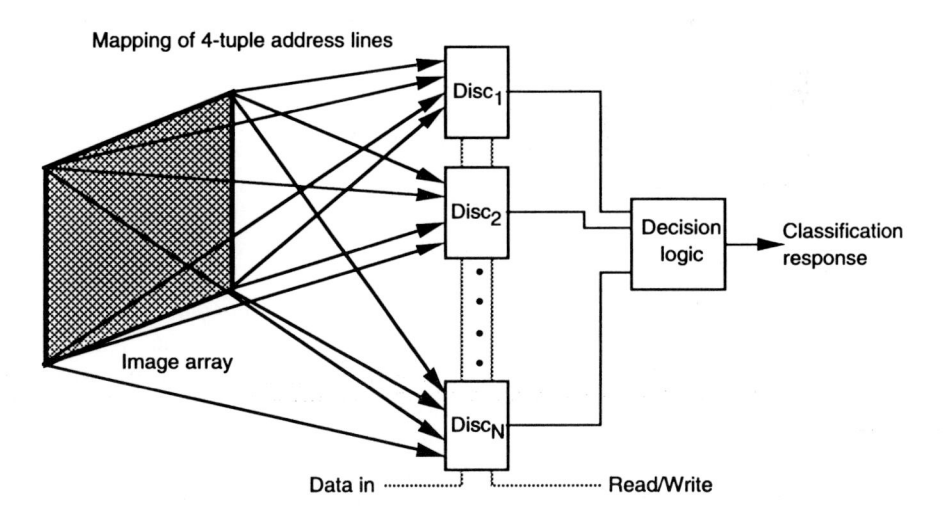

Figure 7.16 Multi-category discriminator

In this way it is possible to obtain a response which incorporates a degree of confidence in the correctness of the pattern classification result.

Many examples of industrial applications of this memory-based classifier, known commercially as 'WISARD', are cited by Redgers [33]. These range from the checking of Equadorian banknotes to confirming that cherries are in the centres of cherry tarts!

**References**

1. R. Schalkoff, *Pattern Recognition – Statistical, Structural and Neural Approaches*, Part 2, Wiley, Chichester, 1992.
2. R. Gonzalez and R. Woods, *Digital Image Processing*, Chapter 9, Addison-Wesley, London, 1992.
3. *British Standard 5464, Specification for Optical Character Recognition*, Parts I and II, British Standards Institution, London, 1977.
4. J. Mantas, 'An overview of character recognition methodologies', *Pattern Recognition*, **19**, pp. 425–430, 1986.
5. I. Aleksander, *Advanced Digital Information Systems*, Chapter 6, Prentice-Hall, Englewood Cliffs, New Jersey, 1985.
6. R. J. Schalkoff, *Digital Image Processing and Computer Vision*, Chapter 6, Wiley, Chichester, 1989.
7. A. Rosenfield, *Picture Languages*, Academic Press, New York, 1979.
8. A. C. Shaw, 'Parsing of graph representable pictures', *J. Assoc. Computing Machinery*, **17**, pp. 453–481, 1970.
9. D. H. Ballard and M. B. Brown, *Computer Vision*, Chapter 10, Prentice-Hall, Englewood Cliffs, New Jersey, 1982.
10. D. G. Lowe, *Perceptual Organisation and Visual Recognition*, Kluwer Academic, Brentwood, 1985.
11. I. Aleksander and H. Morton, *An Introduction to Neural Computing*, Chapman and Hall, London, 1990.
12. W. S. McCulloch and W. H. Pitts, 'A logical calculus of the ideas imminent in nervous activity', *Bulletin of Mathematical Biophysics*, **5**, pp. 115–133, 1943.
13. F. Rosenblatt, *Principles of Neurodynamics: Perceptrons and the Theory of Brain Mechanisms*, Spartan Books, New York, 1962.
14. M. Minsky and S. Papert, *Perceptrons: An Introduction to Computational Geometry*, MIT Press, Cambridge, Massachusetts, 1969.
15. N. J. Nilsson, *Mathematical Foundations of Learning Machines*, Morgan Kaufmann, Palo Alto, California, 1989.
16. S. Mackie, H. P. Graf and D. B. Schwartz, 'Implementations of neural networks in silicon', in R. Eckmiller and C. Von der Malsburg (Eds), *Neural Computers*, Springer-Verlag, Berlin, 1988.

17. A. F. Murray, 'Silicon implementation of neural networks', *IEE Proc. Part F*, **138**(1), pp. 3–12, 1991.
18. M. A. Maher, S. P. Deweerth, M. A. Mahowald and C. A. Mead, 'Implementing neural architectures using analogue VLSI circuits', *IEEE Trans. Circuits Systems*, **36**(5), pp. 643–652, 1989.
19. J. J. Hopfield, 'Neural networks and physical systems with emergent collective computational abilities', *Proc. Nat. Acad. Sci.*, **79** (Biophysics), April, pp. 2554–2558, 1982.
20. J. J. Hopfield, 'Neurons with graded response have collective computational properties like those of two-state neurons', *Proc. Nat. Acad. Sci.*, **81** (Biophysics), May, pp. 3088–3092, 1984.
21. J. J. Hopfield and D. W. Tank, 'Computing with neural circuits: a model', *Science*, **233**, August, pp. 625–633, 1986.
22. G. E. Hinton and T. L. Sejnowski, 'Learning and relearning in Boltzmann machines', in *Parallel Distributed Processing, Vols 1 and 2*, MIT Press, Cambridge, Massachusetts, 1986.
23. S. Kirkpatrick, C. D. Gelett and M. P. Vecchi, 'Optimization by simulated annealing', *Science*, **220**, pp. 671–680, 1983.
24. K. Fukushima, 'Visual pattern recognition with neural networks', *Proc. 2nd Int. Conf. Parallel Image Analysis (ICPIA '92)*, pp. 16–31, Springer-Verlag, Berlin, 1992.
25. R. Linggard, D. Myers and C. Nightingale (Eds), *Neural Networks for Vision, Speech and Natural Language*, Part 1, BT telecommunications series 1, Chapman and Hall, London, 1992.
26. D. E. Rumelhart, G. E. Hinton and R. J. Williams, 'Learning internal representations by error propogation', in *Parallel Distributed Processing, Vols 1* and *2*, MIT Press, Cambridge, Massachusetts, 1986.
27. J. J. Hopfield and D. W. Tank, 'Neural computation of decisions in optimization problems', *Biological Cybernetics*, **52**, pp. 141–152, 1985.
28. T. Kohonen, 'Adaptive, associative and self-organizing functions in neural computing', *Applied Optics*, **26**(3), Dec., pp. 4910–4918, 1987.
29. G. A. Carpenter and S. Grossberg, 'ART2: self organisation of stable category recognition codes for analog input patterns', *Applied Optics*, **26**(3), Dec., pp. 4919–4930, 1987.
30. I. Aleksander, W. Thomas and P. Bowden, 'WISARD, a radical new step forward in image recognition', *Sensor Review*, No. 120-4, pp. 120–124, 1984.
31. I. Aleksander, 'Wisard: a component for image understanding architectures', in J. Kittler and M. J. B. Duff (Eds), *Image Processing System Architectures*, RSP, Letchworth, Herts.,1985.
32. I. Aleksander and H. Morton, *An Introduction to Neural Computing*, Chapman and Hall, London, 1990.
33. A. Redgers, 'Adrian's guide to understanding the WISARD', *Internal Report*, Neural Computing Group, Imperial College, London.

# 8 Image Understanding: Towards Universal Capability

A general-purpose machine vision system must be flexible in the sense that it should be able to operate in virtually unconstrained environments containing ill-defined objects which partially occlude one another. Thus the image analysis descriptions of two-dimensional (2-D) relationships must be enriched and extended to include the three-dimensional (3-D) relationships between objects within a real-world scene.

The human visual system is clearly highly flexible, robust and reliable with an apparent instantaneous response time. Machine vision systems on the other hand offer a comparatively slow response (although implementation in parallel hardware will provide near real-time operation in certain cases), problem-specific solutions and reliability which is governed by appropriate specification of its electronic and mechanical components.

The previous chapters of this text have focused on the processing, analysis and classification of 2-D scenes. Such an approach mirrors the human system which itself offers only a 2-D projection of the real-world onto the retina of the eye. However human 'visual perception' results in the identification of, and relationships between, individual stimuli (edges, points, surfaces) within the image array and hence in the creation of a clear description of the scene. In doing this an understanding of the 3-D relationships within the scene emerge. It is this visual perception which is unique and which is most difficult to capture within a *machine* vision system.

In the first generation vision systems of the early 1970s, features extracted from a captured image were compared and matched with known stored features, the best match being selected for identification. Such systems proved reasonably useful and are still applicable within constrained environments [1]. Once the environment becomes less restricted, i.e. it contains a very large set of objects placed in varying orientations and relative positions, then the machine vision problem becomes much more difficult to solve. Second generation vision systems started to look for solutions to this unconstrained problem from the mid 1970s onwards. Essentially such systems incorporate 3-D acquisition and processing capabilities [2, 3].

## 8.1 Image formation and the visual processes

For a 3-D problem, the recognition process is one which tries to relate representations of objects in an image with representations derived from models stored in the machine memory. It is relatively easy to describe a range of 2-D image features, which result from projections of 3-D objects onto the 2-D image plane. Many industrial visual inspection problems are in themselves essentially 2-D in nature, and in other cases, the depth information available from a 3-D scene can be used to describe a range of salient 2-D features. Such a process can be executed using a 'top-down' or 'knowledge-driven' approach. However, the inverse problem presents extreme difficulty – there is no clear way of transforming a 2-D image to produce a 3-D object model. The depth information has been lost, and any 3-D reconstruction is problematical.

For example, consider a 2-D image derived from a 3-D scene containing a spherical football. From the 2-D image it is appropriate to extract edge features and deduce the diameter and area of the cross-section of the football. However in the 3-D world there is no edge present on the football since its surface is one of continuous curvature. Hence the extraction of an edge from the 2-D image relies upon the precise formation of the image and this will change depending on the lighting conditions present at the time of image capture.

While such an edge extraction operation will produce meaningful results, it does illustrate one of the fundamental problems of reconstructing a 3-D model from 2-D image information without a lot of *a priori* knowledge of the scene. A similar example is that of viewing the moon from earth; the moon is clearly a spherical object but the view (image) from earth changes with its monthly waxing and waning.

The interaction of scene constraints with image acquisition has been discussed in Chapters 2 and 3, but now it is appropriate to return to some aspects in a little more detail. In developing a solution to the problem of interpreting or understanding a 2-D image in order to produce a 3-D model it would be helpful to have an exact mathematical description of the process of 2-D image formation [4]. Two key issues are:

(a) the determination of the geometric correspondence between points in the scene and points in the image;
(b) the determination of brightness at the resulting points in the image.

### 8.1.1 *Monocular images*

A simple model of the operation of the human eye can be obtained by considering it to be a pinhole camera in which the image is obtained from projecting scene points through a single pinhole onto the image plane.

Figure 8.1 illustrates such a system and shows the image plane drawn behind the point of projection, resulting in image reversal.

The optical axis is drawn perpendicular to the 'pinhole plane' which is taken to be the origin of the coordinate system. A point, $P(x,y,z)$, is identified in the 'scene' to the left which is required to be projected onto the 'image plane' drawn to the right at point $P'$. The light ray from P passes through the pinhole and strikes the image plane at a distance $f$ from the pinhole (the image plane being assumed to be parallel with the pinhole plane). The distance from P to the pinhole is taken to be negative in this coordinate system.

From the geometry of the system the image point $P'$ is given by

$$\frac{x'}{f} = \frac{x}{z} \quad \text{and} \quad \frac{y'}{z} = \frac{y}{z}$$

hence

$$x' = \frac{xf}{z} \quad \text{and} \quad y' = \frac{yf}{z}$$

where $x'$ and $y'$ are the $(x,y)$ coordinates of $P'$.

Within the image plane there is clearly no $z$ dimension and hence any $z$-axis information (depth) contained in the original scene is lost. This is an

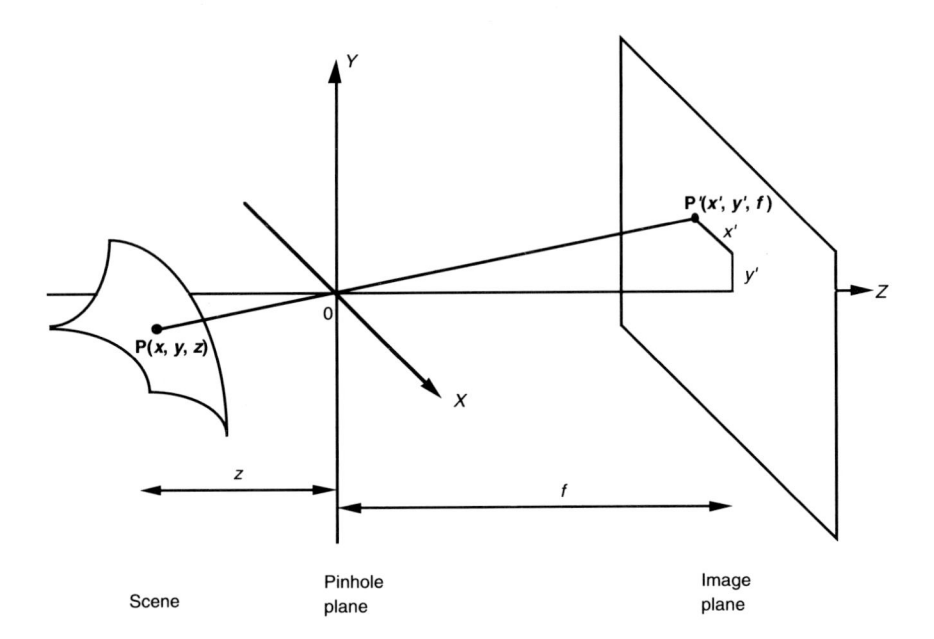

Figure 8.1 Perspective projection

illustration of perspective projection. If the distance from the scene to the pinhole is very large, and hence all depth information in the scene is relatively small in comparison, then a perspective projection can be approximated by an orthographic projection in which the rays between scene and image are simply drawn parallel to the optical axis.

### 8.1.2 Binocular images

Extending the above model, the human vision system can be studied as one which utilises two pinhole cameras set a known distance apart, thus providing a mechanism for deducing depth information from the scene. This technique can also be applied to computer vision and indeed provides one of the depth cues in Marr's computational model (see section 8.2).

In figure 8.2 the two pinholes are separated by a distance of $2d$ (the distance between the optical axes of the two viewpoints). A point $P(x,y,z)$ in the scene produces an image point $P'(x',y',z')$ in the image plane corresponding to pinhole 1 (using optical axis 1), while the same scene point produces an image point $P''(x'', y'',z'')$ corresponding to pinhole 2. The

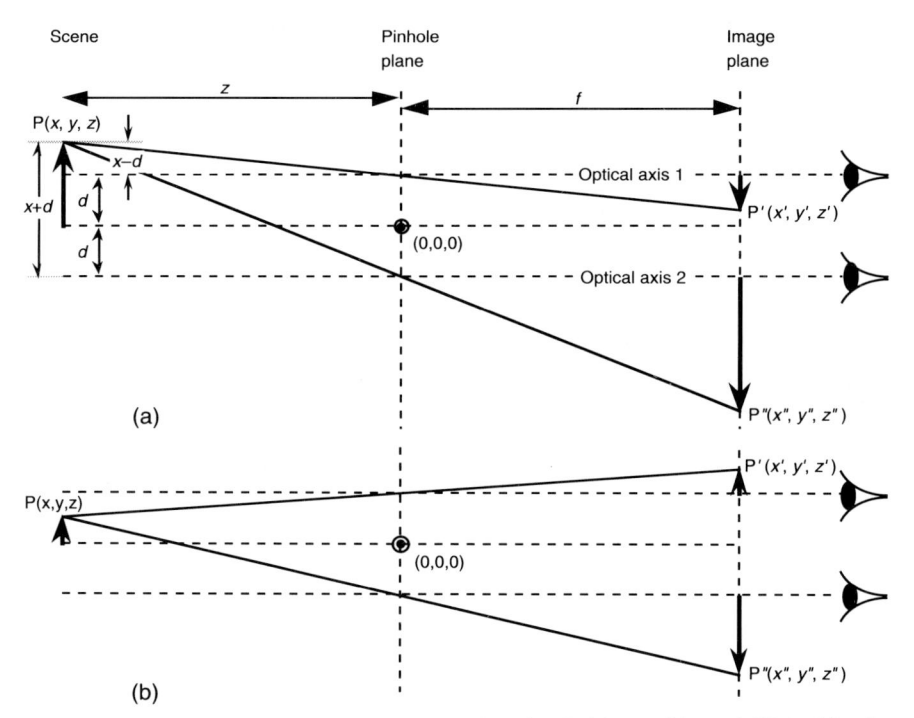

Figure 8.2 Simple binocular geometry: with point P (a) outside and (b) inside, the optical axes

disparity between the two images is related to the distance of the scene object from the camera.

Referring to figure 8.2a, in which the object point falls outside the region bounded by the two parallel optical axes and the $y$ displacement of P is zero, a similar geometrical analysis to the monocular case yields

$$\frac{(x-d)}{z} = \frac{x'}{f} \quad \text{and} \quad \frac{(x+d)}{z} = \frac{x''}{f}$$

or

$$x - d = \frac{x'z}{f} \quad \text{and} \quad x + d = \frac{x''z}{f}$$

Subtracting these produces

$$2d = \left\{\frac{x''z}{f}\right\} - \left\{\frac{x'z}{f}\right\}$$

hence

$$z = \frac{2df}{\{x'' - x'\}}$$

A similar analysis for the illustration of figure 8.2b will yield the same result provided that it is noted that:

(a) the origin of the coordinate system is the origin of the pinhole plane
(b) only one of the images is inverted, hence the signs of $x'$ and $x''$ will be opposite.

Note, however, that a key preliminary step in any practical implementation of stereoscopy is to establish corresponding points in each image so that the elementary trigonometry discussed above can be put to work. This is definitely not trivial in real scenes; points which are visible in one image may be occluded in the other and natural scenes largely made up of foliage or water, for example, can make finding correspondence virtually impossible (see section 8.2.1). Indeed this is precisely why Marr's computational model uses several depth cues rather than relying entirely on one.

### 8.1.3 Visual perception

From the research work reported to date on image understanding, it is apparent that both the organisation and structuring of the acquired image data must be an important part of low level vision operations and that many different low level image representations should be used [5, 6]. The way in which the human visual system structures its low level representations is known as 'perceptual organisation' and involves the detection of perceptually significant groupings and structures within the image. These groupings are used as cues for subsequent analysis and understanding.

For example, consider the images shown in figure 8.3; each one is easy to understand although a detailed look will reveal the level of perception being utilised. In figure 8.3a the reader will perceive a (solid) disc overlapping a square rather than simple outlines which are actually drawn, and in figure 8.3b a circle will be perceived at the centre of the radial lines although no circle has been drawn. Similarly figure 8.3c implies a white square positioned on top of four black discs although none of these objects is actually present. Finally in figure 8.3d the 'wire-frame' drawing in the centre can be interpreted as a cube with two possible orientations, i.e. with the front and back surfaces apparently changing position, as illustrated in the drawings on either side which have 'hidden lines' removed. It is this ability to perceive objects within low level image representations which is at the heart of the human visual system.

The capability of humans to organise low level image features into larger perceptual groupings was first identified by the German Gestalt school of psychologists in the 1920s. They believed that it was the perception of the

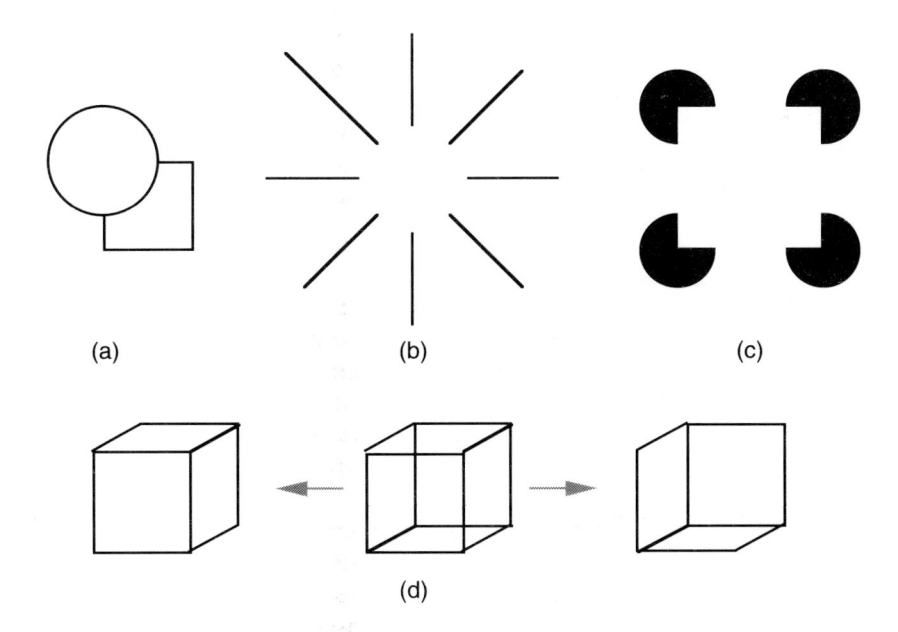

Figure 8.3   Examples of perceived organisation and structure

'whole' rather than the individual 'parts' of the image which was fundamentally important to the way in which humans perceive their environment. They also derived a number of perceptual grouping categories which sought to explain how certain perceptions dominate other possibilities. Demonstrations were also designed to prove that the distinction between objects and background became ambiguous in the absence of any of these dominant groupings.

Figure 8.4 summarises the principal perceptual grouping categories originally developed by the Gestalt school which may be explained as follows;

*Proximity*. When objects are close together they tend to be grouped. For example, the arrays shown in figures 8.4a and 8.4b both contain the same number of elements, however the first array has a smaller horizontal spacing and hence is perceived as rows of black spots while in the second there is a smaller vertical spacing giving the perception of a column of spots.

*Similarity*. Objects which appear to be similar in size, orientation or colour tend to be grouped together. In figure 8.4c the array of spots will be grouped into four regions, two white and two black.

*Closure*. A collection of objects or curves will tend to be perceived as closed rather than open figures. For example in figure 8.4d, incomplete squares, rather than back-to-back brackets, will generally be perceived because of closure.

*Continuation*. Figure 8.4e is perceived as two crossing lines rather than two non-crossing shapes, indicating that perception of continuous figures is preferred to abrupt changes.

*Symmetry*. Perceptual significance is attached to figures which are reflected around an axis as in figure 8.4f.

*Background separation*. In general the larger of two areas will be perceived as the background while the smaller will be perceived as the object or figure where the larger area surrounds the smaller. Thus the right-hand illustration in figure 8.4g shows a black diamond on a white background square but the left-hand illustration has equal areas and can be interpreted in other ways.

Although these grouping categories appeared to provide useful assistance in understanding the basic pixel data within images, the approach is severely limited when it comes to quantifying the degree of significance of the various categories in order that appropriate computer evaluation can take place. This quantification process is extremely difficult and the operation to select the best and simplest possible grouping organisation, inherent in Gastaltist psychology, becomes purely subjective.

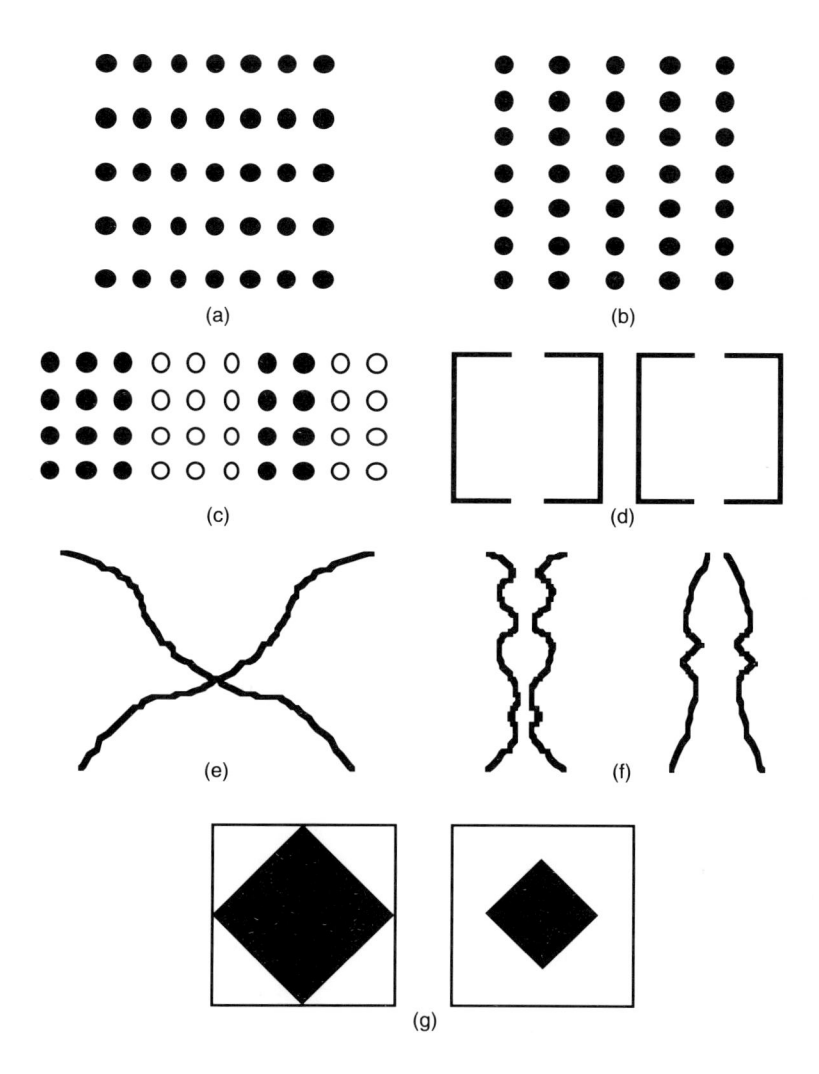

*Image Understanding: Towards Universal Capability* 211

Figure 8.4 Perceptual grouping categories: (a) and (b) proximity; (c) similarity; (d) closure; (e) continuation; (f) symmetry; (g) background separation

An alternative approach, based on the principle of non-accidentalness, has been developed which offers much promise [7]. In essence this says that relations between image features which are unlikely to have occurred through an accident of viewpoint, provide important cues about surfaces and objects. In particular, image features that exhibit parallelism and collinearity will do so over a large range of viewpoints and are therefore unlikely to have arisen by accident. By calculating the probability of a relation not occurring by accident it is possible to focus on the most probable relationships and hence reduce the computational effort needed to search through memory for appropriate model matches.

## 8.2 Marr's computational theory of vision

The previous chapters of the text have discussed a modular approach to machine vision. The low level processes revolved around image acquisition and preprocessing while the high level processes encompassed segmentation and classification. In general the modular system is designed to recover a significant amount of information about physical properties of the scene. The underlying philosophy of this empirical 'bottom-up' approach is that the recovery and combination of these properties can lead to a single coherent and robust description of objects within the scene and hence be used for comparison with stored models for unambiguous identification.

A major step forward in image understanding and visual perception came through the work of Marr who attempted to develop a mathematical analysis of the human visual process [8]. The central tenet of this approach was that the vision process should be seen as an information processing task which must be analysed and thoroughly understood before any attempt should be made to implement an artificial system.

This information processing task can be broken down into three distinct levels:

(a) computational theory;
(b) representation and algorithm;
(c) implementation.

At the computational theory level the investigator must determine what is the goal of the process, why it is necessary and what is the logic of the strategy by which it may be carried out. This *must* be the starting point of any investigation into a process.

The representation and algorithm level is concerned with the way that the computational theory can be implemented. The investigator must determine:

(a) the best form of representation for both the input and the output of the process;
(b) the algorithm that performs the transformation.

The implementation level is concerned with the means by which the algorithm and representations can best be physically realised. It should only be considered *after* the other levels have been thoroughly investigated.

*Image Understanding: Towards Universal Capability* 213

Figure 8.5 illustrates Marr's vision paradigm where perception is realised through the recovery of physical properties. The initial raw image data are first transformed into the primary representation or 'primal sketch'. This consists of edge, termination, bar and blob primitives which are determined from intensity or texture discontinuities within the raw image. These are then grouped to form tokens, at various scales, which make up the full primal sketch.

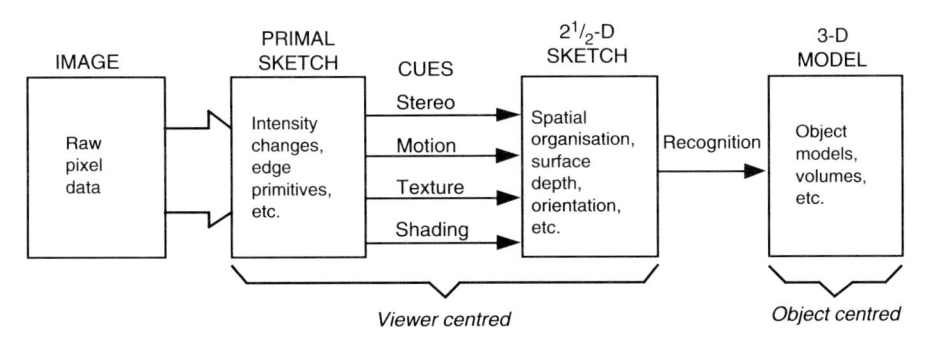

Figure 8.5 Marr's vision paradigm

This complicated array of data must then be decoded to form the next representation which is called the '$2\frac{1}{2}$-D sketch'. It is concerned with making explicit the existence of, and properties of, surfaces in the scene. The decoding is performed with the aid of a set of low level ('early') image analysis modules which operate on various visual cues within the raw image. They offer information about depth and orientation within the scene by determining 3-D shape from cues such as stereo, motion, texture and shading. Firm evidence for the existence of these cues within biological vision systems has been provided by a range of psychophysical experiments [9].

The $2\frac{1}{2}$-D sketch conveys spatial organisation, surface discontinuities, local surface orientation and local distance from the observer. However, it is unable to offer consistent descriptions of objects independent of their position with respect to the viewer. Thus it is desirable to transform the $2\frac{1}{2}$-D sketch into a representation that uses an object-centred coordinate system to explicitly model the scene in 3-D. This can be achieved by modelling volumes which make up the object, rather than its surfaces whose characteristics gave rise to the raw pixel data (see section 8.3).

## 8.2.1 *Depth from stereo*

Perhaps the most obvious cue is concerned with the recovery of depth from stereo [10]. It has been shown above in section 8.1 that the difference, or disparity, between two views of an image can be used to determine the depth from the observer to an object. However there is one major problem to overcome – for every object point there will be two image points corresponding to the left and right views.

In order to measure the disparity between the two views, corresponding points in both images have to be properly matched; this is termed 'the stereo correspondence problem'. Typically, initial feature extraction will be performed to produce corresponding sets of features independently in both images. However, some features cannot be matched because of obscuration by foreground objects and the whole process tends to be time-consuming and unreliable.

Figure 8.6a illustrates the stereoscopic observation of two black objects, but figure 8.6b illustrates the ambiguity that will occur in the recovery process if the correspondence problem is not adequately solved. It shows two possible 'phantom' matching points in the image which could be falsely identified as objects A and B.

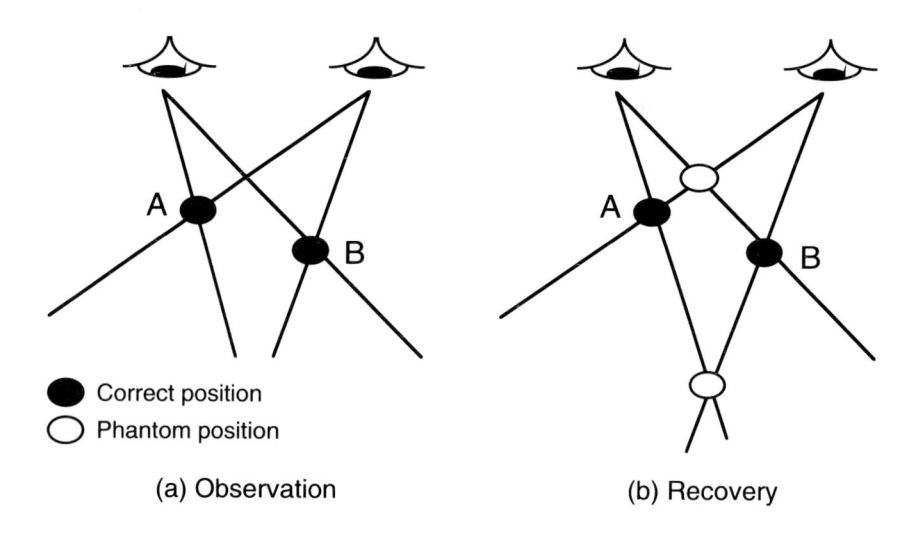

Figure 8.6 Stereo correspondence problem

## 8.2.2 Depth from motion

Motion makes a significant contribution to human visual perception, indeed in the natural world motion is often used by predators to detect prey and the lack of motion used as a method of camouflage. It is also common for one-eyed humans to drive cars, play tennis and perform other tasks which require real-time depth perception.

*Image Understanding: Towards Universal Capability* 215

A real-world scene is generally constantly changing as the objects in view, or the observer, or both change their relative positions. This gives rise to 'optical flow' which associates a two-dimensional velocity with each image point in the 2-D image plane. Shapes from motion algorithms initially determine optical flow then subsequently derive depth and surface information [11].

As with the above stereo approach, difficulties arise when attempting to match local features in an image recorded at two discrete points in time. If the object motion is small, or the time interval between the image samples is small, then a simple process of differencing the images will be sufficient to detect the moving points. However, if the relative movement is large, motion correspondence can be attempted by matching edge segments within the two images.

### 8.2.3 Depth from texture

Many surfaces are composed of uniform patterns of intensity variations which are collectively observed as texture. If these patterns comprise a constant and uniform texture then they can form the basis for image segmentation. Alternatively the apparent variation of a uniformly textured surface results in a texture gradient which can be utilised to recover surface depth and orientation [12].

Figure 8.7a illustrates this effect; the gradual changes in size and density of the ellipses towards the top of the figure give the impression of a surface stretching away from the viewer. In texture analysis, texture gradients (the

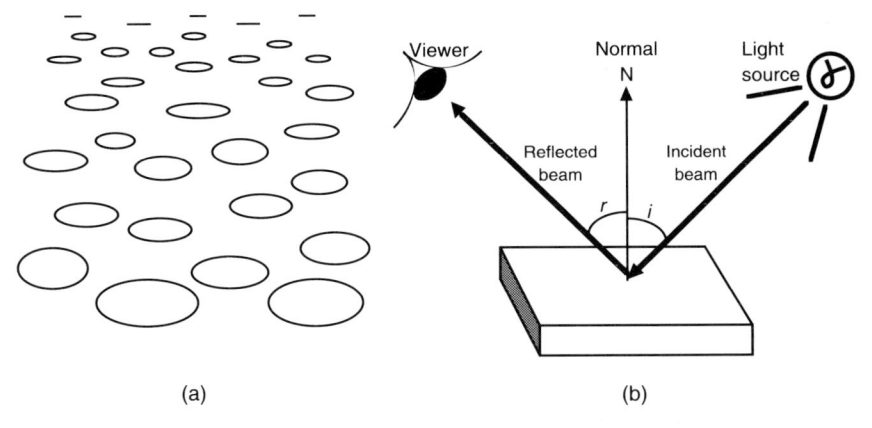

Figure 8.7   Depth from texture and shading: (a) a perceived textured surface; (b) single point surface illumination

direction of maximum rate of change of a specific texture primitive) are found to be the result of viewing, or scaling distance, and surface orientation.

Surface orientation can be further divided into slant (the angle with which the surface is oriented to the viewer) and tilt (the direction of the surface). In the figure, the circle texture primitives appear as ellipses on the tilted surface. The orientation of the principal axes defines rotation with respect to the observer, and the ratio of minor to major axes defines the tilt.

### *8.2.4 Depth from shading*

Chapter 2 highlighted the importance of lighting on the quality of the resultant image and discussed how its control use could be exploited as an effective scene constraint. By contrast the shadows cast on an object surface by a *simple* lighting (e.g. a single point source) can be used to obtain information about the structure of the object.

The important parameter of such an approach is the reflectivity function for the specified surface. For example, a mirror-like surface reflects light only at an angle of reflection which is equal to the angle of incidence. Alternatively, for a matt surface the reflected light varies as a cosine function of the angle of incidence and so most light will be reflected when the light source is normal to the surface.

Figure 8.7b indicates the appropriate parameters for a single point surface illumination; the incident light ray strikes the surface at an angle $i$ to the surface normal N, while the reflected ray leaves the surface at an angle $r$ to the normal. If the reflectivity function of a surface is known, together with light source distribution and intensity values, then surface orientation can be recovered [13]. This process is not trivial, as it involves the solution of complex partial differential equations, but ultimately surface gradient contours can be obtained from which surface normals, and hence surface shape, can be deduced.

### 8.3 Image representations

As stated above it is desirable to transform the $2\frac{1}{2}$-D sketch into a representation that uses an object-centred coordinate system to explicitly model the scene in 3-D. This allows objects to be uniquely described in a way which is stable with respect to variations in viewpoint.

For accurate results it is important to ensure that the object-centred system is fully calibrated. The pinhole analysis discussed above shows that it is possible to determine mathematically the coordinates of specified object points in the resultant image plane. Thus to calibrate the object-centred system it is necessary to extend this concept by identifying a set of reference points to establish the relationship between image coordinates and the corresponding real-world coordinates. This may be done by presenting the imaging system with a calibration grid, then identifying points in the image which correspond to the known real-world positions of grid intersections.

For actually 'modelling' the 3-D scene there are three basic approaches. The first and most popular is the 'generalised cylinders' approach which is based on a skeletal three-dimensional object representation. The second is the 'volumetric' representation which utilises a volumetric element (voxel) primitive and works on the basis of spatial occupancy. The third is the 'surface' representation which uses surface primitives to construct the 3-D model.

Each approach will have strengths and weaknesses depending on the particular problem under consideration, but it should be noted that in all cases there will be a degree of approximation in the resultant 3-D models and hence they will inevitably be imperfect representations [14–17].

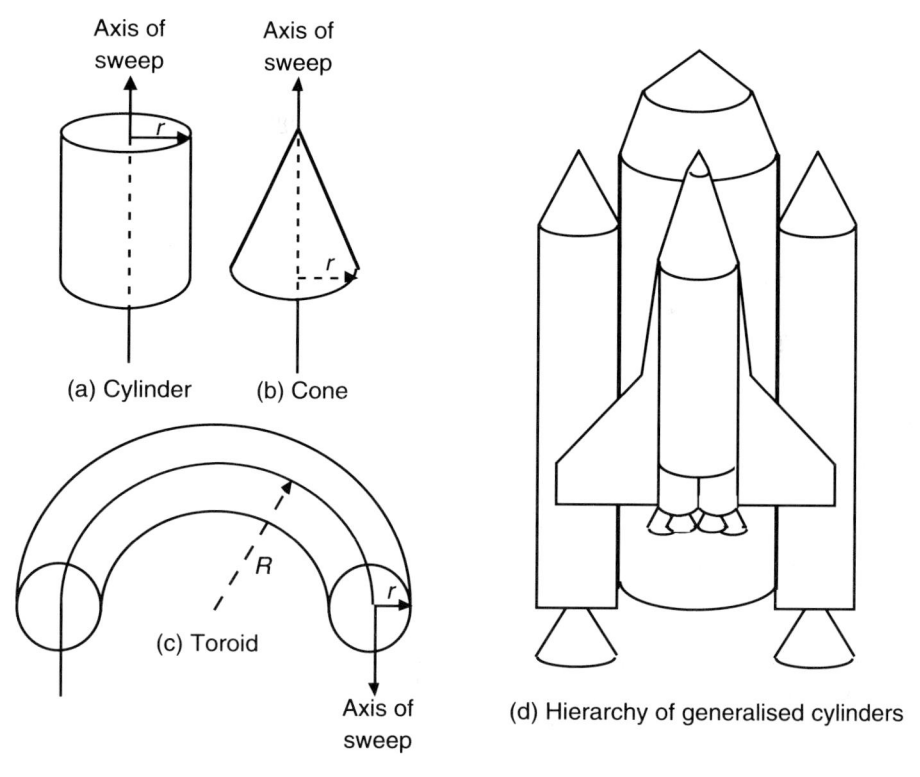

Figure 8.8  Generalised cylinders/cones: (a) a cylinder; (b) a cone; (c) a partial torus; (d) a familiar object modelled as a hierarchy of generalised cylinders

### 8.3.1 Generalised cylinders

The generalised cylinders approach, also termed 'generalised cones', is based on the definition that a cylinder is the surface swept out by a circle whose centre is travelling along a straight line normal to the plane of the circle. This is seen in figure 8.8 where a circle of radius $r$ and its axis of sweep are clearly marked for both a cylinder and cone. The cone is regarded as the surface swept out by a circle whose radius changes linearly with distance travelled along the axis of sweep. The axis of sweep may also follow some defined function and so produce other volumes – e.g. the toroidal shape in the figure is obtained by allowing the circle, radius $r$, to sweep along the axis of another larger circle of radius $R$.

These generalised cylinders are specified by a skeletal spine, a cross-section (which does not have to be a circle), and a sweeping rule. A general 3-D model comprises several generalised cylinders organised hierarchically, with each component subsequently comprising its own generalised cylinder based model – see figure 8.8d.

### 8.3.2 Volumetric elements

The volumetric approach produces a 3-D model by construction from a set of volumetric elements or primitives. These primitives may be cubes, cylinders, spheres or cones which are scaled, positioned and combined together using standard set operators. This approach is also referred to as 'constructive solid geometry (CSG)'.

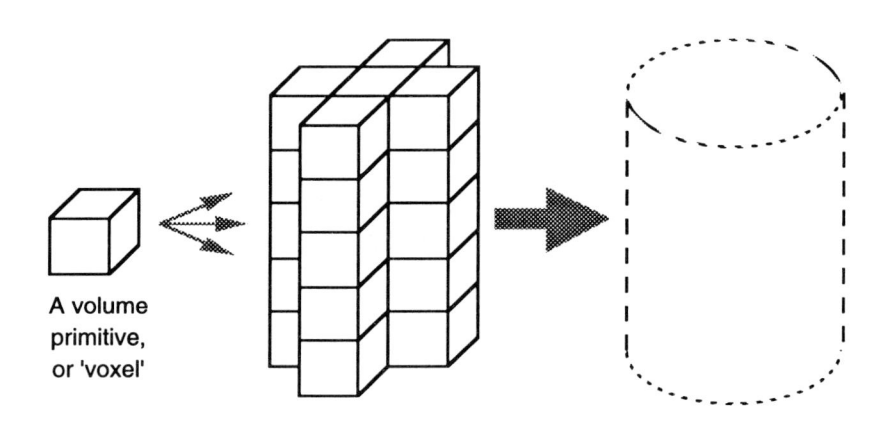

Figure 8.9 Volumetric representation of a cylinder by uniform voxels

Figure 8.9 shows how a cube primitive (e.g. 1 cm$^3$) may be used to model a cylinder and hence produce a three-dimensional array of such cubes. Clearly the only cube primitives included in the model are those that lie totally within the 3-D cylinder and so there will be approximation errors near the curved surface of the cylinder. Such errors will be reduced by decreasing the size of the cube primitive (shown relatively large in the figure for emphasis) according to the Nyquist sampling criterion.

The use of small voxels to obtain accurate modelling requires a relatively large amount of computer memory to store each representation, but this requirement can be considerably reduced if non-uniform voxels can be used. This is achieved in the 'oct-tree' representation method which is a 3-D extension of the quad-tree data structure discussed in section 5.3.1. In this case each parent has eight possible children, but if *all* the children would be contained within the object, then the (large) parent voxel is stored rather than all of the (smaller) children. Thus the construction of the tree ensures that the object is modelled with the largest volumes that will fit entirely within it, but accuracy is maintained thanks to the small 'leaf' voxels.

### 8.3.3 Surface representation

The surface representation approach, also referred to as a 'boundary representation (or B-rep)', is similar to the previous one except that the model primitives are surface elements. A variety of surface elements or polygons may be used and the resultant model is a polyhedral approximation to the 3-D object.

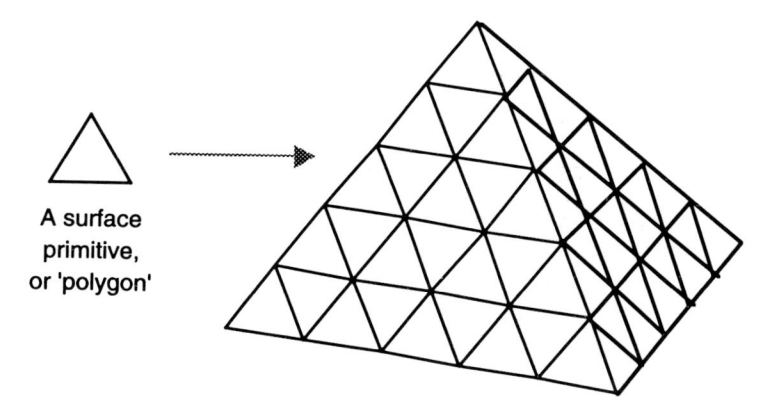

Figure 8.10 Surface representation of a tetrahedron by triangles

Most surface primitives are taken to be planar although it is possible to adopt curved ones. The simplest planar polygon primitive will be the triangle which can then be used to construct a representation of any complex 3-D object. Figure 8.10 illustrates the construction of a tetrahedron from the triangular polygon shown.

This approach tends to be favoured for representing industrially generated components which, in the main, are constructed from planar surfaces, but has limitations when applied to naturally occurring objects. The principal advantage of this technique is that it produces accurate surface area measurements for the modelled objects.

### 8.3.4 Wire-frame representation

Many engineering applications still favour line drawing or 'wire-frame' models of the real-world. Here 3-D objects are represented by a list of vertices and edges which join them. Unfortunately such a list does not necessarily produce an unambiguous representation although to the trained operator they may always be meaningful. Interpretation of line drawings requires a formalised labelling system as illustrated in figure 8.11.

Lines labelled '+' are caused by a convex edge, those labelled with a '−' are caused by a concave edge, while those labelled with arrows are caused by part of the object occluding a surface behind it. The occluding object is to the right of the line, looking along the axis of the arrow, while the occluded surface is to the left. For example, figure 8.11a shows a cube with the nearest

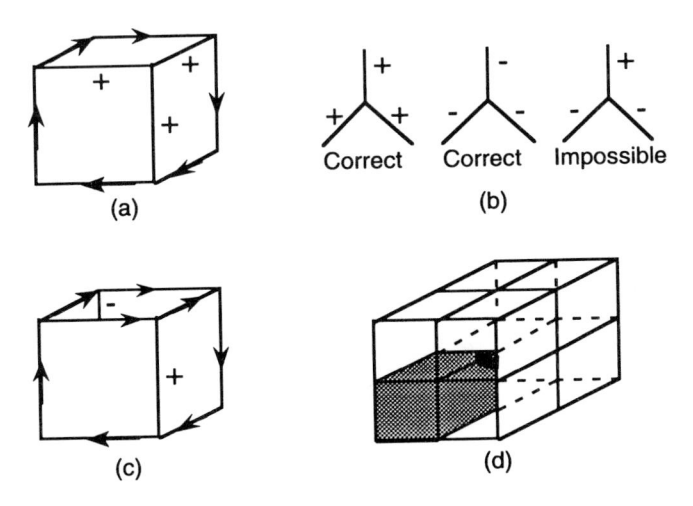

Figure 8.11   Wire-frame representation

vertex formed by three convex edges, the leftmost vertical arrow indicating that the front face of the cube (object to the right of the arrow) is occluding the rear left side of the cube (object surface to the left of the arrow). All other edge labels can be interpreted in a similar way.

The ambiguity problem is illustrated in figure 8.11b where three possible vertices are labelled, two are correct and can be interpreted as a convex and concave respectively, but the third labelling cannot be interpreted as a real-world vertex. The *box* drawing in figure 8.11c is correctly interpreted because of the concave edge identifying the rear inside edge (notice the similarity and difference to the *cube* drawing).

Within any three line vertex each line can be labelled with one of the four defined labels, thus giving a total of 64 possible vertex labels although only 3 vertices can exist in the real-world. In studying this problem, use is made of the number of octants of space filled by the object around the vertex. For example, in figure 8.11d the cube vertex identified has 7 other neighbouring octants that could be occupied with object material.

## 8.4 Pragmatic modelling and matching strategies

Regardless of how the models are organised and what particular model representations are chosen, the recognition process is essentially one of matching data derived from an image with that of a stored model. The 3-D nature of the problem means that the search space becomes impossibly large for complex natural scenes and a dramatic reduction in its size is necessary before practical 3-D industrial machine vision systems become widely realisable.

The model descriptions may be viewed as 'graphs' (networks or trees) stored in memory and the matching process is one of evaluating the similarity between two graphs. In simple terms the graph consists of a set of nodes representing parts of an object, each node contains a set of object properties, and there are fixed relationships between nodes which are represented by the arcs which link them. Searching or traversing the network may be carried out in a number of ways.

The earliest methods employed exhaustive search techniques. The 'depth-first' search assumes that one solution is as good as another and so at each decision point in the graph it makes an arbitrary decision and continues its evaluation. The 'breadth-first' technique considers all alternatives at a given level in the graph before evaluating any one.

Pruning the search space is essential to overcome the combinatorial explosion so techniques such as 'hill-climbing', 'best-first search' and 'branch and bound' have been developed and exploited. Other approaches derived from game playing problem domains, namely 'state-space search', 'means-end analysis', etc., can be applied.

The topic of graph or tree searching is beyond the scope of this text but is a key discipline in the domain of mathematics and artificial intelligence.

Returning to the 3-D image understanding problem, the reduction in search space may be assisted by making use of *a priori* contextual knowledge. This is undoubtedly used by humans and enables search space reduction by an initial consideration of only the most probable regions in which objects are likely to occur. The use of contextual knowledge also leads the system designer to consider knowledge-based system architectures which combine perceptual organisational techniques with a multiresolutional methodology in which a low resolution image is used for focusing attention to a particular area of interest.

Multiresolution representations employ both a 'top-down' and 'bottom-up' processing methodology. Bottom-up control is used to determine candidate objects from a reduced resolution image while a top-down methodology is used to process a restricted high resolution area. Such a multiresolution or pyramidal processing approach has been explored by researchers in medical imaging and aerial reconnaissance applications [18, 19]. In essence this approach utilises a 'hypothesis and test strategy', where hypothesis generation is concerned with identifying potential object locations from within a low resolution image which merits further analysis, while hypothesis verification focuses on the subsequent analysis to evaluate the final matching.

Three general approaches to model-based recognition can be identified [20].

### 8.4.1 Two-dimensional modelling and matching

In the first approach 2-D models are matched with 2-D image data. Systems that utilise this approach assume that the third dimension is relatively small in relation to the other two. In other words, 2-D modelling is most applicable to objects that appear flat and are viewed in a plane that is parallel to the image plane of the camera.

HYPER (HYpothesis Predicted and Evaluated Recursively) [21] uses 2-D models of industrial parts to match against 2-D image data. The system has been coupled to a robot arm, in order that picking and repositioning of overlapped industrial parts can be performed. In order to achieve this objective the system had to be designed with good robustness to partial occlusions, shadows, touching and overlapping objects.

Even though HYPER assumes its input data to be two dimensional, it serves very well to illustrate the contrast between the hypothesis and test approach, which is common in 3-D matching, with the bottom-up approach of continually enriched description that is typical of the 2-D image analysis methodology described in the earlier chapters. HYPER begins by generating a scene description using tools which are also used, off-line, to generate object models.

Object boundaries are detected and a list of connected boundary points is determined in order to generate polygonal object border approximations. Model to scene matching is then initiated by considering a number of salient, or privileged, model features, selected on the basis of line length. Objects are identified and located by initially generating hypotheses.

By matching the privileged model segments to the scene description, a few hundred hypotheses are usually generated and ranked on the basis of a local criterion of merit. The best of these hypotheses are used to predict object locations within the scene, and appropriate model transformations are calculated. These may be matched to the scene description and used to iteratively refine the model transformations.

The matching process terminates either when a sufficient number of the high ranking hypotheses have been evaluated, or when a high degree of match between lengths of the identified scene segments and the relative lengths of model segments has been obtained. The best hypothesis is finally re-examined before being accepted or rejected.

*8.4.2 Three-dimensional modelling and matching*

In the second approach 3-D models are matched with 3-D image data. This is most appropriate when an arbitrary viewing angle is a desirable feature of the end application.

An example of a recently developed system that matches 3-D models against 3-D image data is known as TINA [22]. This is a 3-D vision system designed for an industrial 'pick and place' robot. It is designed to support model-based recognition, and location of objects in a cluttered working environment, enabling a 'grasp plan' for the scene to be computed.

Object location begins by recovering a depth map from two cameras positioned to one side of the working area. A 'Canny' edge detector is used on both images before the edge based depth map is recovered, by using the 'PMF' stereo algorithm [23]. The edge-based depth map is then used to recover a description of the scene in terms of straight lines and circular arcs. Instead of discarding any matches between scene descriptions, they are used to form a composite 2-D and 3-D scene description, thereby forming a more complete description than using the stereo data alone.

Object models are matched to the scene description, by initially choosing a 'focus feature' from the model, and selecting a number of neighbouring salient features, the saliency of a feature being based on line length. From these matches, various model transforms are determined before the best match is selected.

In addition to object identification and location, TINA is also able to generate its own object models, by modelling objects in terms of straight lines and circular arcs, from a sequence of views. As the view is updated, features from the previous view are matched to ones from the current view. Novel features from the current view are added to the model. Only features that have appeared in more than one view are retained in the final model.

### 8.4.3 Three-dimensional modelling and matching to two-dimensional image data

The third approach is a hybrid system, in which 3-D objects are represented as a number of 2-D models, or 3-D models are matched with 2-D image data. Difficulties associated with the stereo correspondence problem are alleviated, but exact object location can be problematical.

ACRONYM [15] was designed to interpret 2-D scenes using 3-D models, and was applied to aerial image scenes of airports. Models of commercial aircraft were defined by using generalised cylinders. Object identification was performed in ACRONYM, by matching object models to special low level shape primitives from the 2-D image data.

Matching is performed by predicting possible object appearances in the image and computing associated constraints on the model. These constraints are then used for object prediction and interpretation of the shape primitives. As objects are modelled in terms of object class, the most general set of constraints concerning the class of object (e.g. wide-bodied aircraft in the case of airport scenes) are initially applied when carrying out prediction and interpretation.

Interpretation is achieved by using both geometric and symbolic reasoning. A consistent or partial match with a model initiates a subclass check (e.g. '747' or 'DC-10'), by applying the extra subclass constraints. Throughout the prediction and interpretation process, ACRONYM does not utilise any special knowledge of airport scenes – all its reasoning is based on purely *geometric* reasoning.

## 8.5 The beginning of the end . . .?

At this point it should be noted that much research still has to be carried out in the area of 3-D scene description before machines can even begin to compare favourably with the human visual system. The very essence of the human visual capability is the perceptual organisation of line features into meaningful groupings which allows robust and reliable recognition and understanding to be performed. The identification of shape from stereo, motion, shading and texture may be seen as a prerequisite to recognition and understanding by machines, but humans are able to perform these tasks even in situations when no such cues are clearly recoverable.

Marr's vision paradigm has rightly been identified as a landmark in understanding the visual process, but it does not indicate how the shape recovery processes can be reliably combined into robust deductions about the real-world. For example, the assumption that information derived from one shape cue will *reinforce* that derived from others is not always true. It is quite possible for conflicting information to be recovered from the scene and appropriate strategies for resolving these conflicts have to be developed.

The few attempts that there have been to recreate a true $2\frac{1}{2}$ representation from multiple information sources have invariably been computationally complex and therefore very slow – offering performance several magnitudes away from real-time operation even on sophisticated computer hardware. Therefore most attempts at producing real-world 3-D machine vision solutions still rely upon a finely tuned and fragile collection of *ad hoc* techniques rather than the universal computer vision ideal. Nevertheless, fundamental research into the visual processes continues in order that there might one day be a solid foundation for machines that can understand their interaction with their environment through the means of visual perception.

## References

1. R. T. Chin and C. R. Dyer, 'Model based recognition in robot vision', *ACM Computing Surveys*, **18**(1), pp. 67–108, 1986.
2. H. G. Barrow and J. M. Tenenbaum, 'Recovering intrinsic scene characteristics from images', in A. R. Hanson and E. M. Riseman (Eds.), *Computer Vision Systems*, pp. 3–26, Academic Press, New York, 1978.
3. D. Marr, *Vision*, Freeman, Oxford, 1982.
4. B. K. P. Horn, 'Artificial intelligence and the science of image understanding', in G. G. Dodds and L. Rossol (Eds), *Computer Vision and Sensor Based Robots*, Plenum Press, New York, 1979.
5. J. Beck, *Organisation and Representation in Perception*, Lawrence Erlbaum Associates, London, 1982.
6. I. Biederman, 'Recognition-by-components: a theory of human image understanding', *Psychological Review*, **94**(2), pp. 115–147, 1987.
7. D. G. Lowe, 'Three dimensional object recognition from two-dimensional images', *Artificial Intelligence*, **31**, pp. 355–395, 1987.
8. D. Marr and H. K. Nishihara, 'Visual information processing: artificial intelligence and the sensorium of sight', *Technology Review*, **81**(1), pp. 2–23, MIT Alumni Assoc., 1978.
9. M. Brady, 'Preface – the changing shape of computer vision', *Artificial Intelligence*, **17**(1–3), pp. 1–15, August 1981.

10. D. Marr and T. Poggio, 'A computational theory of human stereo vision', *Proc. Royal Society*, **B204**, pp. 301–328, 1979.
11. S. Ullman, *The Interpretation of Visual Motion*, MIT Press, Cambridge, Massachusetts, 1979.
12. A. P. Witkin, 'Recovering surface shape and orientation from texture', *Artificial Intelligence*, **17**, pp. 17–45, August 1981.
13. A. P. Pentland, 'Local shading analysis', *IEEE Trans. Pattern Analysis and Machine Intelligence*, **6**(2), pp. 179–187, 1984.
14. T. O. Binford, 'Visual perception by a computer', *IEEE Conf. Systems and Control*, Miami, 1971.
15. R. A. Brooks, 'Model-based three-dimensional interpretations of two-dimensional images', *IEEE Trans. Pattern Analysis and Machine Intelligence*, **5**(2), pp. 140–149, 1983.
16. D. T. Morris and P. Quarendon, 'An algorithm for direct display of CSG objects by spatial subdivision', in R. A. Earnshaw (Ed.), *NATO ASI F-17: Fundamental Algorithms for Computer Graphics*, Springer-Verlag, Berlin, 1985.
17. D. H. Ballard and C. M. Brown, *Computer Vision*, Chapter 9, Prentice-Hall, Englewood Cliffs, New Jersey, 1982.
18. T. M. Silberberg, 'Multiresolution aerial image interpretation', *Proc. DARPA Image Understanding Workshop*, Vol. 2, pp. 505–511, Materials Research Society, Pittsburgh, Pennsylvania, 1988.
19. S. J. Cosby and R. Thomas, 'IRS: a hierarchical knowledge based system for aerial image interpretation', *Proc. 3rd Int. Conf. IEA/AIE*, pp. 207–215, 1990.
20. W. E. L. Grimson and T. Lozano-Perez, 'Model based recognition and localization from sparse range or tactile data', *Int. Journal of Robotics Research*, No. 3, pp. 3–35, 1984.
21. N. Ayache and O. D. Faugeras, 'HYPER: a new approach for the recognition and positioning of two-dimensional objects', *IEEE Trans. Pattern Analysis and Machine Intelligence*, **8**, 1986.
22. J. Porril, S. B. Pollard *et al.*, 'TINA: a 3-D vision system for pick and place', *Proc. 3rd AVC*, Cambridge, UK, pp. 65–72, 1987.
23. S. B. Pollard, J. E. W. Mayhew and J. P. Frisby, 'PMF: a stereo correspondence algorithm using a disparity gradient', *Perception*, **14**, pp. 449–470, 1985.

# 9 Image Processing Case Histories

Computers have many potential uses in the manipulation of images to extract, refine and evaluate information. However, it is extremely taxing of computer power in two respects. Firstly the amount of raw data in an image is very high – a single image of 512 × 512 eight-bit pixels requires 256 kbytes of storage. Secondly it is computationally expensive, with typical non-trivial image processing tasks requiring between 1000 and 10 000 machine cycles per pixel [1]. This translates into either very slow execution times on conventional Von-Neumann architectures or very expensive customised hardware.

Thus the history of computer vision as a practical tool is rather short, as most potential applications have had to wait for the arrival of cost-effective memory technology and the achievement of adequate cost/performance ratios in terms of processing power. Despite this, the underlying disciplines identified earlier in this book have been brought towards maturity by the academic community and application domains which had the right motivation and funding to do so.

The foundations of image enhancement and restoration were laid by Dr Robert Nathan and the team that he built at NASA's (National Aeronautics and Space Administration) Jet Propulsion Laboratory (JPL) [2]. A life sciences/biomedical programme was eventually 'spun-off' from this work, thus beginning the development of medical imaging applications. The investigation of image understanding (as well as image processing) has been heavily supported by military and government agencies such as DARPA (Defence Advanced Research Projects Agency), the NSF (National Science Foundation) and NASA in the US [3].

However, now that high density primary and secondary memory technology and powerful personal computers are a reality, the range of applications of image processing is growing extremely rapidly. It is truly a sign of the times that domestic products such as CD-i (Compact Disc Interactive) and other entertainment systems now have the potential to give such sophisticated technology mass-market appeal.

## 9.1 Industrial machine vision applications

It will be clear from the discussion in Chapter 1 that many industrial applications of machine vision involve inspection of products. Estimates vary, but between 60 and 90% of all existing machine vision applications are classed as automated visual inspection (AVI). It has achieved this dominance because it is only concerned with the question 'Is this item well-made?' rather than the open-ended question 'What is it?', so that it involves the minimum 'on-line' understanding of scene content.

In a well run manufacturing system process faults should be well understood, so that even qualitative inspection offering good coverage of a broad fault spectrum can be developed relatively easily. Once the parameters that must be inspected have been elucidated it is only necessary to devise a way of measuring them by non-contact visual means. Such techniques are abundant and well documented, as in Batchelor *et al.* [4]

Inspection is also a tedious task which used to occupy a significant percentage of the workforce. A Bureau of Census report for the USA in 1970 revealed that 4–6.5% of *all* the workers involved in the manufacture of durable goods were involved in inspection operations [5]. Therefore AVI is blessed with a convenient mix of 'technology push' and 'market pull' which has encouraged its development and application. So the future for AVI is blossoming and it looks set to maintain or increase its share of the machine vision market.

However, AVI does not primarily enhance the *flexibility* of a manufacturing system. This comes from the utilisation of vision in process control, parts identification and robotic guidance and control applications. Process control is essentially on-line, real-time AVI incorporated within a control loop in the manufacturing process. As such, there are no significant technological barriers which prevent its adoption. Instead, there are major problems of awareness of its potential, especially within industries such as fibre-board manufacture, which are not traditionally associated with new technology, but which could benefit enormously from it. Other problems include the need to retrofit any new equipment to existing plant and overcoming anxieties about relying *totally* upon machines to control product quality. However, it is reasonable to speculate that it is only a matter of time before vision-assisted process control truly makes its mark.

Parts identification is essentially concerned with classification of industrial components prior to processing by an automaton which is capable of blindly executing a number of predefined programs. As such, it does represent a significant contribution to achieving flexibility in manufacturing, and many examples exist, particularly in the automotive industry. However, this approach is naturally limited to situations where the decision-making processes can be totally divorced from the subsequent actions. Applications such as assembly tasks, where this is not possible, can be classified as being in need of vision-assisted robotic guidance and control.

This is proving to be the most complicated area to address because the automaton must continuously update its understanding of the scene it sees as it interacts with that scene. In general, scene constraints such as structured lighting are hard to impose in such situations, so the well established techniques of 2-D image analysis are often inappropriate. This means that the vision system must derive three-dimensional understanding from a sequence of 2-D images, bringing the level of complexity up towards that of a universal vision system, albeit operating in a familiar and predictable environment.

Successful existing applications either allow the task to be effectively constrained to two dimensions and/or allow structured light to simplify assessment of the third dimension. Alternatively, more sophisticated attempts remain in the research laboratory for want of more processing power to permit 'real-time' operation. Despite the difficulties, each new development in the field of computer vision brings robots a step nearer to fulfilling their long-predicted potential in manufacturing.

### 9.1.1 Automated visual inspection

The old attitude to inspection was that it was to be performed as a separate function *after* the complete manufacturing process. Therefore it was concerned with the retrospective detection and rejection of faulty workmanship. When such faults were detected the only course of action was costly rework or scrapping.

Modern thinking rejects this view and aims to integrate inspection thoroughly at all stages of the manufacturing process. This is part of a 'zero defects policy' where the aim is to identify and eliminate all process faults which cause unsatisfactory goods. Automated visual inspection (AVI) systems can be of great assistance to this aim. In many applications the only problem is explicitly recognising the need for an inspection system in cases where human workers have provided a free, implicit inspection as part of their normal activity.

When performed automatically by a machine vision system, AVI may be included in a process control feedback loop. This allows both quality and throughput to be maintained or improved simultaneously. In other applications, usually known as non-destructive testing (NDT), image processing and analysis techniques may be used to aid humans in their decision making processes.

There are essentially two categories of AVI, as follows.

*Measurement and gauging*

The first sub-class of AVI system is concerned with making accurate measurements of critical object dimensions in applications such as:

(a) gauging of spark plug gap [6];
(b) measurement of belt width [6];
(c) measurement of tool wear [7];
(d) fiducial mark alignment and pick and place component offset [8];
(e) analysis of crack formation and propagation by interference fringes [9].

*Integrity checking and quality control*

The second sub-class of AVI applications is concerned with qualitative and semi-quantitative assessment. In this case the system emulates a human inspector who assesses manufactured articles for integrity and completeness, correct labelling, surface finish, etc. For example:

(a) safety-critical inspection of Volkswagen brake assemblies [10];
(b) inspection of automotive valve spring assemblies (General Motors 'KEYSIGHT') [11];
(c) checking correct printing of lot code on pharmaceutical labels [12];
(d) surface quality inspection of automotive sheet metal stampings [13];
(e) paint finish assessment [9];
(f) inspection of machined components for surface roughness, nicks, edge damage, flats, etc. [9];
(g) visual inspection of printed circuit boards, bare and loaded [8, 14].

### 9.1.2 Process control

Vision is applied to process control when that control relies on a parameter which can most easily be assessed visually, but human senses are inadequate. This will normally be in circumstances where high throughput rate, high reliability, high uniformity, critical objective measurements, etc. are important issues in the marketability of the finished product.

This is the type of application which seems most likely to bring use of vision into industries that have not been traditionally associated with high technology automation. Successful applications to date (see figure 9.1) include:

(a) fibre analysis in the wood panel industry [15];
(b) control of flatness in float-glass manufacture [16];
(c) elimination of surface defects in photographic film webs [10].

Investigations into application (a) were initiated within the Applied Image Processing Research Unit at the University of Brighton and were ultimately taken to market by a company formed by ex-students, called 'Advanced Fibre Imaging'. It was discovered that existing methods of assessment of fibre quality were neither accurate, repeatable nor able to provide a direct measurement of fibre size. In particular, the most common method consist-

(a)

(b)  (c)

Figure 9.1   Image-based analysis in the fibre products and paper-making industries: (a) the 'Fibertron' transducer, showing sampling apparatus, pulp presentation cell, lighting, camera and controlling computer; (b) grey image of ink particles distributed in recycled paper pulp; (c) result of automatic image pre-processing to permit true particle size measurement for subsequent statistical analysis *(Photographs courtesy of (a) AFI Ltd; (b) and (c) University of Brighton in collaboration with PIRA International)*

ing of occasional visual and tactile examination of fibre samples by the refiner operator is highly subjective, unrepeatable and statistically irrelevant.

The machine vision fibre quality transducer, known commercially as the 'Fibertron', produces a single measure of fibre quality which is the result of direct automated visual measurement of fibres in samples taken from the process. This differs from other attempts at objective fibre measurement which can only infer fibre quality from properties of the fluid containing the fibres and which do not produce unique results. The measurement produced by the Fibertron is called the 'combined fibre measurement (CFM)', and it is determined from measurements of length, area and perimeter of fibres viewed in binary images taken of the fibre samples.

Extensive testing has shown that the CFM does indeed vary in sympathy with the quality of raw material stock and changes in refiner settings, such as grinding plate gap and feed-screw speed. The refiner operator can now be presented with a continuously updating display of CFM. By maintaining it within a set band, fibre quality of unprecedented uniformity is produced and consistent quality of finished board is assured.

Tests with a pilot installation at Evanite Hardboard, Oregon, USA have shown that use of the Fibertron eliminates over-refining and allows levels of wax and resin additives to be objectively optimised for particular board-strength grades. This has resulted in savings of 5% in energy costs and 17% in chemical costs. The latter reduction saves Evanite more than US$90 000 per annum [17] – this alone would ensure that the Fibertron system pays for itself in approximately six months. Other tangible benefits include a better understanding of the mechanisms that cause variable board quality, enabling changes to be made as appropriate at all stages of raw material preparation, storage and processing. This insight is a natural by-product from the existence of an explicit, objective assessment of fibre quality.

At the present time the Fibertron family of transducers relies upon a human operative to control the refiner settings in response to the continuously updating CFM display, because that is what the existing customers prefer. However, it is perfectly possible to develop full automatic control of the fibre refining process, should it be required. News of the success of the Fibertron is travelling fast and orders for two more systems have already been placed by mills in Montana and Carolina.

Meanwhile, research continues into the application of machine vision techniques into other aspects of the wood products industry such as de-inking of recycled paper (in collaboration with PIRA International).

## 9.1.3 *Parts identification*

Identification of parts finds a number of applications in industry, particularly in flexible manufacturing environments. General Motors'

'CONSIGHT' is typical of the class of 'look-then-move' applications which features part identification as a tightly integrated component in an otherwise 'blind' robot guidance system. Such applications generally require the position and orientation of the object to be determined, in addition to a preliminary classification. Bin-picking in its simpler forms may be considered a special case of part recognition.

In other applications, such as (g), (h), (i) and (j) below, a sorting task is a necessary precursor to processing by a first generation robot or other manufacturing system.

Some examples are given below:

(a) sorting of automotive castings ('CONSIGHT') [18];
(b) automotive wheel-to-hub assembly [16];
(c) automotive tyre-to-wheel assembly [16];
(d) precision application of car body seam-sealant [19];
(e) unloading of automotive crankshafts from pallets (bin-picking with intermediate spatial organisation and separation) [20];
(f) unloading materials from matrix packaging (bin-picking with high spatial organisation and separation) [21];
(g) identification of car bodies by outline [10];
(h) classification of automotive axles prior to spray painting [16];
(i) automatic decoration of chocolates [22];
(j) sorting of fish by species, size determination and inspection [23].

### 9.1.4 Robotic guidance and control

In the context of industrial machine vision this activity is concerned with the provision of visual feedback in part handling, assembly and welding, etc. It is not to be confused with the navigation of autonomous vehicles or missile guidance.

In the guidance role, machine vision is often used to eliminate the need for costly and inflexible fixturing and jigging in precision assembly tasks. Control applications are difficult to separate from guidance, the best known ones being seam-following for the purpose of welding etc. However, they indicate the first steps along the path towards 'visual servoing' which promises to radically simplify and cheapen robot construction by relaxing the requirement for high precision mechanics on a large scale.

Some existing applications of robotic guidance are as follows:

(a) automotive windscreen alignment and placement [10];
(b) seam location and following for welding car chassis members [16];
(c) acquisition of cylindrical objects from bins (bin-picking with random spatial organisation) [24];

234    *Applied Image Processing*

(d) vision guided nuclear fuel sub-assembly dismantling [9];
(e) high precision part-mating in aerospace applications [25];
(f) pattern-correct sewing in textile manufacture [16].

In addition to the ultimate need for universal vision capability, development of applications in this area suffers from a lack of appropriate sensor technology. In particular there is a lack of visual and tactile transducers that are well adapted for manufacturing applications. Work is underway at the University of Brighton to develop sensors which can be easily and intimately integrated into the control of robots and other processing machinery in the short to medium timescale [26]. The aim is to

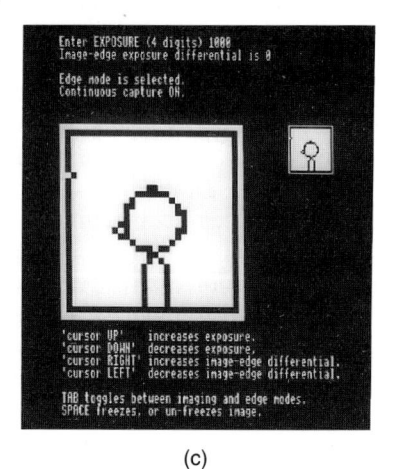

Figure 9.2    Prototype 'smart' custom image sensing array and typical images: (a) the 32 × 32 binary edge-detecting image sensor chip; (b) a plain binary image; (c) a corresponding edge-map, produced from the chip without any further processing. *(Photographs courtesy the University of Brighton)*

produce 'smart' image sensors that provide the machine with visual feedback regarding position and/or status, but without involving the host processor in excessive computational load (see figure 9.2).

In terms of the generic model identified earlier, there are clear advantages to be gained if smart sensors could be developed which could process image data to the point where spatial correlation between neighbouring pixels is lost, i.e. up to and including feature extraction. This would allow the best use to be made of parallel processing architectures within the sensor itself and yet would ensure that the host controller was only required to deal with relatively high level information from its sensors, such as the number of objects in view, their position and orientation, etc. This, in turn, would ensure that visual sensing could be realistically incorporated into position control of the machine itself – this is 'visual servoing' [27, 28]. Ultimately machines that were so equipped would not have to rely upon high mechanical precision to achieve repeatability and accuracy in their tasks and so could be made lighter, smaller, faster and/or cheaper [29, 30].

Custom VLSI design strategies are being investigated to ensure that such sensors are well matched to the needs of machine, rather than human perception. It is also necessary to ensure that the cost of (necessarily) customisable smart sensors is not prohibitive to automation machine manufacturers. This could be achieved by designing field-programmability into generic smart sensing architectures. The first product of this work is a 32 × 32 binary image sensor that performs real-time edge-map production by exploiting the inherent parallelism that exists within the focal plane of the photosites themselves. This has shown itself to be able to operate at over 500 frames per second [31]. Although far from ideal, this prototype 'smart' sensor clearly indicates the potential achievable by *custom* sensor development.

## 9.2 Space exploration

While it is clear that image processing was not invented there, NASA's JPL certainly provided a focus for the development of the field. Dr Robert Nathan, originally a researcher in X-ray crystallography, joined NASA's JPL in 1959. He began to utilise image processing techniques to obtain the maximum information from images sent back by unmanned lunar and planetary exploration spacecraft [32].

Originally, image processing was used on the Ranger missions for the restoration of images suffering geometric and photometric distortions, using pre-flight calibration data. Other applications included contrast enhancement and the removal of periodic noise patterns and linear smearing.

At the time the US space programme was under severe threat of cancellation owing to a series of unfortunate failures. In 1963 the objectives of the Ranger missions were switched from lunar science to lunar surveying in order to support the Apollo programme. This meant that all non-visual experiments were deleted from the craft. Therefore the role of image processing became more central to the successful achievement of the mission objectives. Deconvolution techniques were developed to remove blur introduced by the camera optics and motion of the spacecraft, and these had impressive effects upon image clarity.

The use of image processing went from strength to strength. Its contribution to the success of each space-flight became greater as the probes were sent farther afield because of the increasing intrinsic value of the data contained within the returned images.

*Mariner 4* flew past Mars in November 1964 and returned its video data in digital form, thus signalling the birth of a new branch of image processing for the support of image transmission. *Mariner Mars* 1969 was the first mission to use image encoding techniques embedded within its video data transmission system. This was used to improve the efficiency of the transmission of the low contrast images expected from that planet [33].

By the time of the 1971 *Mariner Mars* orbiter mission, developments in digital compression and transmission techniques ensured that images could be returned to Earth faster than they could be processed. Since this time, increases in available computer power have generally been more than matched by more ambitious and demanding applications. So this processing 'bottleneck' remains a fundamental issue to the present day.

Activity in space exploration declined from the mid-1970s onwards but image processing continues to play a vital part in the work that goes on. For example, it allows the maximum knowledge to be gleaned from the 1970s technology aboard the *Voyager 2* spacecraft even as it nears the edge of the solar system. The European *Giotto* spacecraft encountered Halley's comet and sent back the first close-up pictures of its core. The *Galileo* spacecraft, currently en-route to Jupiter, carries a CCD camera as a primary means of data acquisition. There is also an expanding role for image processing in space-based astronomy.

## 9.3 Astronomy

In recent times the best known contribution to astronomy made by image processing has been its critical role in salvaging the ill-fated Hubble Space Telescope (HST). This was necessitated by a manufacturing error in the primary mirror of the telescope which, although only of the order of two microns, causes severe spherical aberration in the visible wavelengths. The resultant blurring means that the images obtained without the aid of image processing are worse than those from the earth-bound telescopes that HST was designed to supersede.

Fortunately however, de-convolution can go some way towards restoring the lost resolution in areas of the sky containing well known objects. These are used as a template to calibrate the process. Ultimately the optical errors are to be properly corrected by fitting additional lenses and mirrors, but in the meantime the 2.5 billion dollar HST programme can continue to make a worthwhile contribution to the exploration of the universe.

Despite this recent high-profile activity, image processing has been steadily growing in its importance to astronomers over a number of years. In fact the de-convolution techniques used to salvage HST data were initially pioneered by radio astronomers. Furthermore, there has been a rapid increase in the uptake of solid-state image sensors as an alternative to photographic plates in capturing data for subsequent analysis. Charge-coupled device (CCD) sensors combine the advantages of greater sensitivity and wider spectral range over conventional photographic emulsions. Their inherent compatibility with computerised processing is also a considerable advantage.

The CCDs used for astronomy are specialised types with far higher resolution than those used in conventional television applications. Loral Fairchild recently introduced a $4096 \times 4096$ pixel sensor which is among the largest active area imaging devices available (see section 3.2.3). The sensor must be cooled by liquid nitrogen to minimise the thermally induced 'dark current' as very lengthy exposures are often used to maximise the light gathered from faint objects. The parallel/serial architecture is used to maximise the sensitive area of each pixel and minimise the effects of aliasing. Mechanical shutters are used to eliminate smearing that might otherwise blur the image as it is read out from such an architecture.

Simple radiometric de-calibration techniques are used to eliminate the pixel–pixel sensitivity variations across the sensor's field of view. A subtractive technique is also used to minimise the effects of the sensor's 'electronic signature' brought about by amplifier noise, for example.

In order to improve their short wavelength performance some CCDs are fabricated on specially thinned substrates and then illuminated from the rear. This has the unfortunate side-effect of superimposing interference fringes on the image, but these can be removed by a process known as 'adaptive modal filtering' [34].

After correction for known defects in the image acquisition apparatus, image enhancement, filtering, pseudo-colouring, etc. techniques may be used at the astronomer's discretion, depending upon the final application of the data.

There is also much interest in astronomy in the infra-red wavelengths. This is because the attenuation due to interstellar dust and the Earth's atmosphere is dramatically reduced at wavelengths even as short as 2.2 μm. Another factor is the 'Doppler effect' which causes visible wavelength emissions from rapidly expanding parts of the universe to shift into the IR region.

Much of the work at these wavelengths is done by scanning a single, cryogenic detector across the sky, but area imaging devices made from new materials such as indium antimonide (InSb) and mercury–cadmium–telluride (HgCdTe) are under investigation. These materials, originally developed for military applications, are notoriously difficult to fabricate reliably into high resolution arrays at this time, but even relatively low resolution arrays are likely to be of interest to astronomers.

## 9.4 Diagnostic medical imaging

As noted earlier, medical imaging was an early beneficiary of the interest in image processing, with some of the first work being undertaken at JPL. In this field the major motivating factor is to eliminate the necessity for invasive surgery as far as possible, since this always involves some trauma to the patient as well as an inevitable element of risk. See figure 9.3.

### 9.4.1 Medical image processing

The first applications of image processing to the medical field involved restoration and enhancement of conventional radiograms and were thus direct spin-offs from the space exploration programme.

When converted to digital form it is possible to remove noise elements from X-ray images, to enhance their contrast in order to aid interpretation and to remove blurring caused by unwanted movement of the patient. This form of representation also makes it very easy to provide physicians with tools allowing them to accurately measure the extent of tumours, and other significant features.

A slightly more sophisticated technique, which illustrates clearly the benefits of systemic exploitation of scene constraints, is called 'digital subtraction angiography'. This involves taking two radiograms of an area, one as a reference image and the other after injection of an X-ray-opaque dye into the relevant blood-vessels. When one image is subtracted from the other the result is a high contrast image of the blood-vessels, with all other features suppressed. This is an excellent tool for detecting thromboses, aneurysms and other defects of the vascular system. It is also increasingly being used in real-time to assist surgical procedures carried out entirely via the vascular system – e.g. unblocking of cardiac arteries by means of inflatable 'balloons'.

Another technique that involves taking several images of the same area of the body allows unwanted objects to be eliminated without the use of dyes. In this case the images are taken from different viewpoints and then averaged to produce the final result. This is a very simple forerunner of the

Image Processing Case Histories 239

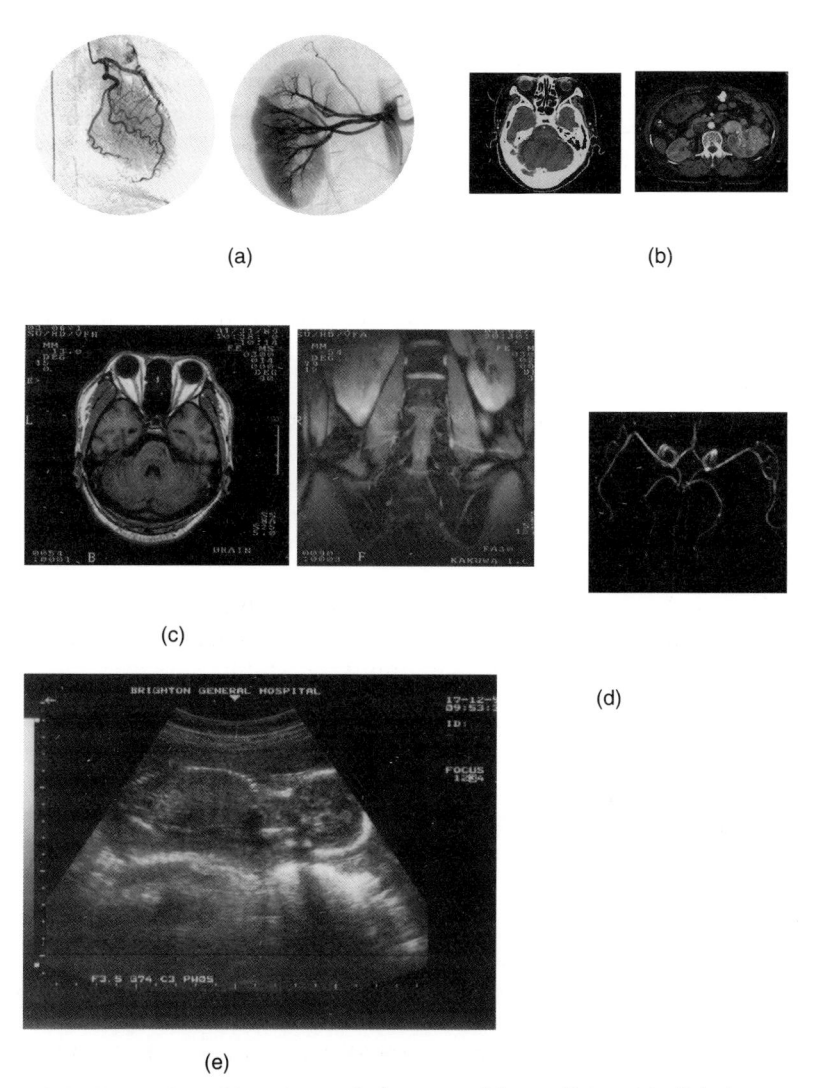

Figure 9.3 Examples of imaging techniques used in medicine: (a) digital subtraction angiograph of the vascular systems of the heart and kidney; (b) X-Ray CT images of skull (posterior fossa, 2 mm slice thickness) and abdomen (through kidney, 10 mm slice thickness); (c) magnetic resonance images of head (T1-weighted, sagittal section) and spinal nerve root (MR myelography, coronal section); (d) 3-D MRI image of the vascular system of the brain (Fine MIP); (e) ultrasound image of an approximately 10-week old foetus (B-scan).*(Pictures (a)–(d) courtesy of Toshiba Corporation, Medical Systems Division; picture (e) courtesy of Hannah Awcock.)*

sophisticated image reconstruction techniques which have now become such powerful diagnostic tools.

### 9.4.2 *Medical image reconstruction*

This second class of applications involves the use of computers to reconstruct two-dimensional images which could not be realised in any other way. In particular they are concerned with 'tomography', which means the generation of images of a slice through the body. This field relies heavily upon the Radon transform, named after the Austrian mathematician who proved that a complete set of 1-D projections of continuous 2-D functions contains all of the information present in the original function [35].

All computerised tomographic (CT) techniques rely upon making measurements of some form of radiation along a circular scan path around the outside of the patient. These data must be combined in a powerful computer which collates the information into a cross-sectional slice ready for display as a conventional image. However, there are several forms of radiation that may be employed to acquire these data, each with its own advantages and disadvantages:

*X-rays*

A thin, collimated beam of X-rays is emitted on one side of the patient and a single detector is placed diametrically opposite. The beam travels in a straight line and is differentially attenuated by bone, soft tissue, etc. in the normal way.

The technique is straightforward and well established, but the radiation is harmful, so doses must be kept low. X-rays give excellent rendition of bone, but images of soft tissue suffer from poor contrast.

*Magnetic resonance imaging (MRI)*

The patient is placed in a very strong magnetic field which aligns all of the hydrogen nuclei in the body. A second magnetic field is applied at right angles to the first, offsetting the alignment of the nuclei. When the second field is removed they revert to their original alignment giving off a characteristic radio-frequency signal depending upon the type of tissue they are in. Billions of these signals are detected and processed to produce images of stunning clarity.

This is a complex operation involving very powerful and relatively expensive computers, but the magnetic field is believed to be completely harmless to humans. As the technique relies upon the hydrogen atoms in water molecules it gives excellent rendition of soft tissue, but poor bone visibility.

The non-invasive nature of this technology coupled with its incredible effectiveness looks set to ensure that it will eclipse all other forms of CT scanning. It has even been suggested that it should be used for routine screening in public health programmes. However this kind of application will not become viable until computing power improves to the point where it allows the duration of a typical scanning session to be reduced from the current 45 minutes, to just a few minutes.

*Ultrasound*

Pulses of sonic energy waves are coupled into the patient's body from an array of transducers via a gel. The waves propagate with characteristics similar to those of light but the technique relies upon reflection rather than transmission. At each interface between tissue of different density some of the sound energy is reflected. The reflections are collected by the same transducers that emitted the pulses and the delay encodes the depth of the interface, while the strength characterises the nature of it. The transducer array is often manually scanned to produce an interactive, real-time image of the patient.

The equipment is relatively cheap and the radiation is considered harmless enough to be used routinely on pregnant women. However, the long wavelength of the radiation limits the ultimate resolution that is achievable. It also has a niche application in the use of Doppler shift measurement techniques to image blood flow, cheaply and in real-time.

*Gamma-rays*

In this case the source of radiation is arranged to be *inside* the patient. A carrier dye 'tagged' with a low-energy radioactive source is either injected into, or ingested by, the patient. The nature of the dye depends upon the organ under investigation. After the dye has travelled to the desired site, a detector consisting of a collimated scintillation crystal and photo-multiplier detector counts the gamma particles emitted along its axis. The detector is rotated around the patient to build up the 2-D cross-section.

Careful choice of the carrier allows good selectivity of the organs under investigation and the technique can be used to study the patient's metabolism. However gamma-rays are potentially dangerous and so they must be used with care.

*Three-dimensional reconstruction*

Any tomographic imaging process can be used to build up a series of cross-sections taken at intervals along the axis of the patient. If these are combined in a suitable database the physician is offered the unprecedented opportunity to visually examine the internal tissue and bone of the patient in 3-D [36, 37]. Even though the data originate from a living patient, all of the tools of modern computer graphics can be brought to bear in order to aid the diagnosis of disease. For example, each organ or structure can be shaded in a different colour. Alternatively, a notional internal light source can be applied to shade the structures and reveal minute textural differences. Such a 'volume-rendering technique' has been developed jointly by the Johns Hopkins Medical Institutions, Philips Medical Systems and Pixar in the USA [38].

New techniques in MRI allow major blood vessels in the brain and the rest of the body to be imaged in 3-D, without being obscured by other tissue. This relies upon the flow of blood inside these vessels acting as a self-contrasting medium and can even be configured to image blood flowing only in a chosen direction. Thus the venous system can be distinguished from the arterial system, making certain types of diagnosis utterly unambiguous. This technique can completely replace the use of invasive angiograms.

Other recent advances include the capability to produce 3-D 'movies' of any part of the body. This looks especially valuable for studying disease of the heart, which can be seen pumping in real-time (although the generation of the 'movie' is an off-line, non-real-time process at present).

*Medical thermography*

The skin is the largest organ in the body and it plays an important role in maintaining thermal equilibrium through heat energy exchange at its surface by radiation, convection, conduction and evaporation. Studies of the process by which the thermal energy is transferred from deeper tissues to the skin for subsequent exchange with the environment reveal a lot about physiological mechanisms in health and disease.

The recent developments in thermal imaging, achieved mainly through military research programmes, have provided tools to investigate the skin in this way. Real-time 2-D thermal imaging arrays are now being made available. This is achieved by laminating specialised sensing materials (pyroelectric ceramics) onto silicon substrates containing processing electronics [39].

Although the resolution of such sensors is limited at this time, they offer advantages over scanned sensing elements. In particular, a 2-D array can view the whole region of interest all the time and can thus be used for dynamic studies in which the spatial distribution of heat must be recorded as a function of time. Since such a camera can also acquire static images, it can be thought of as a more flexible instrument than a scanner, provided that high spatial resolution is not of primary importance.

Similar arguments can also be applied to 2-D gamma camera systems which have been used to study dynamically the pumping action of the heart.

## 9.5 Scientific analysis

X-ray crystallography was among the earliest applications of image processing. Computers were used to determine the structure of crystals by numerical analysis of X-ray diffraction patterns. Since that time there have been many areas of scientific research and application that have benefited from computerised analysis of data presented as a 2-D image.

It is now possible to buy 2-D image analysis workstations from a number of suppliers which feature a wide range of general-purpose image analysis tools for use by scientists who have no previous experience of image processing. A cross-section of existing applications covers many scientific fields and includes:

- chromatography
- chromosome analysis (and abnormal chromosome detection)
- analysis of electrophoresis gels
- intracellular ionic analysis
- autoradiographic analysis
- bacterial colony counting
- real-time tracking of flagellated bacteria
- analysis of wear particle size distribution in lubricating oils
- asbestos fibre analysis
- cytology
- pathology
- metallography
- petrology
- scanning electron microscopy.

As a specific example, intracellular ionic analysis has provided biological and medical researchers with an invaluable tool – the ability to study the chemical function of *living* cells in real-time. This is achieved by 'loading' special dyes into cells which then fluoresce to reveal the minute concentrations of important intracellular ions such as calcium, sodium and chlorine [40].

Another dye called BCECF allows intracellular pH to be measured. It is excited at 440 nm and 490 nm wavelengths and then fluoresces in the range 510–520 nm. However, the fluorescent response to 490 nm excitation is affected by local pH, while that to 440 nm excitation is not. Accurate estimates of pH can be therefore be derived, independent of variable dye concentration, by taking the ratio of the output resulting from the two forms of excitation. This principle is known as 'fluorescence ratio imaging', and is vital to the successful exploitation of optical intracellular 'probes'.

Intensifiers are used with CCD video cameras to acquire 2-D images of the faintly fluorescent scenes. Real-time 2-D analysis of these images is a valuable addition to traditional photometric techniques because some of the most important messages about cellular function are coded in spatial as well as temporal form. In all, this range of applications represents an inspiring example of the critically beneficial interaction of carefully contrived scene constraints with appropriate image acquisition technology.

The necessity to automate the analysis of scientific data has also had a significant role to play in the development of image interpretation techniques. For example, Ledley [41] applied syntactic pattern recognition to his study of chromosomes as early as 1964.

An even more extreme example concerns the development of the range of structure analysis techniques known as 'mathematical morphology'. This arose out of work on the petrography of iron ores and porous media by Serra and Matheron at the Paris School of Mines. This was necessary because they found that the notion of structure within images was highly dependent upon the purpose to which the interpreted image was to be put. Mathematical morphology addresses this issue by introducing the concept of 'structuring elements' which can be interactively selected by the investigator to suit the intended interpretation – see section 6.5.

## 9.6 Military guidance and reconnaissance

Surely there are few who can fail to have been impressed, however reluctantly, by the uncanny accuracy of cruise missile technology. This was demonstrated so graphically by the sight of these craft flying through the streets of Baghdad during 'Operation Desert Storm' in 1991, and is achieved largely with the aid of image processing.

No practical system could achieve such accuracy by dead-reckoning alone, so cruise missiles actively navigate themselves using landmarks and way points that they observe along their route. As the missile flies along it uses radar to gather images of the local terrain. These are processed and interpreted so that they can be compared to maps of the area preprogrammed into the missile's guidance system. These maps, in turn, are often gathered from hostile territory by military reconnaissance satellites.

Radar is used by the missile as a source of imagery because, being 'active', it is unaffected by darkness and can penetrate adverse weather conditions. It is also relatively unaffected by the changes that take place in the landscape with the seasons. For example, snowfall can make terrain look totally different under normal visible light, but obviously does not change the basic underlying structure of the area. Therefore, the use of radar to acquire images can be seen as yet another particularly effective exploitation of interaction between *a priori* knowledge of the scene, imposed structured 'illumination' and image acquisition technology.

Cruise missiles are considered cost-effective use of technology because the type of deep-penetration mission on which they are employed is dangerous in terms of the vulnerability of manned aircraft, which would obviously be expected to complete the return flight! Manned aircraft also suffer the disadvantage that they must be relatively large to accommodate the crew and this further reduces the probability of mission success. Similar problems are faced by land forces who have many jobs which would be better suited to unmanned vehicles which can be unarmoured and therefore smaller, faster and cheaper than their manned equivalents. Typical applications include:

- nuclear, biological and chemical contamination monitoring
- minefield reconnaissance
- decoys
- front-line radio relay, electronic counter-measure and surveillance duties
- anti-tank weapon platform

Some of these jobs are currently being addressed by the UK Defence Research Association's (DRA) MARDI (Mobile Advanced Robotics Defence Initiative) and ROVA (Road Vehicle Autonomous) research programmes [42].

MARDI is concerned essentially with 'telerobotic' operation of a tracked, cross-country vehicle and so does not require considerable interpretation of any visual data on-board. However, the bandwidth of communication links is a problem, as always for the transmission of video images, so image compression is a major feature. The programme also ultimately aims to tackle limited autonomy, in the form of 'blind' driving between designated way points.

ROVA is a much more ambitious programme which has already succeeded in autonomously navigating a 3.5 tonne test vehicle at 40 km/h on simple roads, free of obstacles. It uses a model-based system with two independent models of the vehicle dynamics and the road geometry. These are fed from a camera and interpretation system which identifies the boundaries of the roadway, both near to the vehicle and in the far distance. Steering commands are updated at 12 Hz and are based upon 'Kalman-filtered' observations of the nearby road boundaries supplemented by 'anticipatory control' derived from observations of the distant boundaries. The image processing system consists of an array of transputers, currently fifteen in number, although modular design allows this to be extended up to 160, as necessary.

Future investigations aim to improve boundary detection algorithms and develop multiple edge trackers and obstacle detection. Clearly however, there is some considerable way to go before such a vehicle could navigate unaided across a battle-torn landscape, without the aid of man-made roads.

*Thermal imaging*

The military have been prime movers in the development of thermal imaging technology. This is because of its innate ability to acquire useful images at night and under atmospheric conditions such as fog and smoke that would blind imaging systems based upon normal visible light. See figure 9.4.

The Earth's atmosphere features 'windows' of especially low attenuation to certain wavelengths. The medium IR window passes wavelengths from 3 to 5 μm and the long IR window passes from 8 to 13 μm (there are also windows at 1000 μm and 3000 μm which are exploited by radar signals). Natural sources of thermal IR are found to peak at approximately 10 μm, although useful passive imaging can be achieved in the medium IR window of surfaces at temperatures above 30°C (303°K). Passive imaging is concerned with detection of naturally occurring IR that is either generated or re-radiated by the objects of interest. Active imaging relies upon illumination of the scene using suitable IR sources which are then reflected by objects.

Sensitivity to heat energy also brings interesting new capabilities which make life more difficult for any unwitting adversary. Thermal imagers can easily pick out the heat of human bodies from the ambience of jungle cover and other vegetation, rendering traditional camouflage techniques ineffective. The heat from the exhaust of a tank engine, etc. can be easily detected for a considerable time after the vehicle has been halted and camouflaged. Sophisticated systems are even sensitive enough to be able to detect the recent passage of vehicles, thanks to friction-heating of the ground by their wheels or tracks!

*Airborne reconnaissance*

Reconnaissance has an important role in military strategy. The Joint Surveillance, Target Acquisition and Reconnaissance System (JSTARS) uses radar to determine movements of troops and armaments during times of high tension between adversaries. Any sudden movements towards a friendly border can trigger an appropriate defensive reaction. However, during times of conflict it is often necessary to get detailed reconnaissance from suspicious locations within enemy territory.

Very often, the best way to do this is through the use of high-performance fighter aircraft specially adapted to carry reconnaissance equipment. They can fly fast and low over the region of interest in order to acquire high

Image Processing Case Histories 247

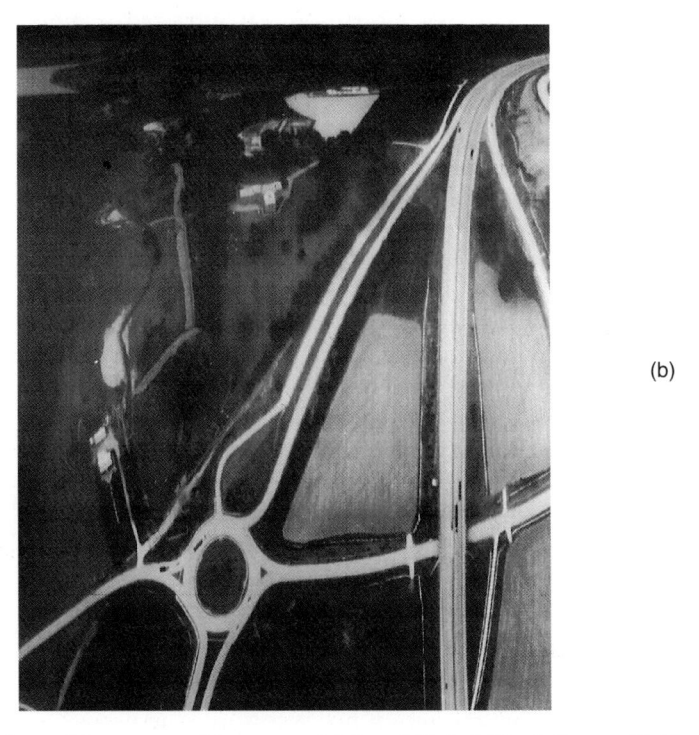

Figure 9.4   Infra-red imagers and imagery: (a) Loral Fairchild platinum silicide staring array camera; (b) infra-red linescan imagery of a motorway complex taken at low altitude (< 1000 feet) *(picture (a) courtesy of Loral Fairchild Image Sensors (USA) and Optimum Vision; picture (b) courtesy of Computing Devices Company Ltd (UK))*

resolution images which allow the movements of the enemy to be determined. This kind of mission has a number of important characteristics. Firstly, since it involves flying close to enemy held territory, it is dangerous. Secondly, the information gathered has a very short 'shelf-life' because of the dynamic nature of modern warfare. Therefore the imagery gathered must be interpreted and acted upon as quickly as possible.

Traditionally this role was fulfilled by cameras that formed a conventional photographic image that needed traditional 'wet film' processing before specialist 'photo interpreters' could do their job. Obviously this has the disadvantage that imagery cannot be acted upon until the aircraft has flown home and the film has been unloaded and processed. Until this time it is impossible to know whether the mission has successfully acquired images of the region of interest but even if it has, the information derived will inevitably be somewhat 'stale'.

In the early 1980s this problem was tackled by Computing Devices Company Ltd who developed the world's first all-electronic tactical reconnaissance system for the Tornado multi-role combat aircraft. The system uses a number of electro-optical infra-red scanning systems and records the resulting imagery onto specially modified videotape recorders. Since the Tornado has a crew of two, the navigator can be provided with the facility to view the imagery recorded from the target to check that it carries appropriately useful information and to identify interesting events which should be brought to the attention of the interpreters. In principle it would be possible for immediate decisions to be based upon the navigator's observations within the aircraft, but a mission over hostile territory does not typically make for a cockpit environment compatible with cool deliberation! However, it is feasible to transmit interesting sequences of imagery back to base via data links *while* the aircraft is returning home.

The Tornado reconnaissance aircraft carries three scanners, an IRLS (IR linescan) slung under the fuselage and two sideways-looking IR (SLIR). The IRLS scans across the track of the aircraft from horizon to horizon and thus allows a two-dimensional swath to be captured as the aircraft flies forward. The imagery is corrected for the geometric 'bow-tie' distortion which results from the variable 'footprint' of a sensor photosite as it is scanned from directly beneath the aircraft's track to the horizon. Therefore a pixel captured from near the horizon should not be displayed with an area equal to one captured from near the aircraft track. This is particularly important because it is often impossible (or unwise) to fly directly over the suspected hostile territory, meaning that much of the useful information is contained in the relatively low-resolution pixels towards the horizon.

The SLIRs are framing sensors that look out from either side of the aircraft. Their imagery is arranged to scroll from left to right on the navigator's monitor while the IRLS imagery scrolls from top to bottom. All imagery is electronically annotated with the aircraft's height, velocity, position and heading at regular intervals. When the data are returned to the ground either by radio link, or physically, they can be further enhanced by a ground-based image processing workstation. This reconnaissance system proved itself to be a great asset to the allies during the Gulf War of 1991.

A third type of IR scanner is used by other aircraft, helicopters and civilian agencies. It is called FLIR (forward-looking IR). Like the SLIR it is a framing sensor but it may be mounted in the nose of an aircraft or slung on a gimbal mount underneath a helicopter or in ground-based surveillance applications. The AGEMA 'Thermovision 1000' is a FLIR operating in the 8–12 µm band. It has a 12-bit 'luminance' resolution and a spatial resolution that allows it to detect a human on land at 3 km, a 30-foot boat at 10 km or an aircraft at 60 km [43]. It employs an electromechanical scanning system to focus radiation onto a five element SPRITE (signal processing in the element) array [44]. This type of detector allows very good signal-to-noise performance to be achieved from a scanning system.

So-called 'staring' arrays (basically a 2-D dimensional detector matrix laid out like a convention CCD TV sensor) generally offer better signal-to-noise ratios, but at the present state of development cannot compete on spatial resolution with scanning mechanisms. GEC–Marconi Material Technology Group have developed a $100 \times 100$ element pyroelectric staring array camera which also operates in the longwave IR window. Loral Fairchild have released a commercial range of platinum silicide video sensors based upon their standard CCD manufacturing technology, which offer up to $488 \times 512$ spatial resolution. However, these operate in the less ideal medium IR window and require cryogenic cooling down to 88 K [45].

*Civilian spin-offs*

Recent developments in 'thermography' have resulted in affordable civilian versions of thermal cameras with 0.1°C temperature resolution and spatial resolutions up to 140 pixels per line. They are used by police and rescue services in searching for missing persons in open country, earthquake victims buried in rubble, etc. Firemen also use simple thermal cameras to identify the heart of fires inside buildings that are filled with dense smoke. Drug enforcement, border control, oil-spill and forest-fire detection and other general surveillance tasks are also among the growing range of applications for this spin-off of military research.

In the electricity supply industry, maintenance engineers use thermography to detect overheated components prior to failure. Since a supply disconnection, or 'outage', is a source of major cost and inconvenience they must be avoided if at all possible. Being a non-contact inspection technique, thermography allows engineers quickly, safely and cheaply to detect and diagnose sources of excessive heat loss caused by problems such as loose connections, oxidation and corrosion. Helicopter-borne imagers can be used to perform quick annual surveys of long runs of power cable and distribution equipment. Any suspicious components can then be further investigated by ground-based maintenance teams using hand-held imagers. These devices require no specialised training to allow engineers quickly to assess likely sources of trouble. If required, an infra-red spot measurement device may then be used to obtain an accurate digital readout of component temperature.

## 9.7 Remote sensing

Early applications of remote sensing were largely concerned with either meteorology or military intelligence gathering. Although processed weather satellite imagery is already beginning to be taken for granted by the public, governments and environmental protection agencies are beginning to recognise the wide ranging benefits that its other civilian applications can bring to mankind.

The 1980s saw an expansion of the availability of data for such applications with the steady growth of the US LANDSAT programme and the launch of the French SPOT satellite. This expansion is set to continue into the 1990s and many more launches are planned. Despite the vast amount of data contained in a typical image, the relentless improvement in computer cost/performance ratios means that these data can be exploited by a rapidly expanding user base.

### 9.7.1 *The nature of remote sensing*

Remote sensing is concerned with the gathering of data about the Earth and its environment by means of visible and non-visible electromagnetic radiation. The data are acquired by photographic means or electronic scanners carried aloft by satellites, specialised high-altitude reconnaissance aircraft and conventional low-altitude aircraft.

Multi-spectral data are gathered, with as many as seven bands being acquired simultaneously. Ultra-violet, visible, infra-red and thermal wavelengths are collected by passive sensors. Active sensors exploit microwave radiation in the form of synthetic aperture radar (SAR). This can detect features which are invisible to optical cameras, including wind patterns [46].

Electronic scanners are becoming increasingly dominant over photographic techniques because they are better suited to the acquisition of radiation outside the visible and IR bands. Although the resolution achievable is generally not as good as that of photographic film, the large volumes of data involved can be collected, stored, transmitted and processed more easily.

Development trends in electronic sensors are moving towards an expansion of the number of spectral bands such that each pixel in the image will effectively yield a full spectral analysis of the radiation emitted from the region imaged. Existing devices are already capable of intensity resolution of greater than eight bits in each of the spectral bands.

Typical resolution is of the order of 1–2 km per pixel side but the SPOT satellite offers the best resolution currently available in the civilian domain; each pixel is a mere 10 × 10 metres. SAR images from satellites such as the European Space Agency's ERS-1 allow even higher resolutions to be achieved, such that wave activity and shipping movements can be readily resolved.

Aircraft may obviously fly in essentially random paths over the ground but satellites operate in one of two modes: 'geostationary' or 'polar orbiting'. Geostationary satellites orbit synchronously with the rotation of the Earth at an altitude of 35 860 km above its surface. Therefore they always look down at the same region of the Earth's surface, but the altitude limits the resolution achievable.

If the whole of the planet is to be covered by one satellite then a polar orbit is required such that a 'raster scan' of the surface is effectively performed. The orbit of the satellite scans roughly in the longitude axis and a combination of linescan sensing in the satellite and the rotation of the earth achieves scanning in the latitude axis. Thus the satellite observes a 'swath' of land as it orbits and images it line by line. The swaths from subsequent orbits can be arranged to overlap slightly so that total coverage is achieved.

Such satellites orbit at a height of approximately 1000 km and can therefore achieve significantly better resolution than geostationary satellites. In practice the orbit does not pass exactly over the poles so complete and contiguous coverage in often impossible to achieve during a single rotation of the earth. Data can also be lost owing to cloud cover, transmission noise, etc.

### 9.7.2 *Applications of remote sensing*

As already noted there are many fields beyond military intelligence which can benefit from remote sensing technology. Major application areas include the following.

*Meteorology*

The application of satellite data to short-term weather forecasting is explicitly obvious and well known. However, satellite data are important in the study of long-term climatic changes such as global warming, etc. Much of this work is concerned with the study of the currents in the oceans and fluxes of heat and water vapour in the ocean/atmosphere boundary. This is important because more heat enters the atmosphere from the oceans than by direct solar heating and the top five metres of water has the same thermal capacity as the whole atmosphere.

*Natural resource location*

Remote sensing is beginning to play a significant role in the development of the economies of third world countries. For example, the government of India is using its IRS I and II satellites to locate new sources of drinking water. The satellite can easily identify the lush vegetation which thrives over underground springs. These data are correlated with maps of the area and are used to plan the development of new towns and villages which will then have a reliable water supply. This is just one aspect of a geographical information system (GIS) which helps to plan development of India's resources. It relies heavily upon satellite imagery as the only cost-effective source of data about the huge Indian sub-continent.

The capability to survey large areas of inaccessible and even unmapped land efficiently is also being exploited by major oil and mining companies to identify new sources of mineral wealth. Supervised learning allows the imagery obtained to be accurately correlated with particular geological features that are characteristic of regions likely to have desirable properties. These can be automatically segmented and then used to direct the activities of ground-based survey teams far more effectively.

*Environmental monitoring*

Satellites find many other applications in monitoring large or remote tracts of land to determine its existing use or future prospects, for example:

- the European Community (EC) is using satellites to check that its system of farming subsidies is not being abused now that it is sponsoring farmers to let their fields lie fallow;
- the Thai Narcotics Control Bureau is using images derived from the SPOT satellite to identify and pinpoint poppy fields within the so-called 'golden triangle' area of northern Thailand [47];
- in the USA, the infringement of National Park boundaries by cultivated forestry is being carefully monitored, as is the destruction of the rain forests in the Amazon and elsewhere;
- regular checks are being made on the rate and extent of the expansion of the Sahara Desert which is resulting from land mis-management and climatic changes.

When natural or man-made disasters strike, such as floods in Bangladesh or the oil spillages following the Gulf War, satellites can help to assess the extent of the damage quickly. More subtle environmental damage such as that caused to marine wildlife by long-term exposure to domestic sewage or cooling water discharged from power stations is also being studied from space.

*Cartography*

It may be surprising to learn that the world is still not adequately mapped. There are large tracts of land such as central Yemen which are only sparsely mapped. Modern satellites allow such surveying to be made affordable, but land relief is hard to quantify from a two-dimensional image. However, it is now possible to use stereo imaging to determine accurate height information remotely. At present the SPOT satellite is the only one that can achieve this and it combines images from two passes in order to do so. Unfortunately the swaths used are seldom consecutive in time and can be many months apart. Therefore the land surveyed can look very different in each view because of seasonal changes and this makes automatic interpretation difficult.

However, British Aerospace has proposed an Optical Mapping Instrument satellite which will improve on the resolution achieved by SPOT and gather both stereo views simultaneously [48]. The acquisition of such data will make it possible to automatically generate 'digital elevation maps (DEMs)' and thence computer visualisations of local terrain from any ground-based location. This will allow the impact of contentious issues such as major civil engineering projects and re-forestation of mountainous scenery to be evaluated *before* work begins. The military have other uses for this technology. They use it to allow visualisation of low-level attacks on hostile terrain from the pilot's viewpoint [49]. It is also possible to incorporate intelligence regarding missile sites and radars so that protection afforded by hills and valleys from ground-based anti-aircraft fire can be fully exploited.

Some areas need to be continually re-mapped because they contain volatile features such as sand-banks or sand-dunes. Sand-bank movements can determine the viability of estuarine projects such as tidal power barrages, but conventional hydrographic surveying is very costly and time consuming. By contrast, satellites can perform this task quickly and efficiently.

### 9.7.3 Image processing for remotely sensed data

At its most basic level, processing of remotely sensed imagery simply consists of assigning a grey level to a pixel intensity value so that the sensed radiation can be visualised. However, raw data in this form will require restoration and enhancement to maximise its application potential.

*Geometric transformation*

It is a natural requirement for users of satellite data to want the images that they use to be centred on *their* particular region of interest. However, it is not economic to achieve this by waiting until the track of the satellite passes immediately over the centre of the region before acquiring an image. Therefore it is a common requirement to process images to change the apparent viewpoint.

The curvature and rotation of the Earth, combined with the elevated viewpoint of the satellite, will cause geometric distortion in the images obtained. Since the mechanism of this distortion is known it can be removed by reverse geometric transformations. However, it is not economically viable totally to eliminate all distortion and thus positional errors of up to several kilometres can still remain. At present, these have to be eliminated by manually identifying key points in the image and then using computerised facilities to 'warp' the image, mapping these control points onto their proper coordinates (see section 4.5).

The provision of accurate height data from stereo imaging satellites such as SPOT will improve the accuracy of remote cartography by allowing the generation of 'orthoimages'. These are 2-D maps, geometrically corrected for height-induced distortions [50].

*Contrast stretching and false colouration*

The amount of data that can be displayed by any conventional means for interpretation by a human observer is generally only a small proportion of that recorded from the scene. Limitations in both display media and the human visual system itself give rise to several problems in this area.

The gamut of contrast that a human can detect over a *whole* image easily exceeds the capability of even the best display monitors and image printing technologies. Despite this, neurophysiological testing has shown that humans can distinguish surprisingly few grey levels within contiguous regions of an image. This means that various forms of 'contrast stretching' are necessary (see section 4.2.2). This is concerned with selecting a subset of the grey levels available which convey information that the human finds difficult to discern. This subset is then mapped onto the full range of grey levels made available by the display medium, thus enhancing the image perceived.

Although poor contrast images should normally never be tolerated in circumstances where constraints may be effectively applied, there are cases where severe compression of the contrast range is unavoidable. A good example comes from an atypical 'inner space' application of remote sensing – the discovery of the wreck of the *Titanic* by a team led by Dr R. Ballard. The remotely controlled underwater submersibles *Argo* and *ANGUS* were used to explore the sea-bed at great depths. On-board cameras relayed the scene to the survey vessel at the surface, but the image contrast was poor owing to particulate matter suspended in the water. Contrast stretching proved to be a very effective form of image restoration in this case [51].

An alternative to contrast stretching is 'density slicing'. This overcomes the limitations of the human eye in distinguishing tonal contrast by exploiting its greater sensitivity to colour variations. Therefore a range of intensity values in the data is mapped to a particular colour, and another range to a second easily distinguishable colour, and so on. The intensity ranges can be chosen so that each colour represents a particular type of land use, e.g. forestation, tarmac, crops, etc.

This 'false colouring' is obviously a potent form of image enhancement, but it can be made even more useful in cases where multiple spectra are acquired from the scene. In such cases radiation normally invisible to the human eye can be used to distinguish more reliably between particular types of ground cover. For example, infra-red bands provide valuable information about the lushness of vegetation. If each spectral band is used to drive a separate channel in a colour display system then a 'colour composite' will result which uses colour in an unreal, but helpful way.

*Classification*

In most applications of remote sensing it is necessary to classify different parts of the image because these represent different land usages, terrain types, minerals, etc. This amounts to a very sophisticated form of segmentation, operating as it does on multi-spectral data.

At the present time this technique usually relies upon 'supervised training'. This means that particular 'spectral signatures' perceived by the remote sensor are correlated manually with features evident in archetypal specimen regions found on the ground. When this has been done, a computer can be asked to search the whole image, or set of images, for occurrences of all of the spectral signatures that are of interest.

A recent application illustrates this approach well. The Royal Society for the Protection of Birds (RSPB) in the UK wanted to monitor the estuarine habitats of many species of wildfowl which are under continual threat from pollution and development. Previously this could only be achieved by employing teams of experienced observers over long periods, and then the statistical relevance of such data was in question. So a pilot study was initiated in the estuary of the river Stour to use satellite data to automate the data collection and classification.

Samples were taken of the types of mud particularly favoured as food sources by each species of bird – shelducks, oystercatchers, etc. The coordinates of these mud samples were then correlated with the spectral signature obtained from that region by the satellite. Now it is possible to survey wide areas of estuary automatically to see how the food source of each species varies with time and thus the threat to particular species can be quickly assessed.

Despite the increasing exploitation of satellite data it is worth noting that it is an application which is still generally under-resourced in terms of processing power. Any satellite worth its salt can generate data at a much faster rate than it can be interpreted. This is why there is considerable interest in the *fully* automated interpretation of such imagery. This involves the use of 'unsupervised training' where the data processing system has sufficient intelligence to identify features from their spectral signatures alone.

The aim is to classify with enough confidence to make ground-based cross-checking unnecessary, at least for preliminary, wide-ranging investigations. Paradoxically, it seems likely that the trend for increasing the spectrum of data obtained (and thereby the processing workload) will ultimately aid the viability of fully automated classification, by increasing its reliability.

## 9.8 Information technology systems

Despite the explosive growth in the amount of information in circulation, brought about by what is now known as the 'information technology' revolution, it is estimated that only 5% is held in electronic form. Computers have certainly not yet yielded the paper-less office, despite the high hopes of a decade ago. However, dramatic reductions in equipment costs over the last 2–3 years have finally begun to make document image processing (DIP) an increasingly attractive solution.

### 9.8.1 *The nature of DIP*

DIP can be divided into two major classes of application:

(a) document archive and retrieval;
(b) document work-flow automation.

Archive and retrieval DIP is concerned with simply replacing paper storage and filing by electronic movement of records. It is effective under the following conditions:

- the volume of information is high
- filing space is expensive (e.g. prime office space), yet frequent access to large archives is necessary.
- multiple access to single documents is necessary.

Essentially, efficiency is greatly improved by bringing information to the staff rather than vice versa. Staff reductions of up to 30% have been reported, as well as improved productivity [52]. The costs of photocopying and repeated re-filing are also eliminated. This type of application is therefore the easiest to cost-justify and can have a very short 'pay-back period', and so it is the most common form of DIP at present.

Work-flow automation adds the capability to interpret information contained within the documents, rather than just storing them. This works well when the following application constraints are added to those noted for document archive and retrieval:

- well established (and well understood) administration procedures exist
- information is derived from a variety of sources
- processing is multi-stage (i.e. more than eight staff involved) and the transaction time is long.

Such applications are much more difficult to set up and justify and generally have a longer pay-back period (of the order of 2–3 years). Therefore they are generally restricted to very large businesses. However, in the right environment productivity improvements of as much as 40% have been reported.

### 9.8.2 Applications of DIP

Insurance companies have been pioneers in the field of DIP, attracted by its ability to assist in case processing and address their problems of storing policies, etc. Banks and building societies have also been quick to spot the benefits, especially in the area of improved service to the customer. DIP offers the highly prized ability to achieve 'product differentiation' in a field where many organisations are able to provide a similar range of services. A good example is provided by a private health insurance company which handles 2 million clients with a workload of 1 million claims annually. Much of their interaction with clients takes place over the telephone and instant access to relevant documentation allows them to give a distinctively superior service [53].

The ability of DIP to offer an 'edge' in business sectors where competition is particularly fierce has also been recognised by some companies offering express parcel delivery services. In this industry better customer service means knowing where every parcel is, when it is delivered, who receives it and then making this information available to the customer immediately upon enquiry. This can be achieved by scanning all parcel 'manifests' into a centralised database at the end of every working day. An image of the manifest, together with extracted bar-code and other information, is stored on-line for a period of 10 weeks. The image can then be called up by an enquiry operator within 15 seconds so that copies of signatures, etc. can be verified or faxed to the client as necessary. Eventually the manifest image is moved off-line to a support facility while the database information is kept on-line at the enquiry office. Images can then be retrieved as necessary, but only after the appropriate optical disc is loaded manually at the support office [54].

The engineering industry uses DIP for the management of large collections of pre-CAD engineering drawings. It is also being actively developed as a way of integrating all of the instructions necessary to manufacture complex assemblies in the military and aerospace industries. Engineering drawings can be supplemented with overlays, text and video images to guide workers through the assembly of items which are complex and manufactured infrequently in small batches [55].

Several local government authorities in the UK turned to DIP to help them cope with the huge increase in administration associated with the switch away from property-based 'rates' to the individual 'community charge' as a means of raising revenue. For example, Lancaster City Council had only 50 000 ratepayers, but 100 000 community charge payers [56]. Allied to the problem of generating the community charge documentation itself was the even greater task of coping with the $\approx 30\%$ of the population who were entitled to claim relief from paying the full charge. No wonder the community charge has been rapidly dropped in favour of a new property-based tax! However, now that DIP has found its way into local government administration and experience has been gained of its advantages when dealing with queries from clients, its use seems likely to spread.

Another big potential application area is medical imaging. Despite the advances of computer reconstructed imagery, projection X-ray images are still a major source of diagnostic data. An average hospital produces between 500 and 1000 X-ray images per working day and it is desirable that these be kept for a period of ten years. Not only is there a huge potential to reduce the space required to store such an archive, but also patient care could be improved through superior reliability and speed of retrieval. It also makes sense to be able to combine this 'analogue' data with the digital data derived from CT, MRI, etc. This application is being addressed by a programme known as 'picture archiving and communication systems (PACS)'. It is very demanding of technology because a typical X-ray image occupies around 5 Mbytes when represented digitally. Therefore the typical hospital output suggested above implies an archive capacity of around 10 Terabytes of data over the ten year period.

Storage is not the only challenge – all aspects of the PACS require high performance. Scanners must digitise images to at least $2000 \times 2000$ pixels with an intensity resolution of better than 10 bits. Pilot studies of PACS-based radiology departments have suggested the networks linking the archive to the diagnostic workstations must deliver performance of the order of 140 Mbits/s simultaneously on *all* of the possible routes [57]. The diagnostic workstation itself requires a display resolution of the order of $2560 \times 2048$ with real-time contrast manipulation to overcome the limitations of CRT technologies. In order to perform other image enhancement tasks within an acceptable time, say one third of a second, processing power of the order of 12.5 Mpixels/s is required.

Rapid development of all parts of DIP technology is ensuring that it is becoming a cost-effective investment even for small companies. Small-scale PC-based archive and retrieval systems can now be bought for less than £10 000 while still giving on-line access to around 25 000 pages – the equivalent of five traditional four-drawer filing cabinets [58]. This downward trend in cost is timely in view of a recent EC directive, which seeks to extend the protection of personal data held by organisations to manual files (i.e. paper and microfilm records) as well as computer records [59]. Elimination of paper through DIP might then become a legal requirement to avoid the loss or theft of personal data. The advantage of DIP is that documents cannot be casually viewed without specialised equipment and that its compact storage capability reduces the cost of securing the archive contents. This kind of legislation would significantly affect the legal profession itself, and it is now showing considerable interest in DIP systems. Indeed, it is now being suggested [60] that:

'... document imaging is the first significant advance in IT for lawyers since word processing'.

### 9.8.3 *Image processing for DIP*

As noted above, applications of DIP are critically reliant upon many technologies other than image processing for practical performance. So much so that there is a school of thought which suggests that the name is misleading and that something like 'document image management' or 'electronic image management' is more appropriate. However, there are several image processing techniques that play important roles in this field.

*Image data compression*

A fundamental characteristic of image data files is their large size. It has been noted that this presents problems with storage and transmission of image-based data. Although there is not as much redundancy in natural images as there is in textual information, they can be compressed, but the amount of compression achievable depends upon whether or not loss of quality is acceptable. Archiving may require totally loss-free compression (i.e. other than losses introduced by quantisation) and this limits compression ratios to about 2 : 1. However, when loss in quality is acceptable, such that the reproduced image merely has to be recognisable, ratios of 32 : 1 are feasible (compression ratios given here are for static images).

Until recently, image compression has been dogged by a lack of standardisation, but this situation was dramatically improved in 1986 by the formation of the joint photographic experts group (JPEG). This is a collaborative exercise between ISO (International Organisation for Standardisation) and CCITT (Comité Consultatif International de Téléphonie et Télégraphie) [61]. JPEG concentrates on 'intraframe coding' where each image is considered to be the only one, so redundancy must be detected within this single picture and eliminated to achieve compression. Typical applications for JPEG standard compression include photographic videotex, still picture teleconferencing, slow-scan pictures for security, medical images, newspaper pictures, computer graphics archiving and satellite weather and surveillance pictures (see figure 9.5).

JPEG has been working on standards for image compression based heavily upon the 'discrete cosine transform' (DCT). This technique relies upon splitting the image into blocks (typically 8 × 8) and then transforming the block contents into the spatial frequency domain [62]. There is much less correlation between coefficients in the transform domain than there is between pixels in the original spatial domain. Most of the image 'energy' is concentrated in the lower order coefficients so the higher order coefficients have low 'visual impact'. They can therefore be quantised more crudely than the low-order coefficients, or even omitted altogether, thus achieving compression of the image data content. After this stage the coefficients that are left are further compressed using 'arithmetic' or 'Huffman' coding with no further loss of information.

For high resolution colour images each pixel is normally represented for display purposes by 8-bit words for each of the red, green and blue primary components of the image. However, techniques already used in the transmission of colour television may be exploited to reduce this to 16 bits/pixel. Firstly the *RGB* data are converted to luminance and colour difference signals according to the equations:

Luminance, $Y = 0.3R + 0.59G + 0.11B$

Colour difference (blue), $C_B$ (also known as $U$) $= B - Y$

Colour difference (red), $C_R$ (also known as $V$) $= R - Y$

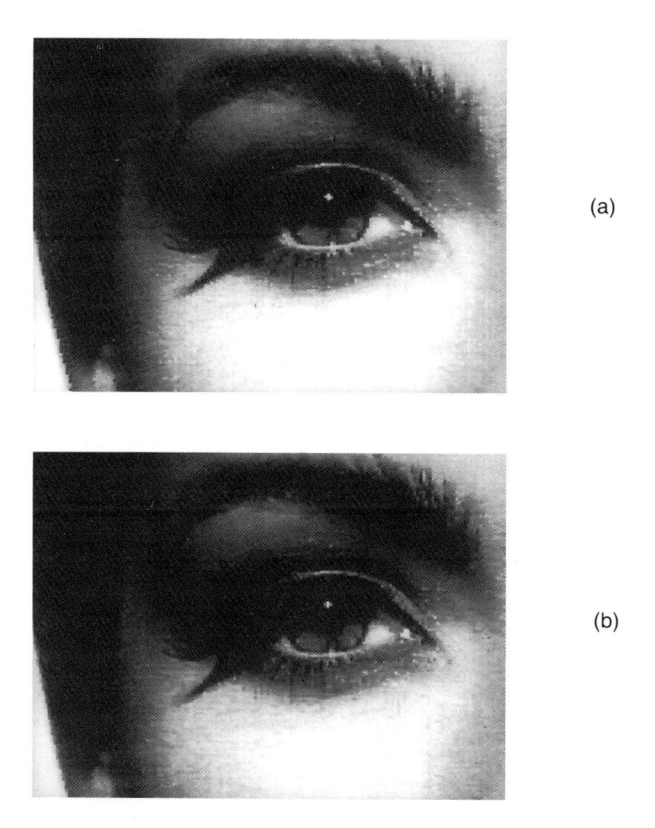

Figure 9.5 JPEG compression applied to a 12-bit (colour) computer digitised image: (a) 160 × 253 section of raw image (12-bits/pixel) magnified 2×; (b) same section of image after compression to ≈3 bits/pixel and then de-compression. Can you see the difference?

This exploits the fact that the human visual system is rather more sensitive to luminance detail than colour detail, making it possible to allocate more bandwidth to the luminance signal at the expense of the colour difference signals. Thus apparent image quality can be maintained with less overall signal bandwidth. This is achieved in the JPEG compression standards by sub-sampling the colour difference signals to reduce their data content. JPEG uses a 4 : 1 : 1 scheme which means that there are four samples of $Y$ to one each for $C_B$ and $C_R$.

Any size of image, with up to 16 bits per pixel, monochrome or colour, can be compressed under the JPEG standard and the degree of compression can be compromised according to the level of acceptable quality required. Pictures that are subjectively indistinguishable from the original can be achieved with DCT compression to 2 bits/pixel, while 0.1 bit/pixel yields a poor but recognisable image. The standard also accommodates loss-free compression by using a combination of 'differential pulse-code modulation (DPCM)' and Huffman coding, but a compression ratio of only 2 : 1 is typical.

Binary images are common in DIP applications. Standards for compression of such images had already been developed for fax transmissions by CCITT and are embodied in the Group III and Group IV fax standards (see section 9.9). Such techniques are effective for compressing text-only documents such as parcel manifests, etc.

One of the problems with compression is the time taken to process the image. In DIP applications this may not be too serious a problem when encoding the data, but when the data are required for immediate response to a client enquiry the decoding time can become critical. Any image decoding time must be added to the time taken for retrieval from an optical 'jukebox' storage system which may be significant because of the need physically to change discs. Therefore many image compression techniques are 'asymmetric' in that they are designed to be relatively easy (and cheap) to decompress, even if this means that the compression process is lengthy and complex.

Decompression can be performed by software, general-purpose digital signal processors (DSPs) or application specific ICs (ASICs). Software does the job relatively slowly, but is supremely flexible and portable to a wide range of applications and hardware platforms. It is therefore correspondingly cheap. ASICs offer the best performance but require high volume to reduce costs and are inherently inflexible. DSPs represent a compromise in terms of speed, cost and flexibility. Since they are programmable they can be upgraded and their off-the-shelf nature makes them relatively cheap. Boards may also be purchased that allow a suitable number of DSPs to be operated in parallel so that the performance of the system can be tailored to the user's application requirements. One manufacturer's product allows up to five TMS 320C25 DSPs to be used to achieve up to thirty times performance enhancement compared with the best software technique available from the same company.

Another exciting new compression technique that seems well suited to image archiving is the 'fractal transform'. This uses simple mathematical expressions to define infinitely complex levels of detail in an image [63]. It is a highly asymmetric technique because encoding is difficult and computationally expensive, but decompression is quick and can provide astonishing detail. At present it is being championed virtually exclusively by British mathematician Dr Michael Barnsley and is available only through his company, Iterated Systems. It is regarded with some suspicion by members of the more traditional image compression community and is not yet thought to have bettered the performance of JPEG standard compression.

*Analysis of document content*

If document image processing, or whatever it is called, is to proceed beyond mere archiving and retrieval, then some form of artificial intelligence must be applied to the interpretation of the contents of the documents processed. This may be as 'dumb' as determining the presence or absence of a tick in a box in a predetermined position on a specially designed form. Such an approach, coupled with careful design and layout of forms, has been shown to yield dramatic productivity improvements in business sectors such as insurance, banking, local government, education, etc.

However, greater flexibility is achieved by increasing the level of intelligence available to the machine. There is an obvious advantage in giving it the ability to read the document without human intervention. It can then automatically absorb, memorise and accurately index a far greater amount of information than a human could be reliably expected to cope with.

The first step in machine reading was optical character recognition (OCR) – it was one of the earliest successful applications of pattern recognition. It relies upon specially designed character sets (fonts) such as OCR-A (in the USA) and OCR-B (in Europe). These are designed specifically to make it easy for the machines to distinguish all of the necessary characters reliably, and as such represents a good example of the effective imposition of scene constraints.

OCR uses a template matching scheme to recognise the strictly limited range of characters that are allowable. Therefore it can offer very reliable recognition of the predetermined fonts, provided that the text is not badly corrupted by noise, or tilted, or scaled in any way. Given that these constraints are acceptable, the task of OCR can be achieved relatively simply without recourse to exceptionally powerful computing facilities. However, the degree of flexibility achievable is very limited. It tends to be most successful in 'closed loop' applications, such as bank cheque clearance, where an organisation can guarantee that all of the relevant information is presented in an acceptable font and in a controlled manner.

This issue has been tackled to some degree by the development of 'omnifont' recognition which copes with a range of standard fonts, such as those commonly used in typewriters. However, there will always be cases where useful documents contain non-standard, proportionally spaced, emboldened, underlined, italic or mixed fonts and thus require human intervention. Another problem is that many documents contain mixed text and diagrams where both types of information are of interest.

The answer to each of these problems is, of course, to endow the machine with yet more 'intelligence'. In 1974 Ray Kurzweil produced a reading machine for the blind, and was the first to apply artificial intelligence techniques successfully to character recognition [64]. Thus the new field of 'intelligent character recognition' (ICR) was born. This uses an algorithmic approach instead of simple template matching to identify the topological features of characters in much the same way as humans do. For example a 'b' consists of a roughly circular component with an ascending vertical stroke attached to its left side while a 'd' is similar but the vertical stroke is on the right. In the case of a 'p' the vertical stroke *descends* on the left, but if it is on the right it is a 'q' and the addition of a 'tail' changes that into a 'g', etc.

This kind of understanding allows humans and similarly equipped machines to recognise characters in *any* font without difficulty. This in turn confers much greater flexibility upon machine reading systems and improves the reliability of recognition of degraded images of documents. Confidence can be further enhanced by the exploitation of contextual constraints – i.e. when a candidate character identification is produced, its relationship with preceding characters can be checked for lexical integrity to ensure that is a viable suggestion. ICR systems have no problem with proportionally spaced text and can recognise font attributes such as emboldening, italics, underlining and variable point size. Diagrams can be distinguished from dirt and scribbles and can then be partitioned into a separate file suitably linked to the text file.

The greatest challenge for ICR is to enable it to deal with hand-written, rather than printed, characters. In many cases it is simple and therefore wise to eliminate the need for this altogether, but it can be achieved, albeit at lower speed and reliability and greater cost. One current high performance ICR system can recognise 10 000 type-written characters per second but only 2000 characters per second for hand *printed* characters (i.e. *not* flowing script) [65]. By contrast, the post office in the UK has recently commissioned vision-based letter sorting machines which cannot deal with hand-written envelopes at all. However, since type-written envelopes represent approximately 75% of the total workload and the machines can sort 20 times faster than a human (at 35 000 envelopes per hour), this seems to be a very pragmatic division of labour between men and machines.

## 9.9 Telecommunications

Telecommunications is concerned with the transmission of meaningful human communication from one individual or business to another sited at a remote location. The most powerful form of human intercommunication is undoubtedly the image (*'a picture paints ten thousand words'* [Anon.]), but of course this is the most challenging form of data for communication networks to carry. Therefore, although we have become used to 'making do' with cost-effective telephony (audio-only) and telex (text-only) services, there has always been a desire to communicate through images.

*Facsimile*

Facsimile image transmission was first patented by Alexander Bain in 1843, several decades before Bell invented telephony. This first system relied upon the use of a communications channel to synchronise two pendulums, one being used to drag a probe over an electrically conductive image. The second pendulum used a stylus to pass current through paper soaked in potassium ferricyanide, which then selectively turns black. The pendulums were soon replaced by a 'cylinder and screw' raster-scanning arrangement and 1902 saw the invention of photo-electric scanning by Arthur Korn. As early as 1910 there was an international facsimile service between London, Paris and Berlin using Korn equipment, but the transmission of a single picture could take as much as a week!

This problem was tackled by what must be recognised as one of the earliest applications of image processing: the Bartlane cable picture transmission service, established in the early 1920s. This used the submarine cable between London and New York to achieve a transmission time of just one day per picture, by means of coding. The final image was reproduced by a modified telegraph printer using a special typeface to produce halftones yielding just five grey levels. This technology must have been impressive at the time and surely produced some real 'scoop' pictures for the newspapers that could afford to use it. However it did not go far towards making facsimile transmission accessible to all.

Newspapers remained the principal users of facsimile until the late 1960s when the CCITT produced a standard which allowed facsimiles to be transmitted over the 'public switched telephone network (PSTN)'. This Group I facsimile (fax) recommendation continued to allow grey levels to be transmitted using a frequency modulated analogue system. This allowed an A4 page to be transmitted over the PSTN in 6 minutes with a vertical resolution of 3.85 scanlines per millimetre. In the mid-1970s the transmission time was halved by the use of a new modulation technique, but it was not until the arrival of the *digital* Group III standard in 1980 that fax really began to catch on.

This standard was the first to acknowledge that most business documents really only require binary representation and that high spatial resolution was more important in this application than grey-scale resolution. Group III fax offers 1728 binary pixels per line (8.03 dots/mm) and either 3.85 or 7.7 lines/mm. The data are compressed using 'run-length' coding and this allows a page to be transmitted via a modem in one minute at 4800 bits/s, although 9600, 7200 and 2400 bits/s can also be accommodated. The CCITT have designated a set of eight typical documents to test the performance of Group III fax machines and these reveal that run-length coding achieves compression ratios in the range from 5.6 : 1 to 21.4 : 1. The variation is due to the amount of solid black or white in the image.

The Group III fax terminal has proved to be one of the most successful telecommunications innovations ever, with over ten million terminals in use world-wide in 1989, a growth rate of over 30% *per annum* and no imminent signs of saturation [66]. Overall, this means that fax is now second only to telephony in terms of the volume of traffic generated over the PSTN system. It is clear from the application statistics that fax comes into its own in the business community and especially over international routes where office hours do not overlap.

During the early-1970s the operators of PSTN systems (PTTs or post, telegraph and telephone organisations) were actively engaged in converting their systems to all-digital exchanges. They soon realised that something was needed to generate more revenue from the spare capacity created by this heavy investment programme. In response, the CCITT drew up proposals for the Integrated Services Digital Network (ISDN). This would extend 64 kbit/s digital links directly to individual subscribers and allow them to utilise a wide range of digital services over a common link. This was firm recognition of the spectacular growth of *data* traffic over the existing analogue and hybrid PSTNs.

Predictably, the promise of widely available facilities for greater capacity and superior reliability has created a demand for more sophisticated services, rather than merely improving the performance of existing ones. The new Group IV fax standard defined in 1984 is one example. This provides higher resolution images with up to 15.8 dots/mm (400 dots/inch) in both horizontal and vertical directions, which is better than most laser printers can achieve! It is also compatible with the earlier Telex and Teletex character-based services and can support mixed character and facsimile image blocks on a single page.

Despite the improved performance of ISDN connections it is necessary to utilise superior data compression techniques for Group IV fax in order to allow such high resolution pictures to be transmitted in times as short as five seconds. The coding strategy adopted by CCITT is a two-dimensional code called 'modified modified Reed (MMR)' or 'modified relative element address designate (READ)' coding [67]. This is an advanced form of run-length coding which compares the line being coded with the previous line of the document which therefore reduces redundancy in two dimensions. After run-lengths have been obtained they are further compressed using Huffman entropy coding to allocate the shortest code words to the most frequently occurring run-lengths. For the highest resolution images two-dimensional coding yields compression ratios between 40 and 60% better than simple one-dimensional methods.

*Videotex*

Videotex is the generic name given to a range of services offered over the PSTN which allow textual and low-resolution graphical information to be accessed from a central database facility using a low cost terminal. It was initiated in 1979 by British Telecom as 'Prestel' and is now widely used in Europe to source information in travel agents, building societies, showrooms of multi-outlet shopping chains, etc. In 1983 the Conference of European Postal and Telecommunication Administrations (CEPT) established a comprehensive standard which also allowed for the provision of photographic quality pictures. This would make the system attractive to applications such as estate agencies (realtors) who could benefit from having high resolution images as well as text in their real estate database.

Until that time both graphics and text was 'character coded' – i.e. the screen area was divided up into small rectangular blocks and the contents of each block was defined by a 7- or 8-bit code word. This was an extension to the way that ASCII code represents pure textual data. For display, the code is converted to a block of pixels representing either a textual character or a block-graphic by reference to a simple 'look-up table' or 'character generator'. This yields an efficient system, since a full screen of 24 rows by 40 blocks requires only 960 bytes of data which takes just 6.4 seconds to transmit reliably over the PSTN (less than a second over ISDN).

However, for photographic quality images each pixel must be individually defined and, unlike fax, it is necessary to do this at many intensity levels and colours to achieve life-like picture quality. Therefore it is usual to use 8-bit words to represent each of the red, green and blue primary components of the image, allowing 256 levels of intensity and 16.7 million colours to be portrayed. At a resolution of $512 \times 512$ such a system requires 786 kbytes of data per image, which would take approximately 100 seconds to transmit even at 64 kbits/s over ISDN. In order to get this down to an acceptable delay of 5 seconds it is necessary to compress each pixel so that it is described by 1 bit (on average) rather than the 24 required in the raw data. Such compression ratios are achievable at very acceptable levels of picture degradation using the JPEG DCT-based standards (see section 9.8.3).

In fact the DCT can achieve recognisable pictures at 0.1 bit/pixel so that an image is compressed to 3.3 kbytes and transmits over ISDN in under half a second. This level of compression is best used in a 'progressive' picture build-up system where a low quality, but recognisable, image is transmitted quickly and detail is added by subsequent transmissions. In this way the picture quality is progressively refined until it reaches the desired level. This is ideal for applications where users wish to browse through an image database and therefore want a quick response. When the desired image has been chosen the user will usually be prepared to wait rather longer for a more faithful representation.

*Video-conferencing and video-phones*

Despite being even more demanding of transmission bandwidth, moving pictures convey much more information than static ones. Therefore the ability to carry moving pictures over general-purpose communication networks is perceived as highly desirable, even if that network has to be ISDN.

For example, businessmen would welcome the ability to carry out detailed negotiations without having physically to travel great distances to meet one another face-to-face. Since facial expressions and 'body language' have a large part to play in such transactions, any suitable remote communications must allow the parties to see one another. This is 'video-conferencing', a technology which has been in use for some time, but only over expensive private leased lines with capacities up to 2 Mbit/s, thus seriously limiting its appeal. Increasing availability of switched digital telephone services like ISDN has begun to revolutionise this field and prices for video-conferencing facilities look set to fall dramatically. For example, first generation video-conferencing suites cost £40 000 ($60 000) and above, while a new generation of low-end equipment using ISDN is available for between £7500–£15 000 ($11 250–$22 500) [68]. It is now also possible to buy cards which plug into a PC to provide a desktop video-conferencing capability over ISDN lines for around £3500 ($5250)!

Another application for such technology is in 'distance learning'. In 1992 France Telecom established a £2 million pilot 'viseocentre' in Lille and another in Lyon, with others to follow. This allows knowledge to be disseminated from centres of learning such as Universities to any community or business which has invested in suitable equipment. France Telecom sees this as a major advance in the cost-effective provision of continuing education within the community. Businesses can tap into video-based courses offering the very best training for its staff without incurring the direct or indirect costs involved in travel, or the overheads involved in one-off on-site courses.

Lectures can be delivered using a wide range of video- and computer-based techniques under the control of the lecturer himself (see figure 9.6). Students can pose questions by pressing a button in front of them which registers on the remote lecturer's control panel. Whenever he or she is ready to answer these questions a camera automatically homes in on the student(s) who originated the question, and a microphone is activated to establish a two-way dialogue between teacher and pupil. Such facilities can be realistically contemplated at a national, and even multi-national, level thanks to the existence of ISDN and the emergence of low-end video-conferencing standard products which exploit it.

For users who cannot afford such exotic facilities there has been considerable progress towards the provision of video-phones. These would

Figure 9.6  An experimental live video-conferencing lecture link between the University of Brighton and ENIC (Ecole Nouvelle d'Ingenieurs en Communication) at Lille, Northern France, conducted in October 1993

allow suitably equipped subscribers to see a moving, colour picture of their correspondent, albeit with low resolution and jerky motion. A video-phone offering a refresh rate of 8 frames per second over the normal analogue PSTN went on sale in Britain in early 1993 at a cost of about £400 ($600) [69]. At the same time, European PTTs and some of their customers are participating in the European Video Telephony Experiment (EVE-2) trials. These will evaluate a 25 frame per second digital system operating over ISDN. Among the British customers involved in these trials are the Royal National Institute for the Deaf, who have obvious reasons to be interested in the use of such technology.

Obviously the success of any of these services is going to be critically dependent upon the use of image compression according to an internationally agreed standard. The CCITT has produced recommendation H.261 for this purpose. It is aimed at video-conferencing and video-telephony over links offering bit rates in the range 40 kbits/s to 2 Mbits/s. In order to ease problems of different television standards around the world a Common Intermediate Format (CIF) has been selected which specifies images of 352 × 288 pixels at 30 frames per second. Without any form of compression such images would require transmission channels with a capacity of 36.5 Mbits/s when using the 4 : 1 : 1 colour sampling scheme specified by H.261. Clearly, compression ratios of at least 18 : 1 are required for successful transmission over the specified links.

Like JPEG, H.261 standard uses the DCT and other techniques to minimise *intra*frame redundancy, but since moving images are involved it is also possible to detect and eliminate *inter*frame redundancy. This means that if there is no motion taking place in a part of the image there is no need to transmit data describing it after the first frame where it appears. This can be achieved by 'frame differencing'. A second technique is 'motion compensation' which aims to detect spatially correlated groups of pixels which are moving at a fixed velocity within an image. If these can be found, their position in subsequent images can be predicted from previously transmitted data. Then it is only necessary to transmit a signal corresponding to the error between the predicted data and the actual data. The error signal is required because object motion reveals background pixels that were invisible in the initial frame used for prediction.

Care has to be taken with interframe encoding to avoid the build-up of effects due to quantisation noise and the initialisation of predictors, especially in areas of the picture which are static for long periods. In the working standard this is achieved by ensuring that purely intraframe-coded pictures are transmitted at regular intervals. Motion compensation of randomly occurring *objects* is too difficult computationally, so a compromise is achieved by dividing the picture up into 16 × 16 blocks and checking each of these for motion within a 7 pixel 'radius'. Even this task involves 684 288 000 pixel comparisons per second for a CIF image, necessitating the use of dedicated hardware array processors.

The performance achieved by H.261 codecs (coder/decoders) means that picture quality over 2 Mbit/s links compares favourably with that offered by home video recorders. However, at 64 kbits/s a fixed camera position is really essential to minimise unnecessary movement in the background, and large amounts of foreground movement can cause the picture to become noticeably jerky. The outcome of the EVE-2 trials is eagerly awaited to gauge customer perception of the viability of this technology.

## 9.10 Security, surveillance and law enforcement

Machines have great potential to replace humans where continual vigilance is required over very long periods, especially when the rate of occurrence of noteworthy events is relatively low. They also excel where decisions must be based on objective rather than subjective visual measures. These characteristics are often present in security and surveillance operations and this makes them suitable applications for machine vision systems.

## Verification of identity

There is a common requirement automatically to validate the right of a person to enter a particular zone or gain access to classified data, etc. Facial recognition is complex and unreliable, speech recognition suffers from long term changes due to colds, etc. and signatures are too easily forged. However, the human body has several features which are unique to the individual, constant over time, relatively easy to quantify and yet virtually impossible to forge.

The best known feature in this category is the fingerprint, which has long since been associated with criminal investigations. Each fingerprint contains a pattern of lines, intersections, loops and whorls which uniquely identifies the individual. The British Home Office has for some time been investigating the use of multiple transputer arrays for automatically cross-matching fingerprints with those in its criminal records database, but such equipment is too complex for routine use. However, workers at the University of Edinburgh have recently developed a system intended to perform on-line fingerprint verification. The prototype device features a $258 \times 258$ image sensor and 40 000 gate signal processing array integrated into a single chip, supported only by an external microcontroller and a RAM chip [70]. The image from the sensor is preprocessed and binarised before being passed to a 64-cell correlator array running at 2 billion operations per second. Decision logic then determines the likelihood of a satisfactory match. Such a system is capable of verifying a fingerprint against a stored reference in just one second.

Although it is generally recognised that the pattern of blood vessels in the retina is unique to the individual, methods for accessing this characteristic feature may not be widely acceptable to the public. However, a new system under investigation by the British Technology Group and Cambridge Consultants performs a similar check on the veins in the back of the hand [71]. In this case the individual requiring recognition would merely have to place a clenched fist inside the machine while inserting a 'smart' card containing his personal details and recorded vein pattern. The system uses an infra-red sensing camera to highlight the pattern of veins on the back of the fist, while rejecting the arteries. The image of the veins is firstly corrected for the variable position of the hand and then the pattern is compared with that stored on the card. Comparison times of the order of a 'few seconds' and error rates of less than 1% have been predicted from a working demonstrator model.

## Monitoring and surveillance

Closed-circuit television (CCTV) systems are commonly used for monitoring company premises and warehouses outside working hours, but until recently the output, often from many cameras, had to be interpreted exclusively by human eyes. This is a tedious task which would tax the concentration of any individual, leading to poor performance and job dissatisfaction. Consequently there is keen interest in developing automated systems that can continuously keep watch with total reliability. Potential application areas include alarm verification, 'smart' movement detection, event recording, door-entry security and even domestic security and baby watching duties.

Alarm verification is very important because typically more than 95% of automatic alarm indications turn out to be false, consequently wasting much time and reducing the 'believability' of any given alarm indication. One system recently developed uses transputers to analyse scenes for moving objects or characteristic shapes [72]. It stores four images at $512 \times 512$ resolution and processes them continuously on-line. When an alarm condition is detected a manned control room is automatically notified. The operatives can replay the last four images captured by the unit so that they can evaluate the event which triggered the alarm. This allows them to decide on appropriate investigative action or to eliminate false alarms, etc.

In an attempt to raise the utility of such systems still further, workers at the University of Edinburgh have developed a miniature $156 \times 100$ camera system using 1.5 µm CMOS ASIC technology [70]. This is coupled to a fixed-focus 90-degree field-of-view lens and a hybrid signal processing module to form a complete standalone unit capable of verifying intruder alarms produced by passive infra-red detectors. When an alarm condition is detected the camera captures a short sequence of video and sends it in compressed form to a central monitoring facility via a modem. Thus, within 24 seconds of an alarm condition being detected, the operatives of the system can determine the likelihood of a false alarm. The high level of integration also means that the units should be cheap enough (at suitable production volumes) and small enough ($40 \times 20 \times 10$ mm) to be liberally dispersed around a site needing protection.

Another common application for CCTV surveillance is in monitoring and control of traffic flow. Once again this application is easily capable of overloading the capabilities of unaided human operatives. Traffic control centres are often required to handle images from up to 60 cameras, each equipped with pan, tilt and zoom controls. While trying effectively to monitor all of this information for accidents, breakdowns and tailbacks, operatives must handle calls from roadside emergency telephones and respond appropriately. Clearly this is a job which could benefit from some automation of routine tasks.

Roke Manor Research laboratories in the UK have begun to tackle this problem with the Automatic Road Traffic Event Monitoring Information System (ARTEMIS) [73]. The system is based around a Sun SPARC Unix workstation which is configured to perform real-time automated video monitoring. The system can perform traffic flow monitoring, vehicle classification and number plate *recording* (note: *not* recognition) without recourse to dedicated real-time image processing hardware. This is achieved by adopting a philosophy of initially detecting regions of interest in an image and then concentrating more significant image processing operations only within those 'active zones'.

In its traffic flow mode the ARTEMIS system can automatically generate statistics regarding percentage occupancy, number of vehicles per minute and average vehicle speed, for each lane of a three-lane highway. This operation is far from trivial because of problems such as huge light level variations, motion blur, shadows cast by vehicles in adjacent lanes and obscuration by high-sided lorries. Reference background intensity levels must be adaptively generated to cope with these widely varying conditions. Vehicle presence is detected by noting significant grey-level deviations within active zones and vehicle speed is determined by spatial correlation.

The system can also be used to detect vehicles which are static for a period of more than fifteen seconds while other traffic is flowing freely. This is achieved by temporal filtering and produces an enhanced image for the operator which is free of moving traffic. This allows a swift response to breakdowns and accidents. This is particularly important when they occur within busy sections of road undergoing maintenance or widening, etc.

In surveying traffic conditions for planning new roads and developments of existing networks it is vital that data are accurate, reliable and up-to-date. In addition to gathering the statistics relating to traffic flow discussed above, ARTEMIS can also analyse the types of vehicles using a particular section of road. For example, vehicles can be classified as cars, lorries, buses, vans or unknowns by overlaying the image with a grid of pixel detectors and correlating their responses to achieve an estimate of speed and size of vehicles.

Licence plate extraction is useful for census studies and more particularly for law enforcement, because the vehicle ownership can be uniquely identified from this information. However, the lack of constraints that can be applied to the imaging process make it very difficult to do in real-time under typical operating conditions. The licence plate must first be located at relatively low resolution when working with a practical field-of-view. This is complicated by the vehicle motion, obscuration, varying light conditions, moving shadow patterns and the existence of dirt, defects and other 'clutter'.

However, the licence plate is a relatively high contrast region on any car and this represents the key to locating it. After detecting such regions they are thresholded and subjected to connectivity analysis, a size criterion and a horizontal adjacency criterion. This yields the location of the licence plate and the character blocks within it. At present automated character recognition is not attempted because of the problems of achieving sufficiently reliable performance under such difficult conditions. However, licence plate images can be automatically recorded in response to trigger events such as a speeding offence or a traffic-light violation, ready for subsequent interpretation by a human operator. As such, this represents an effective form of intelligent image data compression.

*Forensic investigations*

Low cost PC-based image processing facilities have recently begun to make a significant impact on the admissibility of forensic evidence. The US internal revenue service (IRS) have successfully used image restoration and enhancement techniques to prove culpability in some cases of income tax fraud [74]. Latent fingerprints can be extracted from tax returns to try to identify who has been involved in their preparation. In one particular case a company principal was proved to have personally handled fraudulent tax documents despite his denials. All conventional evidence pointed towards his executive assistant as being the sole perpetrator of the crime.

Faint latent prints existed in an area of the tax form which featured a shaded background and conventional photographic enhancement and retrieval methods could not extract usable evidence from them. Contrast stretching and histogram equalisation of images of these faint prints showed that they definitely could not belong to the executive assistant, but the pattern of the shading made conclusive identification impossible. However, since the shading was strongly periodic, fast Fourier transformation into the spatial frequency domain allowed it to be eliminated by cutting out the corresponding group of frequency components. When inverse transformed the background pattern was successfully removed and the fingerprint could then be enhanced using edge operators and filters in the conventional manner. Computerised analysis of the mystery print revealed that it belonged to the principal of the company himself and he finally confessed to the fraud when presented with this new evidence.

The validity of computer enhanced forensic imagery as evidence in the US courts was tested during a murder trial in 1990. Once again the frequency domain transformation and filtering technique was used to identify the fingerprint of a key suspect, but this time formed on the cloth weave of a blood-soaked pillow-case. Defence lawyers moved to suppress this evidence at a pre-trial hearing, saying that the technology was [74]

' . . . *not generally accepted by the scientific community*'.

However, the judge was shown the full image enhancement process and ruled that the enhanced print was indeed admissible evidence at the trial.

In order to further the applications of image processing within forensic science a group of federal agencies from the law enforcement, intelligence and defence communities in the US have formed the Federal Forensics

Laboratory Image Processing Work Group (FLIPWG). Members meet regularly to review the progress of the field and exchange information. Image processing technology has allowed forensic scientists to extend their capabilities in the analysis of documents, video tapes, photographs, ballistics, tool marks, etc.

## 9.11 Entertainment and consumer electronics

By now everybody is quite used to seeing the effects of image processing on television, whether they are aware of it or not. Weather forecasters are routinely superimposed over graphics of the local area and weather fronts. These graphics now often contain satellite derived images of cloud or rain-radar images. In television drama, actors are shown performing in impossible places, often at highly reduced or enlarged magnifications. Picture tumbling, bouncing, stretching and squashing effects are now commonplace in all light-entertainment shows and advertisements. Documentary programmes are fond of using coarse pixelation to mask the faces of individuals who do not wish to be identified. All of these techniques are the result of sophisticated real-time image processing hardware produced by companies such as Quantel.

A particularly interesting and unusual application of this technology recently enabled the restoration of several rare episodes of the science fiction TV series *Dr. Who* [75]. When the story called 'The Daemons' was made in 1971 the British Broadcasting Corporation (BBC) did not realise the cult status that it would achieve and therefore paid little attention to proper archiving. Now however, there is considerable marketing potential for every episode, especially the rare or unusual older ones. The only surviving copies of 'The Daemons' were a complete monochrome film copy and a domestic Betamax recording of a US NTSC colour transmission. The quality of the video recording was not sufficient for commercial purposes, although the film copy would be very collectable if colour could be restored.

Since the human visual system is very tolerant of colour degradation, it was possible to lift the colour information from the NTSC version and apply it to luminance signal derived from the film. However, it was not that straightforward because the US transmission had been extensively edited to make enough time for commercials to be shown, so the colour information for many scenes was missing. It was also found that the aspect ratio of the film version was different from that of the video because of geometrical distortions in the film transcription process. Real-time geometrical processing available in standard special-effects units was therefore used to match the dimensions of the luminance and chrominance components of the pictures. The colour information for the missing scenes was painstakingly added by hand using a Quantel Paintbox special effects unit. A single 20-

second scene took 2 hours to re-colour in this way and the whole 125 minute series required over 40 hours of work to reconstruct in total.

*Colourisation*

Despite the time and cost involved, the process of 'colourising' classic black-and-white films such as *Citizen Kane, It's a Wonderful Life*, Laurel and Hardy comedies, etc. has been found to be highly lucrative [76]. When presented in colour such films apparently have a much greater appeal to younger audiences, although the process offends many film critics and aficionados. In order to make the process viable it is necessary to automate the colourising process. The art director in charge of the conversion must choose a palette of 4096 colours for the whole film, from a full range of 16.7 million. Demands upon hardware performance can be further reduced by limiting the number of colours in any one frame to 64.

However, after hand colouring a key frame such systems can automatically colour subsequent frames. Colour saturation is derived from the luminance information and frame-by-frame comparisons detect moving objects and 'track' them, adapting the colouring accordingly. This makes sure that moving features such as folds in clothing, etc. are portrayed realistically. Whenever there is a dramatic change in scene content a new key frame has to be hand coloured and a film like *It's a Wonderful Life* has 1100 such key frames. In 1987, when the technique was first reported, the relatively limited performance of available processing hardware meant that each minute of film took four hours to 'track' automatically. This performance has undoubtedly improved in the intervening years.

*Mass-market picture processing*

Although real-time processing of moving images has been affordable to broadcast television companies for many years, it has only very recently become possible to think of including it inside individual domestic products. Early evidence of a move towards real-time domestic image processing came in the form of the picture-in-picture (PIP) and perfect still-frame facilities that became fashionable in video cassette recorders a few years ago. This was achieved by the inclusion of a framestore and was often accompanied by a range of limited effects such as 'digital zooming', 'mosaic', 'solarisation', 'posterisation' and multi-picture 'strobed' animation (see figure 9.7).

More recently, so-called 'digital television receiver chassis' have become more commonplace as manufacturers move towards multi-standard televisions and attempt to prepare the marketplace for wide-screen and high definition television transmissions. So-called 'flicker-free' televisions achieve 100 Hz field scan rates by storing the incoming 50 fields per second

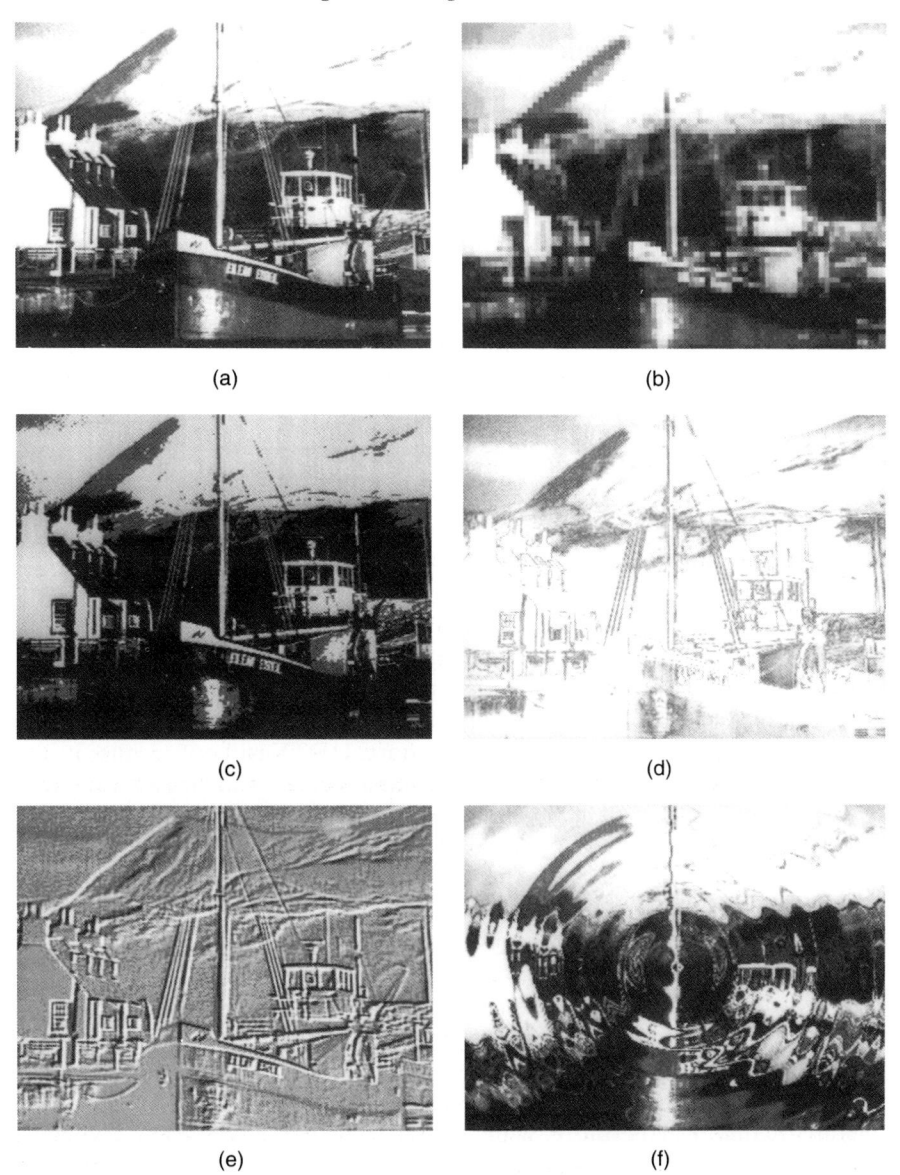

Figure 9.7 Picture processes for entertainment and artistic purposes: (a) original image; (b) 'mosaic'; (c) 'posterisation'; (d) 'solarisation'; (e) 'bas relief' effect; (f) 'water ripple' distortion

in a digital framestore and then reading it out twice to double its scan rate. This reduces the flicker perceived by the human visual system when viewing large screen sizes. This development seems relevant because televisions sales statistics currently indicate a trend towards larger average screen sizes. Unfortunately, early users have reported problems of jittery motion of objects such as rolling balls, presumably due to the frame rate doubling.

An alternative to doubling the scan rate is to double the number of displayed lines, thus making it possible to market televisions that could be upgraded from standard definition to the proposed high definition systems. It is also possible to begin marketing 16 : 9 aspect ratio screens that can be used with the existing 4 : 3 aspect ratio broadcasts because picture height or width can be digitally modified. Wide-screen (16 : 9) TV models currently offer the option to display a standard 4 : 3 aspect ratio picture with blank bars down the sides, or with separate sub-pictures (so-called 'picture-out-of-picture (POP)') in addition to the main one. Their main claim to fame, however, is their ability to expand so called 'letter-box' format wide-screen transmissions to fill the whole screen and thus give the full 'cinematic experience'

*High-definition television*

True high-definition television (HDTV) makes image processing within the receiver an absolute necessity because of the requirement to perform real-time decompression of the broadcast signals. The HDTV standards aim to offer picture spatial resolutions approximately twice that of the current broadcast systems in both axes. Furthermore, HDTV will only realise its full potential on large wide-screen televisions and so the frame update rate must also be raised to eliminate perceived flicker [77].

All in all, this means that the raw bandwidth of an HDTV signal is around 40 MHz. However, in the case of direct-broadcast-satellite (DBS) delivery the bandwidth available to carry these signals is at most only 11 MHz. For terrestrial broadcasting the signals must be fitted within the same bandwidth (approximately 6 MHz) as the existing broadcast channels, unless a corresponding reduction in choice of channels is to be tolerated by the public. Therefore, in order to fit HDTV signals within the strictly controlled frequency spectrum, it is necessary to perform considerable compression of the raw signal without significantly degrading the perceived picture quality. Furthermore, while compression equipment can be complex and expensive (because it is paid for by the broadcasters), the decompression circuitry must be simple enough to make the system available to consumers at the right price.

Competing systems exist to achieve these aims, although they use broadly similar techniques. In Japan the 'multiple sub-Nyquist sample encoding' (MUSE) standard operates at 1125 picture lines and 60 Hz frame rate. It is available commercially and Japanese users are regularly served with 8 hours of broadcasting per day in this format. In Europe the 'high-definition multiplexed analogue components' (HD-MAC) system operates at 1250 picture lines and 50 Hz frame rate. It has been developed with the help of various EC initiatives and was used in 1992 to broadcast both the Winter Olympics from Albertville and the Summer Olympics from Barcelona to the few (non-domestic) audiences equipped to receive it. The USA has not backed either standard and seems likely to go for an all-digital system rather than the all-analogue or 'digitally assisted analogue' approach of the other two systems.

Since the aim is to dramatically reduce the bandwidth of images without *noticeable* degradation, the basic principles of compression rely upon known inadequacies of the human visual system. In particular, we cannot see great detail in objects which are moving and we also have poor spatial resolution of diagonally disposed features. Therefore, the encoding schemes are motion adaptive (that is they reduce the resolution in areas of the screen where motion is detected) and a diagonal spatial filtering scheme is employed. Major bandwidth reduction (approximately 40%) may also be achieved by the use of sub-sampling in conjunction with an anti-aliasing pre-filter in the horizontal, vertical and temporal planes. Further compression can be achieved by eliminating non-essential components after the discrete cosine transform has been applied to blocks of data in the image.

The MUSE system uses simple motion *adaptation*, basically determining if an area of the picture is moving or static and then passing it through an appropriate processing path. HD-MAC uses more complex motion *compensation* with three possible processing routes for blocks of the image which will be updated at 12.5, 25 or 50 Hz. The information on which motion channel should process which block is then transmitted digitally during the vertical blanking period of the TV waveform. This 'digital assistance' considerably reduces the complexity and therefore the potential cost of HD-MAC decoders.

A further constraint that was applied to the design of the HD-MAC system was that the signals it produces should be fully compatible with standard 625-line MAC transmissions. This has been achieved, but unfortunately MAC signals are not widely broadcast in Europe (despite their technical superiority) because of the demise of the businesses planning to offer MAC DBS services.

Digitally assisted enhancements to the existing European 'phase-alternate line' (PAL) broadcast system may ultimately give the consumer the most favoured route to improved picture performance at reasonable cost. This so-called 'PALplus' system offers improved resolution at 16 : 9 aspect ratios, but viewers with conventional receivers would actually get less resolution from such signals.

Ultimately, it seems possible that Europe and Japan will be forced by commercial pressures and compatibility issues to abandon their analogue high-definition systems in favour of an all-new, all-digital system. It is also quite possible that the consumers may never want to pay what is necessary for a full high-definition receiver. Instead, they may prefer to see digital compression techniques used to offer more channels of standard definition within the currently allotted frequency spectrum.

*Interactive video media*

Free from the constraints of compatibility with old established standards, image processing also has a significant role to play in delivering totally new media for home entertainment. A good example of this is the CD-i (compact disc interactive) full-screen, full motion video (FSFMV) enhancement.

CD-i is interesting in its own right as a way of interactively accessing massive amounts of text, graphic and sound data. It is part of the new wave of 'multimedia' technologies. However, without real-time image decompression it cannot deliver FSFMV. This appears to be the 'holy grail' for multimedia developers because it would greatly add to the capabilities of all forms of interactive software.

For example, film scholars would be able to watch a scene from a classic film and interactively dissect its contents by accessing interviews with the director and the leading actors, etc. It is even thought possible that FSFMV CD-i could rival domestic VHS tapes as the medium for delivery of prerecorded film entertainment. Limited capability for interactive video is even now being offered by CD-based computer games consoles at a very reasonable price. New ventures into interactive home entertainment media, such as the recently released Commodore 'Amiga CD$^{32}$' and the Japanese '3DO' system (released in the USA in late 1993), feature processing architectures specifically targeted at exploiting FSFMV.

A big advantage with CD-based video standards is their truly international nature, so that a disc made anywhere in the world will play satisfactorily in any other country. This compares favourably with video or laser-disc storage where the software supplier must manufacture material in NTSC, PAL and SECAM TV standards. Further plus points for CD-i are that it is fully compatible with standard CD music discs and also with the Kodak PhotoCD system that allows domestic snapshots to be archived on CD and viewed on TV at excellent resolution.

Obviously, however, for any of these ideas to reach commercial viability it must be possible to build the necessary sophistication into these domestic products at a price which the individual consumer will find attractive. The sophistication required is far from trivial. Normal video signals require to be represented at a data rate of 200 Megabits/s for broadcast quality, but the maximum data transfer rate of a compact disc is fundamentally limited to only 1.2 Megabit/s. Therefore compression ratios of 167:1 would be required for full broadcast quality. Fortunately, as discussed earlier, it is possible to compromise quality for reduced compression ratio and the deficiencies in the human visual system can mask many such compromises when viewing moving pictures. CD-i uses the MPEG-1 (motion pictures expert group, level 1) standards to achieve a full 72 minutes of FSFMV from a single disc.

The MPEG recommendations are rapidly gaining support as the standard protocol for digital video and are being actively developed. Like JPEG, MPEG is the result of collaboration between the CCITT and ISO, and the proposed standards use the same basic approach to eliminate intraframe redundancy. However it is also possible to employ '*inter*frame coding' because each image is part of a sequence. This allows more opportunity for compression since it is possible to detect and eliminate redundancy owing to the similarity of successive pictures.

MPEG-1 is an elaboration of the H.261 standard for telecommunications which aims to improve picture quality achievable at a given data rate. Primarily, this is achieved by making the coding algorithm highly asymmetric in that the major complexity of the compression algorithms is contained in the encoding process. This suits the essentially 'read-only' nature of systems like CD-i and also dramatically reduces the cost to the consumer. It is assumed that the publishers of such material will be able to pay for sufficiently capable hardware and software to make the encoding process efficient and cost-effective. By contrast, H.261 is concerned with bidirectional image transmission: so the user's apparatus must be able to both encode and decode data and therefore it is not acceptable to raise the cost and complexity of one component in order to simplify the other. Overall, the MPEG approach means that more effective compression of moving image sequences (and of their accompanying audio tracks) can be achieved.

MPEG-1 generates three kinds of frame for transmission: I, P and B. The I (Intra) frames are coded without using any motion compensation at all (just JPEG-like compression). Although these require more bandwidth to convey, they are repeated around twice a second to eliminate artefacts due to prediction and quantisation errors. They also provide essential entry points for a random-access video system such as CD-i.

The P (Predicted) frame is like the standard motion compensated frame in the H.261 standard. However, instead of being predicted solely from the previous frame they are based on P and I frames that have been transmitted several frames earlier. The B (Bidirectionally interpolated) frames are motion compensated using data from both past and future P and I frames. This tackles the problem of background pixels being revealed by object motion, since revealed background is visible in the 'future' frames and can be efficiently coded into a B frame. Since the H.261 standard only supports P-type frames it must transmit blocks of revealed background pixels as uncompressed error data, thus greatly reducing its efficiency and/or compromising its picture quality.

However, this improved performance does not come cheap since many frames must be stored to allow such prediction, and movement at a similar rate must now be detected over a much greater range of pixels. In addition, the range of movement expected in natural images is greater than that allowed for in video-conferencing applications. The processing load for MPEG encoding hardware is typically *thirty-six* times greater than that for H.261-capable hardware. This, plus the fact that some 'intelligence' is necessary to adapt the compression to different types of scene, means that it currently takes approximately 40 minutes to encode one minute of video data for CD-i use. It is expected that custom chips currently in development will achieve 'real-time' encoding in the very near future.

The first CD-i players cost around £500 ($750) without FSFMV but can be upgraded by a plug-in module which is likely to sell for around £200 ($300). This is due to become available in the very near future and may soon be built-in to all CD-i players. To ensure that the market place for domestic digital video does not suffer a 'format war' like that between VHS and Betamax, a June 1993 meeting of major manufacturers agreed extensions to the existing CD standards which incorporate MPEG-1 picture coding into a common 'Video CD' format [78].

Definitions for a second generation of MPEG standards (MPEG-2) have now been approved [79]. These are concerned with any raw data rate above 2 Mbits/s. It seems likely that chips complying to this proposed new standard, offering full 'CCIR601' standard performance (720 × 576 pixel resolution at 25 Hz frame rate) and compressed data rates of less than 10 Mbits/s, could be available in products as early as 1995. This new standard forms the basis of most current proposals for digital television broadcasting in Europe. Furthermore, in June 1993 ten major manufacturers committed themselves to the development of a common standard for domestic digital video recording, using either JPEG or MPEG-2 standards.

A third MPEG group, known confusingly as MPEG-4 (since the activities of MPEG-3 were subsumed by the MPEG-2 committee), was formed in 1992 to consider the use of very low bit rates (tens of kilobits per second). The target date for an MPEG-4 standard is 1997 and it is intended for applications such as cellular video-phones.

*Multimedia and computer video editing*

The compression achieved by standards such as MPEG means that worthwhile segments of video can be stored on, and replayed from, modern high capacity computer hard discs. Coupling this to the processing power of current computers has given rise to a new application field for digital image processing – namely, video editing in the digital domain.

In this application, clips of video are digitised and compressed so that they can be stored on hard disc. Sequences of images can then be recalled in any order and manipulated under computer control with absolute single-frame precision, making it trivial to cut images accurately to music, for example. Edited sequences of video can be combined with computer generated graphics, titles and soundtrack to create impressive 'multimedia' presentations.

Entry level implementations can cost as little as £1000 ($1500) over and above the cost of a high-performance Macintosh or IBM-compatible PC. At this level, images will be limited to perhaps one quarter screen size at normal playback rates and approximately 10 minutes of video material can be stored per 100 MBytes of disc capacity. Their use is likely to be limited to desktop multimedia in applications such as business presentations or training programmes. More powerful systems, which include a hardware 'codec' for compressing and decompressing video, can be bought for around £5000 ($7500). These can achieve image quality equivalent to high-band domestic video formats (e.g. S-VHS).

Even more powerful systems have been used by professional broadcasters in the USA to compose complete television shows, entirely within the computer domain. Indeed, there is a steady development of broadcast-quality digital video-recording systems that rely on image data compression. This seems set to ensure that all programme makers will have to consider using digital video editing in the near future. Some broadcasters see this as a mixed blessing, because although digital video recording apparently allows multi-generation editing without degradation, there is concern about the formation of unwanted 'artefacts' owing to the use of several incompatible stages of compression/decompression. This could obviously be a serious problem when programmes are bought-in and then edited with different equipment from that used originally. However, it seems likely that such problems could eventually be solved by the adoption of suitable standards for broadcast-quality video data compression (note that none of the MPEG family standards is intended to meet *current* TV broadcasting requirements).

## 9.12 Printing and the graphic arts

Applications of image processing to any form of art are generally characterised by the need for unusual image acquisition, storage and processing hardware to deal with atypically high spatial resolutions. Until recently this has held back the development of such applications, but now the necessary tools are sufficiently cheap, powerful and widely understood to allow image processing to begin to make a significant impact.

*Colour fidelity in desktop publishing*

In recent times computers have brought about a revolution in the world of publishing. Most newspapers and magazines are now produced by 'desktop publishing' systems which have eliminated many of the traditional stages in the production of the printed page. The benefits achieved include much lower costs and higher productivity which means a wider choice of material, able to serve even the smallest specialist interests. In addition, shorter 'turn-around times' mean that the contents of journals can be made more topical and responsive to the needs of their readers.

Most of this has been achieved through the development of low-cost computing power and hard-copy technologies but image processing has an important supporting role. For example, a considerable amount of image processing is necessary to ensure that the colours which a designer uses on a desktop publishing workstation actually match those that can be reproduced on the pages of a magazine, printed fabrics, woven materials, etc. In addition, such colour fidelity is becoming increasingly important with the growing preference for 'remote shopping', where items are selected from the pages of a catalogue (or more latterly from images stored on a CD-i disc, perhaps). Problems arise because the totally different technologies used to reproduce the colours in each medium and the limited gamut of colours which each can achieve.

In the case of colour monitors, it is convenient to drive them with separated red, green and blue (RGB) control signals. Colour printing uses an absorption process based upon cyan, magenta, yellow and black inks (CMYK). However, neither of these systems is helpful in describing the way that humans actually perceive colour, especially under various lighting conditions, which is what matters in the colour fidelity issue. Recent work arising from the British 'Alvey' research programme has resulted in a standardised colour model which quantifies the appearance of colour while taking viewing conditions into consideration. This 'Hunt–Alvey colour appearance model (ACAM)' describes colour in terms of lightness, colourfulness and hue (LCH) [80].

In using the model, it is necessary to convert from the source colour specification (say RGB for a monitor) to the long-established CIE international chromaticity standard which uses three practically unobtainable artificial primary colours known as the tristimulus values (XYZ: X = 'red' @ 700 nm, Y = 'green' @ 546.1 nm and Z = 'blue' @ 435.8 nm) [81]. The XYZ image is then transformed via the ACAM model into perceptual LCH coordinates, using the known viewing conditions of the monitor.

This data format is independent of device characteristics or viewing conditions, so it must be modified to take into consideration the difference in colour gamut between the source and destination colour reproduction devices. The 'filtered' LCH image can then be applied to an inverse ACAM transformation to achieve XYZ coordinates by applying the known viewing conditions for the destination medium.

Finally the tristimulus values can be converted to a CMYK specification for output to a colour printing device. Clearly this is a major image processing task when executed on typical high-resolution 24-bit colour images. However, when successfully applied, desktop designers can confidently predict the appearance of their product without the need for time-consuming 'proofing'.

*Art history*

Conservation of unique and fragile art treasures is now a rapidly growing application for image processing. For example, Leonardo da Vinci's *La Gioconda* (Mona Lisa) has been the subject of a ten year investigation by a team from the University of California at San Diego Project for Art/Science Studies [82]. The original aim was to restore a high resolution photographic image of the painting by removing the effects of yellowed and cracked varnish. Initially this was achieved by digitising the photograph into six million pixels in each of three colour bands at NASA's JPL.

A sample of similar varnish was analysed for its spectral transmission properties and then its effect was removed from the image by de-calibration. The result was blue sky where once there was only brown, alabaster skin tone instead of unhealthy yellow, and a gown which revealed greenish tints instead of uniform black.

The resolution achievable with such a digitisation was clearly inadequate for art historians and so a smaller area of the picture, containing the head and neck was digitised to a similar number of pixels using a scanning microdensitometer. A more advanced colour restoration technique was applied and this revealed yet more of the beauty of the original painting, but the result was marred by surface glinting introduced by the 'craquelure' (fine cracks) in the varnish. This effect was minimised by transforming the image into the frequency domain and interactively customising a low pass filter function. After filtering, the inverse transform revealed impressive results with a considerable reduction in glint and craquelure.

This restored image allowed art historians to identify previously invisible artefacts in the image including a possible adornment on *La Gioconda's* neck, modifications to the scenery behind her head and evidence of much earlier (*physical*) restoration work. These findings were investigated using local intensity enhancement, level slicing and edge enhancement. The results fuelled speculation that Raphael, a contemporary of Leonardo's, had gained inspiration for his painting *La Muta* from a viewing of the unfinished *La Gioconda*. *La Muta* features a neck adornment which closely matches the remnants visible in the restored *La Gioconda*. This reinforced suggestions that Leonardo may have chosen to adopt a more simplified style *during* the execution of *La Gioconda* rather than before it, as previously thought. Therefore image processing has made a significant contribution to the understanding of the history of art.

## Art conservation and dissemination

A recent programme supported by the EC's 'ESPRIT II' funding aims to ensure that archiving of art treasures is more accurate than ever before. The programme was called VASARI (Visual Art System for Archiving and Retrieval of Images) [83]. It aimed to prove the feasibility of high resolution colorimetric digital imagery direct from paintings, *in situ*. This avoids undue movement of the delicate works of art.

Digital storage of image archives eliminates the degrading effects that ageing has on photographic archives, which currently require regular updating. It also overcomes the insensitivity of traditional photographic techniques to *gradual* changes in pigment colouration. The technique of microdensitometry proved to be unsuitable for routine archiving of large paintings because a resolution of 10–20 pixels/mm is necessary for adequate reproduction of brushwork and craquelure. This means that up to 400 000 000 pixels must be acquired for each square metre of painting for each spectral band!

The VISARI system uses a Kontron ProgRes 3000 camera which achieves a resolution up to $3000 \times 2320$ pixels by virtue of having a high quality $500 \times 290$ CCD mounted internally on a platform which can be moved by piezo-electric actuators in a $6 \times 8$ grid. The camera itself can be moved over a range of 1.5 m in each of the $X$- and $Y$- planes by a motorised portal in steps of 0.005 mm. 100 mm of movement in the $Z$-plane is also allowed for fine focusing since a fixed focus lens is used for ultimate image quality. The optimal setting for focus is automatically determined in 'real-time' by summing the squares of the differences between neighbouring pixels in a (relatively) low resolution representation of the image.

The colour of each pixel is measured in CIE $XYZ$ tristimulus coordinates, but this cannot be achieved by the successive application of just three filters because of the unobtainable nature of the primaries used by this system. In practice it is necessary to gather images in *seven* spectral bands and combine these to a full colour image using an elaborate calibration technique. Colour filtering is achieved by means of a computer-controlled filter wheel mounted in front of the source used in the multi-point fibre-optic light distribution system. The spectral response of the camera's CCD is accommodated by the use of neutral density filters for some colours plus computer control of the gain in the camera's electronics.

Calibration of the light source colour and spatial distribution is done with the aid of a white target area. It only needs to be done once per painting because the lighting system is carried on the portal and therefore remains stationary with respect to the camera's field-of-view.

Despite the accuracy of portal movement which is achieved, precise alignment of the $3000 \times 2320$ pixel 'frame' from each camera position cannot be guaranteed. Therefore the frames are deliberately arranged to overlap by approximately 100 pixels along their borders. These frames are automatically spliced together to form a mosaic covering the whole area of the painting. This is achieved by identifying 60 high contrast reference points along the border of a frame already in place for use as possible 'tie-points'. The border of the newly acquired frame is searched for points which correlate well with these reference tie-points. The candidate points are fitted iteratively to the reference points and any error is accumulated and evaluated. When a satisfactory match is found the new frame is translated to the appropriate coordinates and the borders are blended together.

All image processing is implemented on a SUN 'SPARCstation 2GS' workstation with 32 MBytes of RAM and 2 GBytes of disc storage. It is assisted by a PC to control the camera and a microcontroller to manage the portal movement and lighting. It takes several hours to capture an image but this compares favourably with the processing time of photographic archive media. Two complete systems exist: one in the National Gallery in London, UK, and the second in the Doerner Institüt in Munich, Germany.

The success of the VASARI project has encouraged the EC to fund more programmes such as MARC (Methodology for Art Reproduction in Colour) which will allow printed art catalogues to faithfully reproduce the colours of the paintings themselves. RAMA (Remote Access to Museum Archives) will allow curators and researchers to have remote access to a database of accurate reproductions of art treasures using ISDN digital communication networks.

## References

1. D. R. Reddy and R. W. Hon, 'Computer architectures in vision', in G. G. Dodd and L. Rossol (Eds), *Computer Vision and Sensor Based Robots*, pp. 169–186, Plenum Press, New York, 1979.
2. K. R. Castleman, *Digital Image Processing*, Appendix I, Prentice-Hall, Englewood Cliffs, New Jersey, 1979.
3. V. Hunt, *Smart Robots*, Chapter 5, Chapman and Hall, London, 1985.
4. B. Batchelor, D. Hill and D. Hodgson, *Automated Visual Inspection*, IFS (Publications), Bedford, 1985.
5. C. A. Rosen, 'Machine vision and robotics: industrial requirements', in G. G. Dodd and L. Rossol (Eds), *Computer Vision and Sensor Based Robots*, pp. 3–22, Plenum Press, New York, 1979.
6. J. A. Mitchell, 'Integrating vision sensors in the manufacturing process', *Proc. 3rd Int. IMS '87 Conf.*, pp. 180–191, Intertec. Commun., 1987.
7. H. H. Cook, *Tool Wear Sensors*, Final Rep., Dept. Mech. Eng., MIT, Cambridge, Massachusetts, 1976.

8. J. Edwards, 'Machine vision and its integration with CIM systems in the electronics manufacturing industry', *Computer-Aided Engineering Journal*, pp. 12–18, Feb. 1990.
9. J. D. Todd, 'Advanced vision systems for computer-integrated manufacture: Part II', *Computer-Integrated Manufacturing Systems*, **1**(4), pp. 235–246, Butterworth, London, 1988.
10. J. Hollingum, *Machine Vision: The Eyes of Automation*, IFS (Publications), Bedford, 1984.
11. L. Rossol, 'Computer vision in industry – the next decade', in A. Pugh (Ed.), *Robot Vision*, pp. 11–18, IFS (Publications), Bedford, 1983.
12. C. Loughlin, 'A hitchhiker's guide to vision systems', *Sensor Review*, **8**(2), pp. 93–99, IFS (Publications), Bedford, 1988.
13. H. Holland and O. Hageniers, 'Pressed metal surface inspection and body assembly gauging at the General Motors of Canada "Autoplex"', in *Proc. 19th Int. Symp. Automotive Tech. & Automation (ISATA)*, Vol. 1, pp. 229–249, Allied Autom., 1988.
14. A. C. M. Gieles, 'Introduction to machine vision for industrial automation', Philips Centre for Manufacturing Technology, Eindhoven, Netherlands, 1990.
15. G. Spaven, An automatic system for measuring and controlling the defibration process in the manufacture of hardboard, PhD Thesis, Brighton Polytechnic, 1992.
16. H. Geisselmann *et al.*, 'Vision systems in industry: application examples', *Robotics*, **2**, pp. 19–30, Elsevier (North-Holland), Amsterdam, 1986.
17. G. Spaven, Paper presented at *Pacific Rim Bio-Based Composites Symposium*, 7–13 Nov., 1992.
18. R. D. Baumann and D. A. Wilmshurst, 'Vision system sorts castings at General Motors Canada', in A. Pugh (Ed.), *Robot Vision*, pp. 255–266, IFS (Publications), Bedford, 1983.
19. C. Loughlin, 'Automatix provides the seal of success for Austin Rover', *The Industrial Robot*, **14**(3), pp. 145–148, IFS (Publications), Bedford, 1987.
20. J. Hermann, 'Pattern recognition in the factory: an example', in *Proc. 12th ISIR*, 1982.
21. J. Henry and C. Preston, 'Implementing machine vision: an IBM case study', *Sensor Review*, **8**(2), pp. 73–78, IFS (Publications), Bedford, 1988.
22. A. J. Cronshaw, 'Automatic chocolate decoration by robot vision', in *Proc. 12th Int. Symp. Ind. Robots (ISIR)*, 1982.
23. J. Shave, 'A fine selection of fish', *Image Processing*, **5**(3), pp. 28–29, European Technology Publishing Ltd, 1993.
24. R. B. Kelley, J. R. Birk *et al.*, 'A robot system which acquires

cylindrical workpieces from bins', *IEEE Transactions on Systems, Man and Cybernetics*, **SMC-12**(2), pp. 204–213, March/April 1982.
25. D. Smith and D. Johnson, 'An application of vision sensing to high-precision part-mating', *Int. J. Robot. Autom.*, **4**(1), pp. 36–42, 1989.
26. G. J. Awcock, Image aquisition technology for industrial automation, PhD Thesis, Brighton Polytechnic, Oct. 1992.
27. G. J. Agin, 'Real time control of a robot with a mobile camera', *Technical Note 179*, SRI International, Feb. 1979.
28. L. E. Weiss, A. C. Sanderson and C. P. Neuman, 'Dynamic sensor-based control of robots with visual feedback', *IEEE J. Robot. Autom.*, **RA-3**(5), pp. 404–417, 1987.
29. P. Regtien, 'Sensor systems for robot control', *Sensors and Actuators*, **17**(1–2), pp. 91–101, May 1989.
30. G. Hirzinger, 'Sensory feedback in robotics-state-of-the-art in research and industry', in *Proc. Automatic Control – World Congress*, Vol. IV, pp. 204–217, Pergamon, Oxford, 1988.
31. G. Awcock, 'An edge detecting VLSI image sensor – towards dedicated sensors for industrial automation', *Proc. 4th Int. Conf. Image Proc. and its Applns*, Maastricht, Netherlands, pp. 546–549, IEE, April 1992.
32. K. R. Castleman, *Digital Image Processing*, Appendix I, Prentice-Hall, Englewood Cliffs, New Jersey, 1979.
33. J. A. Rindfleisch *et al.*, 'Digital image processing of the Mariner 6 and 7 pictures', *J. Geophys. Res.*, **76**, p. 394, 1971.
34. I. McLean, *Electronic and Computer-Aided Astronomy*, Ellis Horwood, Chichester, 1989.
35. J. L. Horner, *Optical Signal Processing*, Chapter 5.2, Academic Press, New York.
36. K. H. Hohne, H. Fuchs and S. M. Pizer (Eds), *3D Imaging in Medicine: Algorithms, Systems, Applications*, NATO ASI Series, Series F, 1990.
37. R. Evans, 'Surgical vision', *Image Processing*, **5**(3), pp. 12–15, European Technology Publishing Ltd, 1993.
38. K. Sheldon, 'From outer space to the inner man', *BYTE*, **12**(3), p. 146, March 1987.
39. 'Uncooled IR thermal imaging camera', in A. J. Walkden, (Ed.), *GEC Journal of Research*, **9**(2), pp. 121–123, 1991.
40. W. Mason and G. Relf, 'Making light work for biology', *Image Processing*, **2**(5), pp. 14–18, Reed Business Publishing Group, London, 1990.
41. R. S. Ledley, 'High-speed automatic analysis of biomedical pictures', *Science*, **146**(3461), pp. 216–223, 1964.
42. C. Blackman, 'The ROVA and MARDI projects', *Advanced Robotic Initiatives in the UK Colloquium*, Digest No. 1991/081, IEE, April 1991.

43. AGEMA Infrared Systems, 'Infrared hotline', *International Newsletter*, Spring 1993.
44. C. Elliot, D. Day and D. Wilson, 'An integrating detector for serial scan thermal imaging', *Infrared Physics*, **22**, pp. 31–42, 1982.
45. Loral Fairchild, *CCD Imaging Data Book*, Milpitas, California, 1991.
46. 'Space radar images show new view of earth', in A. J. Walkden (Ed.), *GEC Journal of Research*, **9**(2), pp. 123–124, 1991.
47. 'Imaging tackles drug barons', *Image Processing*, **4**(1), p. 5, Reed Business Publishing Group, London, 1992.
48. M. Jackson, 'A new dimension in mapping', *Image Processing*, **3**(1), pp. 10–12, Reed Business Publishing Group, London, 1991.
49. D. Braggins, 'A new reality in imaging', *Image Processing*, **3**(5), p. 50, Reed Business Publishing Group, London, 1991.
50. Anon., 'DEMS and orthoimages', *Image Processing*, **4**(4), p. 59, Reed Business Publishing Group, London, 1992.
51. M. Spalding and B. Dawson, 'Finding the Titanic', *BYTE*, pp. 97–110, March 1986.
52. D. Austin, 'Poised to take the DIP', *Banking Technology*, **9**(6), pp. 20–23, Banking Technology Ltd, London, July/Aug. 1992.
53. M. Christian, 'A healthy image', *Image Processing*, **4**(4), pp. 42–43, Reed Business Publishing Group, London, 1992.
54. B. Hamilton, 'DIP on delivery', *Image Processing*, **3**(3), pp. 39–40, Reed Business Publishing Group, London, 1991.
55. D. Braggins, 'Management moves in on DIP', *Image Processing*, **3**(2), p. 58, Reed Business Publishing Group, London, 1991.
56. K. George, 'Planning for the poll tax.', *Image Processing*, **3**(2), pp. 23–24, Reed Business Publishing Group, London, 1991.
57. D. Meyer-Ebrecht, 'The "filmless" radiology department – a challenge for the introduction of image processing into the medical routine work?', *Proc. 4th Int. Conf. Image Processing & its Applications*, Maastricht, Netherlands, pp. 13–20, IEE, 1992.
58. M. Hill, 'DIP for the desktop', *Image Processing*, **4**(1), pp. 36–37, Reed Business Publishing Group, London, 1992.
59. D. Wells, 'Images protect data', *Image Processing*, **4**(4), p. 20, Reed Business Publishing Group, London, 1992.
60. Document imaging in the law office, conference programme advertisement, Unicom Seminars, *Image Processing*, **4**(1), p. 35, 1992.
61. A. Wilson, 'US report: image compression cut down to size', *Image Processing*, **3**(3), pp. 8–9, Reed Business Publishing Group, London, 1991.
62. J. Robinson and F. Coakley, 'Picture coding for photovideotex', *Computer Communications*, **6**(1), pp. 3–13, Butterworth, London, 1983.

63. 'MPEG gets go ahead as fractals emerge', *New Electronics*, p. 9, Dec. 1991.
64. S. Wilmer, 'Scanning technology comes of age', *Image Processing*, **3**(3), pp. 15–17, Reed Business Publishing Group, London, 1991.
65. 'Products: intelligent OCR', *Image Processing*, **4**(4), p. 60, Reed Business Publishing Group, London, 1992.
66. J. Griffiths (Ed.), *ISDN Explained: Worldwide Network and Applications Technology*, 2nd edn, Chapter 8, Wiley, New York, 1992.
67. A. Jain, *Fundamentals of Digital Image Processing*, Chapter 11, Prentice-Hall, Englewood Cliffs, New Jersey, 1989.
68. P. Newton, 'Seeing is believing', *Communications Networks*, pp. 39–42, October 1993.
69. R. Dean, 'Here's looking at you, caller', *What Video*, pp. 56–57, April 1993.
70. P. Denyer et al., 'On-chip CMOS sensors for VLSI imaging systems', *Proc. VLSI '91*, pp. 4b.1.1–10, 1991.
71. 'A firm hand for identification', *Image Processing*, **3**(3), p. 6, Reed Business Publishing Group, London, 1991.
72. 'Security processing', *Image Processing*, **3**(1), p. 47, Reed Business Publishing Group, London, 1991.
73. R. Blissett, 'Eyes on the road', *Image Processing*, **4**(2), pp. 18–21, Reed Business Publishing Group, London, 1992.
74. P. Ringer, 'Forensics fights fraud', *Image Processing*, **3**(5), pp. 17–18, Reed Business Publishing Group, London, 1991.
75. 'Daemon technology', *What Video?*, p. 53, Jan. 1993.
76. K. Sheldon, 'A film of a different color', *BYTE*, **12**(3), pp. 164–165, McGraw-Hill, New York, 1987.
77. I. Childs, 'HDTV: putting you in the picture', *IEE Review*, **34**(7), pp. 261–264, 1988.
78. B. Fox, 'The big squeeze', *What Video?*, pp. 64–69, November 1993.
79. 'Second level digital video standard approved', *What Video?*, p. 9, October 1993.
80. L. MacDonald, 'Colour crunching', *Image Processing*, **3**(1), pp. 15–19, Reed Business Publishing Group, London, 1991.
81. R. Gonzalez and P. Wintz, *Digital Image Processing*, Addison-Wesley, London, 1977.
82. J. Asmus, 'Digital image processing in art conservation', *BYTE*, **12**(3), pp. 151–165, McGraw-Hill, New York, 1987.
83. K. Martinez, D. Saunders and J. Cupitt, 'Paintings as numbers', *Image Processing*, **4**(4), pp. 10–13, Reed Business Publishing Group, London, 1992.

# Appendix: The Video Image Format

The raster scan TV format is defined by the CCIR standard within Europe (or the EIA RS170 standard in the USA and most other countries). It defines fully all parameters necessary to make all equipment broadcast compatible.

In brief, the CCIR Standard says that an image (or frame) is displayed 25 times per second and is built up from two interlaced rasters (or fields). Each field consists of 312.5 lines, giving a hypothetical maximum vertical resolution of 625 lines at an image refresh rate of 25 Hz in a full implementation, figure A.1a.

Figure A.1 Raster scan TV format: (a) interlaced rasters; (b) non-interlaced rasters

The 2 : 1 interlaced field system was designed for the portrayal of moving television pictures at the best possible resolution. However, when static images such as those produced by machine vision systems are displayed, the interlace produces a vertical 'jitter' which can be subjectively very annoying. This may be overcome by the use of the non-interlaced mode, figure A.1b, where the same data are written on both rasters without the half-line offset which is necessary for interlacing. This results in a display refresh rate of 50 Hz but the price to pay is a reduced hypothetical maximum vertical resolution of only 312 lines.

Appendix: The Video Image Format 293

## A.1 The TV line format

Since the image is built up from a number of lines of information, it is necessary to modulate the image data upon waveforms suitable for driving a 'monitor' (which consists of a cathode-ray tube (CRT) plus the interface and support circuitry necessary for raster generation). These waveforms are tightly specified in the full CCIR standard, and are designed to control and synchronise the information, in order to achieve a stable raster display (see figure A.2). Thus each TV line must carry synchronising pulses in addition to the serialised data corresponding to a horizontal slice of the image. All this must be combined into one analogue (continuous and multilevel) waveform, called the 'composite video' waveform.

The portion of the waveform intended for display is called the 'active video' and in conventional broadcast TV it is a fully analogue signal, capable of displaying any number of intensity levels (see figure A.2a). For many computer display applications it is only necessary to represent two levels, corresponding simply to foreground and background. Thus the

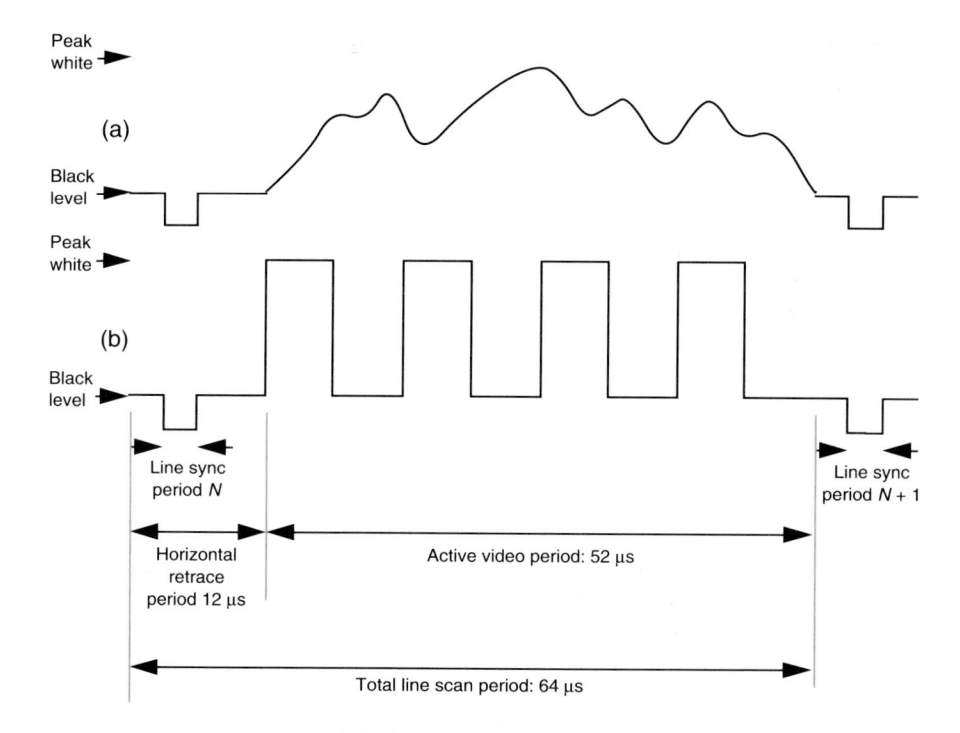

Figure A.2 Illustrations of composite video signals: (a) signal from typical grey-scale image; (b) signal from typical chequerboard pattern image

complexity of the waveform generating circuitry can be considerably reduced by the need to produce only binary active video, with logic '0' corresponding to background (usually pure black) and logic '1' corresponding to foreground (usually white, green or amber depending on the spectral emission properties of the particular CRT phosphor), as seen in figure A.2b.

The minimum and maximum values of the active video waveform are known as 'black level' and 'peak white' respectively (regardless of the actual colour of the final displayed image). Although the active video portion of the waveforms may consist of a simply generated binary waveform, the synchronising pulses require that a third voltage level can be represented in the composite video. In practice, these pulses are simply mixed with the active video by a separate interface circuit under logic control, allowing most of the work of computer video generation to be done by straightforward *digital* circuitry.

## A.2 Textual displays

To illustrate the method of TV format raster generation and the display of digitised images, consider the waveforms necessary for the production of a large format textual display.

The method which is generally adopted is to display each character as a matrix of dots such as $5 \times 7$, $6 \times 8$, $7 \times 7$, etc. similar to that used by dot-matrix LED or LCD displays. The rows of the matrix consist of consecutive TV lines, while the columns consist of time periods along each line governed by a master 'clock' oscillator called the 'dot-clock' or 'pixel clock'. The waveforms depicted in figure A.3 illustrate the display of a selection of alphabet letters made up from a $5 \times 7$ character matrix.

In reality, the composite video waveforms consist of a continuous stream of active video separated by line synchronising pulses. In the schematic shown the synchronisation pulses have been arranged to lie underneath one another. This is precisely the function performed by a monitor's synchronisation circuits in order to produce a stable raster display. Vertical synchronisation is also necessary to prevent the displayed raster from 'rolling' vertically. This is achieved by detecting a sequence of special video lines which convey no active video at all and some of which have considerably lengthened line-synchronisation pulses. These lines occur during the vertical retrace period which occupies around 1.5 ms of the 20 ms period of each field. This is why the number of usable video lines per frame is considerably less than the hypothetical limit of 625.

Figure A.3 shows that the text characters are actually displayed on a $6 \times 8$ matrix to provide separation of the individual characters on a character-row and of the consecutive character-rows. The character matrix is not always

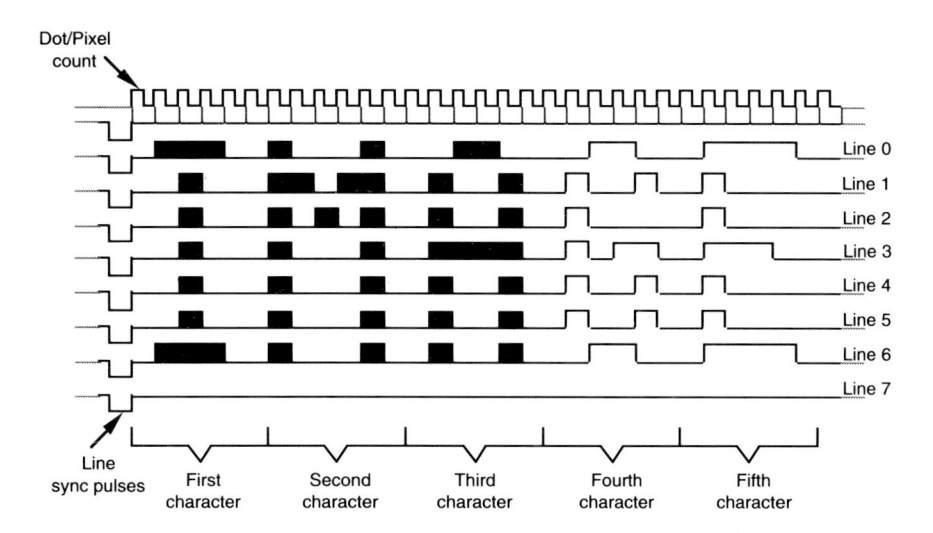

Figure A.3  Portion of textual picture constructed from linescans

augmented in this way, particularly on systems with a mixed text and graphics capability.

The active video portion of each TV line only needs to be a binary waveform for this type of application. It is made up of just one of the matrix-rows for all the horizontally adjacent characters making up a character-row. Thus the full character-row requires, in this case, eight consecutive TV lines to be displayed. The number of characters per character-row is limited by the dot-clock frequency, which in turn is limited by the switching speed of the video output circuitry and the bandwidth of the monitor used for display purposes.

# Index

*a priori* probability   185
*a posteriori* probability   186
aberrations, lens   51
accuracy, of vision system   57
acquisition   8, 59*ff*
actuation   13
aliasing   60
ambient   32, 42, 53
amplitude quantisation   61
analogue-to-digital conversion (ADC)   83
architectures for imaging   72
area arrays, imaging architecture   74
area determination   157
array tessellation   63
art conservation, applications   286*ff*
art history, applications   285*ff*
artificial neural networks   177
aspect ratio   85
associative network   191, 193
astronomy, applications   236
auto-associative networks   195
automated visual inspection (AVI), applications   229
automatic gain control (AGC)   39

background separation, visual perception   210
backlighting   36
 dark field   38
 light field   36
Bayes classifier   184
bimodal histogram   103
binary images   62, 103, 126
binocular images   207
blob analysis   160
boundary   4, 130, 136, 219
boundary-based features   156
B-rep   219
brightness modification   99

calibration   57, 85

Canny edge detector   136
capture of images   83
cartography, applications   253
centroid   27, 163
chain code   152
characteristics, segmentation   129
charge transfer efficiency (CTE)   80
charge-coupled device (CCD)   80
charge-coupled photodiode (CCPD)   80
charge-injection device (CID)   80
class conditional probability   185
classification   12, 255
closing   171
closure   210
colour fidelity, applications   283*ff*
colour temperature   44
colourisation, applications   276*ff*
compass gradient operators   134
compass points   152
computer vision   1
conditional risk   187
confidence level   188
consumer electronics, applications   275*ff*
continuation, visual perception   210
contrast enhancement   101, 254
control, scene constraints   30
convex hull   156
convolution   93
convolution integral   90
correlation   92, 96
crack code   155
cylinders, generalised (cones)   217

2.5-D sketch   213
3-D model   213
3-D reconstruction, applications   241
data compression, applications   259*ff*
decision theoretic   177, 182
'depth from' strategies   213*ff*

depth of focus, depth of field   51
design, system   4
desktop publishing, application   283
detectors
   photoconductive   69
   photoemissive   68
   photovoltaic   69
diagnostic medical imaging, applications   238*ff*
digital convolution   93*ff*
digital-to-analogue conversion (DAC)   86
digitisation   83
dilation   169
directional operators   134
discriminator   198
display adjustment   119
display of image   86
document image processing (DIP), applications   256*ff*
dynamic threshold   145

economic considerations   21
edge detection   131
edge enhancement   114
edge map   155
edge-based segmentation   130
edgels   130
edges   113, 130
electromagnetic spectrum   66
emergent properties, systemic view   5, 14
enabling technology, AIP as a   20
enhancement   90
entertainment, applications   275*ff*
erosion   169
Euler number   161
exploitation, scene constraints   25
external photoeffect   68

facsimile, applications   265*ff*
false colour, applications   254
feature analysis   181
feature extraction   10, 148*ff*
feature space   182
feature vector   148, 176
feedback networks   194
feed-forward networks   194
fiducial marks   31
field of view   29
filtering   97*ff*
fisheye lens   55
focal length   50

forensic investigations, applications   274
Fourier Transform   92, 161
frame buffer   85
frame difference   123
frame transfer, CCD sensor   82
frame-based operations   123
framestore   84
frequency domain   93
Fresnel lens   56
front lighting   34
   directional   35
   omni-directional   34

gamma distortion   77
gamma rays, applications   241
gauging, applications   230
Gaussian filter   110, 186
generalised cylinders   218
generic model   6
geometric operations   99, 118, 254
glare reduction   42
global methods, segmentation   141
global operations   99, 106
global threshold   145
gradient operators   114, 131
graphic arts, applications   283
grey level, grey scale   61
grey-level histogram   99
guidance
   military, applications   233
   robotic, applications   244

heuristics   129
hexagonal tessellation   63
high-definition television (HDTV), applications   278
histogram   99
histogram equalisation   106
histogram specification   108
hit–miss transform   167
Hough Transform   138

illumination
   dark field   38
   directional   35
   light field   36
   omni-directional   34
image acquisition   8, 59*ff*
image analysis   126
image capture   83
image codes   150
image display   86

image enhancement 90
image features 148
image formation 205
image preprocessing 90*ff*
image representation 59, 216
image restoration 90
image segmentation 126*ff*
image sensing 66
image sharpening 113
image smoothing 109*ff*
image transduction 66
image transform 108
image understanding 204*ff*
imaging architectures 72
imposition, scene constraints 30
impulse response 91
industrial machine vision 3, 228*ff*
information format, systemic view 14
information technology (IT), applications 256
inherent features, scene constraints 27
interactive video, applications 280*ff*
interline transfer, CCD sensor 82
internal photoeffect 69
interpolation, grey level 121
invariant moments 164
Inverse Fourier Transform 92
inversion 120

*k*-nearest neighbour filter 111

lamps 44*ff*
   characteristics 46
   fluorescent 47
   quartz halogen 47
   sodium discharge 47
   tungsten filament 47
Laplacian of Gaussian 133, 146
Laplacian operator 114, 117*ff*, 132
law enforcement, applications 270
learning
   supervised 197
   unsupervised 197
lens aberrations 52
lens formulae 48
lenses
   commercial 53
   fisheye 55
   Fresnel 56
light stripe 39
lighting, structured 39
lighting conditions 32
lighting techniques 34

limitations, scene constraints 29
linear arrays, imaging architecture 72
linear imaging device (LID) 72
linear system 90
lines, definition and image features 113, 130
local methods, segmentation 130
local threshold 145
log-polar mapping of photosites 65
lppm 53

machine vision 1*ff*
magnetic resonance imaging (MRI), applications 240
magnification 53, 120, 121
manufacturing
   automation 16
   safety, reliability 18
mapping 99, 121
Marr's theory of vision 212
matched filtering 178
material characteristics, surfaces 25
mathematical morphology 165*ff*
matrix arrays *see* area arrays
measurement, visual inspection, applications 230
medial axis transformation 156, 173
median filter 112
medical imaging, applications 238*ff*
medical thermography, applications 241
memory-based classifier 197
meteorology, applications 251
minimum distance classifier 183
mode filter 111
modelling and matching 221*ff*
moments
   central 162
   invariant 164
monocular images 205
motion, depth from 214
moving window 97
multimedia, applications 282
multi-resolution 135

nearest neighbour 108, 184
negation 101
neighbourhood averaging 109
neighbourhood operations 99, 108
neural network
   classification 177, 191
   nodes 192
   training 196

*n*-tuple   199
Nyquist criterion   60, 94

object features, scene constraints   30
object position, scene constraints   31
opening   171
optical character recognition (OCR)   29, 179
optical flow   215
optics   48, 55
ordered liberty, systemic view   6
orthogonal tessellation   63

parsing   188
parts identification, applications   232
pattern classification   176*ff*
pattern primitives   188
perceptron   192
perceptual organisation   209
performance, system   88
perimeter determination   157
photocell   69
photoconductive detectors   69
photoeffect
   external   68
   internal   69
photoelasticity   43
photoemissive detectors   69
photometric units   45
photoMOS   71
photon flux integration   70
photosensing mechanisms   67
photosite   79
photovoltaic detectors   69
pipeline processor   88
pixel-based methods, segmentation   130
point operations   99
point spread function   95
polarised light   39
positional constraints   31
predicate, uniformity   127
preprocessing   8, 90*ff*
Prewitt operator   115, 134
primal sketch   213
printing, applications   283*ff*
probabilistic classification   177, 184
process control, applications   230
product quality, visual inspection   19
proximity, visual perception   210
pseudo-colour   87

quad-tree data structures   143

quality control, applications   230
quantisation parameters   60, 61
quantum detectors   68
quantum efficiency   68

radiometric units   45
real-time   88
reconnaissance, applications   244, 246*ff*
region-based features   160
region-based methods, segmentation   141
remote sensing, applications   250*ff*
representation of images   59
resolution, amplitude   61
resolution, spatial   60
resolution pyramid   135
restoration   90
Roberts operator   115
robotic guidance, applications   233
robust solutions, achieving   56
rotation of images   120
run code   150

safety critical processes   18
sampling, non-uniform   64
sampling frequency   84
scene constraints   7, 25*ff*
   control   30
   exploitation   25
   inherent features   27
   lighting conditions   32
   limitations   29
   material characteristics   25
   object features   30
   object position   31
   structured lighting   32
scientific analysis, applications   243
security, applications   270
segmentation   9, 126*ff*
self-scanning photodiode (SSPD)   79
sensing   66
shading, depth from   216
shape factor   157
sharpening   113
sigma filter   111
signature, polar radial   155
similarity, visual perception   210
skeleton   155, 172
smoothing   109*ff*
Sobel operator   116
solid-state technology   78
space exploration, applications   235

spatial averaging   109
spatial frequency   94
spatial quantisation   61
spatial resolution   84
spatial transformation   120
spectral distribution   44
split and merge   142, 144
statistical classification   177
stereo, depth from   213
stereo correspondence   214
structured lighting   32, 39
structuring element   149, 167
surface representations   219
surveillance, applications   270
symmetry, visual perception   210
synchronisation, video waveform   84, 86, 294
syntactic classification   188
system accuracy   57
system boundary   4
system calibration   57
system design   4
   hard system   15
   soft system   15
system performance   88
system reliability   57
system survivability   57
systemic   6
systemic considerations   34

telecommunications, applications   264*ff*
template matching   178
templates   116
temporal operations   99, 123
tessellation
   hexagonal   63
   orthogonal   63

texel   161
texture   148, 161, 215
thermal detectors   67
thermal imaging, applications   246
thinning   156
threshold   63, 103, 144, 145, 192
time domain   93
topological features   161
training strategy   196
transduction   66
transforms   108
TV line format   293

ultrasound, applications   241
understanding   1, 204
   'on-line'   3
uniformity predicate   127
units
   photometric   45
   radiometric   45
universal vision   2
unsharp masking   117

vacuum tube technology   75
video conferencing, applications   268
video signal   83, 292
Vidicon   76
visual perception   208
visual sense   1
volumetric elements (voxels)   218

warping   120
weighted mean filter   111
weighting, neural networks   192
wire frame   220
WISARD   202

X-rays, applications   240